# ENGINEERING KARSTOLOGY
# OF DAMS AND RESERVOIRS

# ENGINEERING KARSTOLOGY OF DAMS AND RESERVOIRS

**Petar Milanović**

Professor Emeritus/President of the Serbian IAH Chapter
Belgrade
Serbia

CRC Press
Taylor & Francis Group
Boca Raton London New York

CRC Press is an imprint of the
Taylor & Francis Group, an **informa** business

A SCIENCE PUBLISHERS BOOK

*Cover illustration*: Photograph PB 150010; Klinje Dam, Gacko, Herzegovina, 26.4 m high, arch gravitu, built from 1888 to 1896

CRC Press
Taylor & Francis Group
6000 Broken Sound Parkway NW, Suite 300
Boca Raton, FL 33487-2742

First issued in paperback 2020

© 2018 by Taylor & Francis Group, LLC
CRC Press is an imprint of Taylor & Francis Group, an Informa business

No claim to original U.S. Government works

ISBN-13: 978-1-4987-4807-0 (hbk)
ISBN-13: 978-0-367-78123-1 (pbk)

This book contains information obtained from authentic and highly regarded sources. Reasonable efforts have been made to publish reliable data and information, but the author and publisher cannot assume responsibility for the validity of all materials or the consequences of their use. The authors and publishers have attempted to trace the copyright holders of all material reproduced in this publication and apologize to copyright holders if permission to publish in this form has not been obtained. If any copyright material has not been acknowledged please write and let us know so we may rectify in any future reprint.

**Visit the Taylor & Francis Web site at**
**http://www.taylorandfrancis.com**

**and the CRC Press Web site at**
**http://www.crcpress.com**

# FOREWORD

It is a pleasure and an honour to be invited to write the Foreword for Dr. Petar Milanovic's new volume "Engineering Karstology of Dams and Reservoirs". He and I are of similar age—early eighties. Petar was raised in Serbia in the great Dinaric Karst of Europe, and trained in civil engineering. I began as a schoolboy caver in the southwest of England and went on to study physical geography and geology. Over our lifetimes we have seen our notions of what cave and karst science is, expand and then expand again: now, at meetings and in publications, karst geoscientists and engineers mix together with each other and with hydrologists, hydrogeologists, archeologists, anthropologists, scientists from the many branches of biology and, lately, biochemists seekine exotic bacteria. A broad and fascinating new multi-disciplinary field, "Karstology", has come into being.

In his field engineering practice and in his writings Petar Milanovic has experienced most of karstology. "Engineering Karstology of Dams and Reservoirs" is the third major textbook about the practical problems encountered when designing, building and operating water supply, hydroelectric generating and many other kinds of installations in karst terrains, that he has written since 1981. Published in English and firmly based on his unparalleled engineering field experience in Europe, Asia and the Americas, they are invaluable aids to the world's engineers, scientists and managers. The thirteen chapters of this latest book offer readers a careful, very systematic, review of karstic problems at dams and in reservoirs, the first half describing the initial areal and site survey and exploration techniques that should be undertaken during the design stages, then layout and practice in the grouting and other waterproofing works that will be needed, technical advice on choice of grout mixtures, pressures, etc. The middle chapters consider dams in evaporite rocks (to be avoided if at all possible), and dams and other regulators built underground in existing caves, even those that are submarine. The applications and limitations of computer modeling in the anisotropic and heterogeneous karstic medium are evaluated, and the practical problems of leakage and mechanical failure that may arise in the years following construction and filling are described, and solutions offered. The concluding chapters stress the need to consider the environmental impacts at all stages of the work, review the political problems where these cross major political boundaries, and present a global catalogue of the successes and failures. There are illustrative case studies throughout the book, supported by technical drawings, analytical figures, many site photographs on the surface and down inside the rock. "Expect the unexpected" has been Petar's guiding principle in karst engineering for many years. In this most helpful, sometimes alarming, new book he shows you the reasons why.

Derek Ford, PhD, FRS Canada,
Emeritus Professor of Geography and Earth Sciences
McMaster University

19 December 2017

# PREFACE

From the very beginning design and construction of dams and reservoirs in karstified rocks deal with challenges, difficulties and frequent failures. It was a time when designers and engineers did not entirely understand the role of geology as an important science in dam construction, particularly in karst. During the years, particularly during the 20th century, many of failures were analyzed and experience was presented at a number of scientific and professional conferences and articles. These experiences have also helped in development of new investigation methods and new technologies to minimize risk of failure.

Karstified rock as subject of dam foundation and space for reservoirs is not 'friendly' geological formation. Geological and hydrogeological properties of karstified rock vary extremely from place to place at short distances. Precise detection of exact space position of karst conduits and caverns from surface at depth more than 20 m still is not possible. Investigation works are expensive and time consuming.

Dams are expensive structures. Consequences of failures are expensive also and sometimes disastrous. At many cases expensive remedial works does not justify the money invested. In some cases, after many years of remedial works and enormous money invested, the dam is abandoned. However, each failure is a new experience, very helpful for future dams from investigation and design to the construction phase. Knowledge and experience together are required for construction watertight reservoirs in karst.

In this book is presented an extensive review of number of dam examples, particularly its problems with construction in specific karst environment. In my previous book *Water Resources Engineering in Karst,* published by CRC Press, 2004 is presented wide spectrum of engineering problems in karst including dams but also other large structures. During the past 13 years a number of new dams have ben constructed in karstified rocks, a number of remedial works was done to improve the old dams, new investigation methods and waterproofing techniques were developed. The number of examples is presented at different geological and ICOLD conferences and relevant publications. This is an effort to select crucial data and present it in concise form.

This book is focused on the engineering karstology of dams and reservoirs including a number of examples. This is not officially a second edition of the book published in 2004. However, it was not possible to avoid repetition of some examples including some figures and photos presented in previous book. The idea was to present the reader with as many as possible examples in the relatively low volume book. It is important to emphasize this type of book cannot be never completely finished because the number of new dam sites in karst are permanently under investigations, the number of dams under construction and at a few ten of them running waterproofing works.

I would like to express a lot of thanks to Prof. Derek Ford for reviewing the manuscript in detail, for the excellent language editing and for many of his instructive discussions and suggestions to improve the quality of the book.

# CONTENTS

# CHAPTER 1

# GENERAL INTRODUCTION

## 1.1 KARST AND ENGINEERING

From the practical viewpoint, what does the term *karst* mean? For the people who enjoy landscapes, karst means spectacular and exciting surface morphology; for speleologists it is separate science but also a great challenge and may offer extreme sport; for biologists it is a paradise of exotic endemic species; for archaeologists it is a potential treasure trove of artifacts belonging to ancient humans and civilizations; for geologists, geomorphologists and hydrologists it is a phenomenon that is extremely complicated and problematic for investigations; for local (native) people karst often means suffering and permanent struggle with nature to survive, and for engineers the term can be bad news—its presence may be a threat, implying possible failures.

Why does the term *karst* suggest failures for engineers? During the last decade of the 19th century intensive construction of dams and reservoirs in karst regions began all over the world. The result was expensive failure in many cases. For many of these cases, the water reservoir never filled up despite extensive investigations and remedial works, ending in the site being abandoned. It was a time when designers and engineers did not entirely accept that understanding the role of geology as being important for the building of large structures at the surface or underground. The investigation techniques and technology of any remedial works were also at inadequate levels. As a consequence of these failures, many important lessons were learned when constructing in karst.

However, karst is still far from being accepted as a friendly environment for any construction, particularly the construction of large dams and reservoirs. Almost anybody working on engineering in karst, engineer, contractor or owner, is under pressure is fearful of two things: fear of the risk and/or fear of failure. Generally, fear of risk has a positive effect because it stimulates studies to minimize the hazards. In contrast, fear of failure is negative because it inhibits creative solutions and can lead to conservative and expensive solutions.

Successful solutions in karstified rocks require serious and complex geological and hydrogeological investigations and the close co-operation of a wide spectrum of scientists and engineers. However, the risks of constructing in karst cannot be eliminated completely due to its hydrogeological and geotechnical complexity, even when best engineering practices are followed. Some failures are consequences of long-lasting processes (mechanical erosion of sediments deposited in cavities or dissolution of evaporites) but mostly occur unpredictably and are practically instantaneous. Complete knowledge of the nature of karst together with specific engineering (geotechnical) methods is the only way for a successful struggle with geological environments so complex as karst.

Besides the existence of a good geological map, of primary importance are: full understanding of the local geological structure during the dam site and reservoir selection; understanding of karst aquifer evolution processes; application of non-conventional investigation methods; assessing the risk of seepage, leakage, collapses, instability, and environmental impact; and application of different engineering surface and underground technologies and remedial measures including sealing of deep karst channels under hydrostatic pressure. *Modifications and adaptations of the final design during the execution phase of any construction in karst is not the exception, rather it is rule.*

This book presents a working framework for identifying and evaluating a variety of the processes that present a risk for dams and reservoirs in areas with such complex geological and hydrogeological properties as are found in karstified rock masses. As suggested by the title of the book the focus is on *engineering karstology*, which means identifying the engineering problems of dams and reservoirs in karst. Karstification as a set of specific geological processes and all theoretical and scientific aspects of those processes have been presented in detail in many articles and textbooks over the past 120 years. These aspects are not covered in this book or, in some cases, are presented only in brief but informative summaries.

## 1.2   DISTRIBUTION OF KARST

Karstified rock masses are found all over the world in areas where the rocks are carbonates, evaporites or conglomerates with a carbonate matrix, at all latitudes and altitudes, from a few thousands meters below the sea level up to the elevation of 4000 m or more. Karstified rocks are distributed widely in permafrost areas of Canada and Russia (Ford and Williams, 2007). Karstified limestone is recorded also at elevations of 3000 m in Iran and at 4000 m or more, at Peruvian Andes. Pictures of the south summit ridge of Everest show some small karren features, at 8000+m (pers. comm., Ford, 2017).

According Yuan (1991) a large number of karst phenomena have been discovered on the Tibet Plateau at altitudes of 4000 m, even 6000 m.

Deep weak zones detected by applying remote geophysical methods of investigation, caverns discovered by boreholes at depths between 200 and 400 m or more, large amplitudes in water table fluctuations (100–400 m or more in a few cases), and results of speleological explorations indicate that the extension of highly permeable zones from several hundred meters down to thousands meters in depth is possible. More than 20 shaft caves around the world have been explored by speleologists to depths between 1000 and 2250 m. Extremely deep karst features has been detected by drilling to a depth of 2.236 m below the mean sea level in the Dinaric karst region.

Karstic features are known in all chronostratigraphic units, from the many soluble Precambrian formations up to carbonates of Paleogene and Neogene age. For instance, the Samanalawewa Dam in Sri Lanka is situated in karstified Precambrian marble (metamorphosed limestone).

## 1.3 TYPES OF KARST

There are many classifications of karst types in the literature, based on different geological, hydrogeological and geomorphological criteria. Depending upon which factors are considered to be fundamental, various authors have classified the karst according to a number of different types: regional expressions of morphology (holokarst, merokarst, transitional karst; Cvijić, 1925, Fig. 1.1); according specific morphologic shape (tower and cockpit types); according to morphology and climatic conditions (holo-, fluvio-, glacio-nival, tropical, arid and semi-arid, Sweeting, 1972); according to regional geological structure (platform and geosyncline); according to tectogenetic factors (epi-orogenic or epi-continental and orogenic or geosyncline, Herak, 1977), according depth (deep and shallow, Katzer, 1909); barré (dammed) and covered (Jennings, 1971); subsoil, mantled, buried and interstratal karst, (Quinlan, 1978); alpine or glaciokarst; bare, covered (under unconsolidated deposits) and buried karst (under consolidated insoluble rock, Yuan, 1981); based on lithological properties (bare limestone karst, dolomite karst, marble karst, chalky/marly karst, gypsum-anhydrite and karst in salt rocks, Gvozdecki, 1981). Ford and Williams (2007) list the following types of karst: alpine, barré, buried, coastal, contact, crypto, endo, evaporite, exhumed, exo, fluvio, glacio, holo, hyper, hypo, interstratal, mero, nival, paleo, pseudo, syngenetic, thermo, tower and volcano.

Some karst landscapes from different areas are presented in Fig. 1.2.

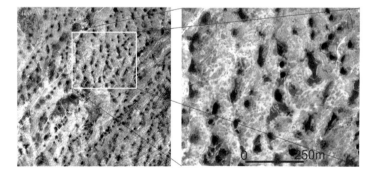

**Figure 1.1.** Typical Dinaric holokarst (bare karst).

## 1.4 INTRODUCTION TO ENGINEERING KARSTOLOGY

The term ***karst*** originates from the geographical name of the region that borders Italy and Slovenia. More than 700 years ago people used the Slavic world ***kras*** and Italian world ***carso*** to characterize this region (toponym). With germanization of those words, the term ***karst*** was created by J. Cvijić and accepted as an international scientific term (e.g., in books such as 'Das Karstphanomen', Cvijić, 1893, Fig. 1.3, and 'Karst', 1895). A few more books and papers using *karst* in the title were published shortly afterwards: 'Die Karsthydrographie, Studien aus Westbosnien' (Grund 1903), 'Der

**Figure 1.2.** Different karst landscapes: (A) platform karst in Spain; (B) geosyncline (Dinaric) karst; (C) tower karst (Li River) China; (D) covered karst (Djerzinsk, Russia); (E) isolated mogote, Malaysia; and (F) karstified salt, Iran.

Geographische Zyklus im Karst' (Grund 1914), 'Karst und Karsthydrographie' (Katzer, 1909), 'Hydrographie souterraine et evolution morphologique du karst' (Cvijić, 1918). In China an old Chinese term 'Yangrong' (*rock to be dissolved*, probably created by Xu Xaike, 1587–1641) was in use and equivalent to *karst*. The first book related to engineering problems in karst was 'Wasserbauten in Bosnien und der Herzegovina, Meliorationsarbeiten und Cisternen im Karstgebeite' (Ballif, 1896).

The first geotechnical conference, on the topic 'Consolidation in Karst', was organized in Split in 1961 by the Yugoslav Society for Geological Investigations and Consolidation. An early comprehensive book, focused to the development and conservation of karst waters, was the 'Handbook of Karst Hydrogeology—with Special Reference to The Carbonate Aquifers of the Mediterranean Region by D.V. Burdon and N. Papakis, published by the FAO (1963, Athens, Greece). A textbook related to dam construction in karstified rocks 'Karst and Dam Engineering' Lykoshin et al. was published in Russian in 1992 ('Geoproyect', Moscow).

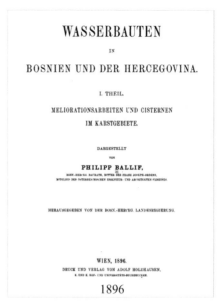

**Figure 1.3.** First books published with the term *karst*: *left*–Cvijić—scientific explanation and *right*–Ballif —the first explanation of engineering works in karst.

Research results of dam and reservoir problems encountered in karst were discussed in many articles presented at a number of ICOLD congresses and conferences. For over 30 years the 'Multidisciplinary Conference on Sinkholes and the Engineering Impacts in Karst' initiated by the late Dr. Barry F. Beck has been one of the premier conferences for all engineering aspects of karst, including problems of dams and reservoirs; there have been a total of 14 Conferences, held in different karst regions of the U.S.A. International Engineering Geology Symposium 'Engineering Geological Problems of Construction on Soluble Rocks' was organized in Istanbul, 1981.

Six karst conferences that include engineering problems have been organized in Turkey (1977, 1979, 1985, 1990, 1995 and 2000), two in Iran (1983 and 1998), and a few in China. Over the past two decades similar conferences were organized in Belgrade (Serbia), Malaga (Spain), Dzershinsk and Ufa (Russia), Besançon (France) and Neuchâtel (Suisse).

Many articles relating to the building of different structures in karst have been presented at a number of international and national conferences all over the world, particularly in different hydrogeological and engineering-geological publications: Dam building, leakages, failures and impacts is presented as a subchapter in 'Karst Hydrogeology and Geomorphology' (Ford and Williams, 2007); Dams and Reservoirs on Karst—is an entry in the 'Encyclopedia of Caves and Karst Sciences' (Editor, Gunn, 2003); and in 'Karst Aquifers—Characterization and Engineering', Chapter 13 (Editor, Stevanović, 2015).

Water resources problems in karst are centuries old. The most frequently encountered are related to water supply, flood control, irrigation, deforestation, hydropower utilization, mining, tailings pond construction and protection of the environment. With increasing demands on water resources and energy in karst regions, an important issue is how to keep the balance between the need for regional socio-economic development and the conservation of nature? This problem is often much more complex in karst than in non-karst regions because each karst region is unique, the nature of environmental changes is unpredictable (often occurring very rapidly), and very similar conditions are seldom found in different regions. Socio-economic development in karst regions, particularly in rural areas, is much slower than in other areas. For instance the average income of farmers in the karst regions of southwest China is only 60% of the average Chinese farm income.

Due to these facts, interdisciplinary cooperation of all those who have common interests in karst is necessary. In spite of different professional backgrounds the problems related to karst phenomena have united engineers and scientists in one specific science known as – ***karstology***.

In the last decades of the 19th and first decades of the 20th century, the general approach in karst investigation was primarily scientific and theoretical. However, because karst terrains produce a variety of resources, intensive construction of different man-made structures began in the second part of the 20th century. Various types of water resources development have been carried out in many karst regions of the world. But the road to success has been paved by numerous problems, sometimes by failures. Dams with seepage beneath the foundations, reservoirs that were impossible to fill, water outbursts into tunnels, mines and other underground excavations, leakage from mine tailings and waste disposal facilities, thousands of induced sinkholes in urban and reservoir areas, and a wide spectrum of unexpected environmental problems have created a specific and important engineering branch – ***engineering karstology***. The philosophy of engineering karstology can be defined with one simple phrase: **expect the unexpected.**

The term ***engineering karstology*** ('inzhenernoe karstovedenie') was used for the first time by G.A. Maximovich at the Molotov (Perm) Karst Conference, Russia, 1947 (Tolmachev and Leonenko, 2011). This term includes karst geology, as the basic science, as well as hydrogeology, hydrology, engineering geology and geotechnique. With the increasing knowledge and understanding of the nature of karst, including development of new investigation methods and geotechnical remedial technologies, the number of failures considerably decreased during the second half of the 20th century. However, karst is still far from being accepted as a friendly environment for the building of large structures, particularly dams and reservoirs.

## 1.5  DAMS AND RESERVOIRS IN KARST

### 1.5.1  Historical Review

From time immemorial the people living in karst regions have had to cope with two contrasting misfortunes: floods and droughts. How to prevent floods in rainy periods and how to keep water available at the surface during dry periods of the year was a key question. Various types of water conservation and protection structures have been constructed in many karst regions of the world, ranging from small and simple reservoirs (Fig. 1.4) up to enormous structures.

**Figure 1.4.** Small ancient reservoir in crushed (pulverized) dolomite, eastern Herzegovina.

A number of ancient engineering structures still exist in the Chinese karst. A water conservation project to serve an extremely karstified area in the province of Guangxi, the Lingqu Canal, was constructed in the period of the Qin Dynasty (221 BC-206 BC) and is still operational for irrigation and shipping (Lu, 2012). The 533 m long Siloam Tunnel in Jerusalem, dated ~ 700 BC, is one the oldest engineering structures built to tap a karst spring (Ghion Spring, mentioned in The Bible; Frumkin and Shirmon, 2006). The main outlet of the Atashkadeh karst spring zone (Firuzabad, Iran) is protected by a 17 m deep circular man-made structure constructed during the Sassanian Empire (229 AD). In the Mediterranean region many such structures date from pre-Roman times. However, building modern engineering structures in karstified rocks is relatively new. In 1436 CE three karst springs were tapped and their water transferred by an 11 km long canal for the water supply of Dubrovnik (Croatia).

As part of a large reclamation project in the Dinaric karst of Herzegovina, a number of karst-specific structures such as the Klinje Dam were constructed in the period 1890–1900. For the water supply of Trieste (Italy) two tapping galleries, 650 m and 250 m long, were constructed between 1898 and 1911 at the large karst Timavo Spring (Galli, 1999). Ancient dams to prevent the clogging of large ponors by flood-born detritus were constructed in many karst areas of the Dinarides, Taurides and Helenides (Fig. 1.5).

Significant hydrogeological investigations, including the development of new exploration methods and treatment of caverns and lesser highly permeable zones, were undertaken by the Tennessee Valley Authority in U.S. The TVA territory covers more than 100.000 km² and includes portions of the seven states along the Tennessee River. The earliest works on this project, including improvements, go back as far as 1922 and possibly even earlier because dam building in the region started in 1910. In the period, 1940 to 1945, extensive grouting and cut-off wall works were performed in the karstified terrains. There were very significant developments in exploration methods, in drilling technology, and in pressure grouting in karstified rocks (TVA-Tennessee Valley Authority projects, 1949).

**Figure 1.5.** Ancient dam to protect a ponor against clogging, Peloponnesus, Greece.

Following World War II and beginning in the early 60s in the Dinaric karst (former Yugoslavia) a number of dams have been successfully constructed in extremely karstified limestone formations.

In the karst of China (3.443 million km²) 32 large reservoirs have been constructed, each with a storage capacity of more than 100 million m³, plus several thousand smaller-sized reservoirs (Chengije, 1994). The Guanting Dam (45 m high) was the first one constructed in the karstified area of Hubei, Northern China (Lu, 2012). According to Kutepov et al. (2004) a substantial number of dams and reservoirs in Russia are also situated on karstified rocks.

Russian engineers first encountered serious karst-related problems at the end of the 19th century during construction and operation of railways in the Volga River basin and in the Urals (Tolmachev and Leonenko, 2011). In the first part of the 20th century modern grouting methods became very important means of protecting structures constructed in karstified terrains. The French and ex-Yugoslavia karst schools earned particular merit for their great contributions to the development of engineering karstology, particularly dams and reservoirs. A number of major engineering structures were also built on karst in Spain, Switzerland, Turkey and Iran (Milanović, 2004, 2015).

In the karst areas, different types of dams have been developed to adapt to particular karst hydrogeological conditions. Due to the hydrogeological singularity of karst, the underground karst water flows through large karst channels (solution conduits), appears at the surface as concentrated points of outflow (spring and estavelles), disappears into the underground in concentrated infiltration points (ponors (or sinkholes) and estavelles), may appear on the sea floor in the form of submarine springs, or discharge at the surface from deep siphoning channels ('phreatic loops'). In response to the local hydrogeological and hydrological conditions in karstified rocks, unconventional barrier structures have been developed: circular and semicircular dams around the openings of ponors and estavelles, and underground dams in the form of plugs, cutoff walls and grout curtains in flowing water karst channels.

Figure 1.6 shows an ancient circular dam that functioned as a mill for maize and grain, constructed around a ponor ($Q > 1$ m³/s) in the banks of the Trebišnjica River, Herzegovina. The purpose of the dam is to use the power of water as it sinks into the ponor to grind maize. The dam is equipped with all of the facilities found at modern dams: water gates, pressure pipes, and wooden turbines to drive the mill stones.

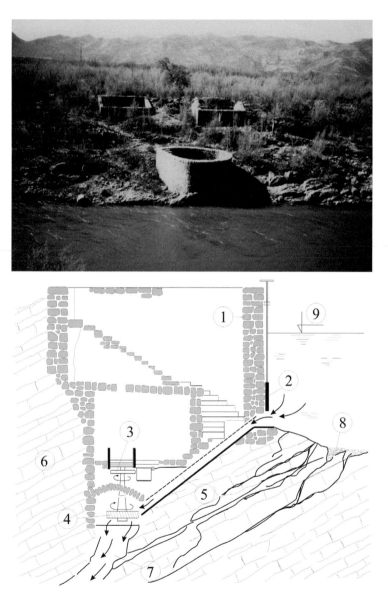

**Figure 1.6.** Herzegovina. Circular mill structure, photo and cross-section. (1) Dam wall, (2) water gate, (3) mill stone, (4) wooden turbine, (5) pipe under pressure, (6) karstified limestone, (7) karst channel (ponor), (8) karstified river bottom, (9) river level.

## 1.5.2 Failures, Risk and Hazard

In many karst regions of the world, deep and narrow canyons have been carved into the carbonate rocks by fluvial entrenchment. For many dam designers this jaws-like morphology was the key property when selecting the dam site. In the first half of the 20th century understanding of karst and karstification was in its infancy from the engineering view point. A significant number of dam and reservoir failures occurred. In many of these cases, the reservoir never filled up despite intensive remedial works, ending in abandonment: examples include the Anchor Dam, Hales Bar, Wolf Creek and Upper Mangum dams (U.S.A.); Liverovići and Vrtac (Montenegro); Montejaque and Maria Christina (Spain) and Apa (Turkey). Hales Bar Dam (1905–1913, USA) was abandoned after more than 30 years of unsuccessful remedial works. The Montejaque Dam has also been abandoned due to similar experiences. Many other dams in karst are operating, but with huge leakage losses. In the karst area of Guangxi Province (China) no less than 644 reservoirs suffer from excess seepage (Yuan, 1991).

Dams and reservoirs are among the most significant and hazardous manmade structures constructed with the intention of improving complex surface and underground water conditions. In general, dams are considered to be dangerous structures. Dam failures can cause massive destruction and loss of life when they occur. However, in the case of dams situated in karst, the risk can take a quite different form. Catastrophic failures, meaning physical collapse of the dam, are very rare, the exception to the rule: destruction of the local population and environment due to dam failure in karst has not been reported. The dam foundations in karst mostly consist of carbonate rocks which have strong mechanical bearing capacity in spite of the fact that the rock mass is intensively karstified.

Instead, dams in karst are risky structures due to the distinctive nature of karst, which is the development of solution channels within the rock. In spite of comprehensive investigations and extensive repair works, many structures built on karst have experienced problems with leakage at the dam and/or losses from the reservoir. This is the reason why some reservoirs can be only partially filled and others fail to fill. Difficulties during dam construction and operation are mostly the consequence of this specific kind of karst porosity, which can be described as a geotechnical problem, but is one with strong impacts on ecosystems and biodiversity in the catchment in which the dam is situated. Technical problems at dams and reservoirs may result not only in the form of empty reservoirs but also in modifications of the water quality and quantity, including impacts on the underground and surface hydrologic regimes. Change in water regime can have great impacts on the flora and fauna, including many consequences of a socio-economic nature, and may possibly cause serious trans-boundary problems.

Undoubtedly dams are constructed with the best of intentions to be positive interventions in nature, which actually happens in most cases; however, the many consequences of such interventions, positive or negative, cannot always be predicted, especially in the karst.

On the other hand this does not mean that dams and reservoirs can't be successfully constructed in areas where karstified rocks are predominant. A number of large dams and huge reservoirs are constructed and successfully operating in Bosnia and

Herzegovina, Montenegro, Croatia, China, France, Honduras, Iran, Spain, Thailand, Turkey, the U.S.A. and many other countries.

In spite of very comprehensive and often time-consuming investigations, including the most up-to-date research methods, acceptance of risk when undertaking engineering works in karst areas is unavoidable. The risk component is impossible to eliminate even by extending the investigation program. It can be reduced to an acceptable level but never totally eliminated. Where any form of failure cannot be tolerated, the resources allocated to the investigation and design stages should be adequate. *Any restriction of funds during investigation phases usually lead to erroneous conclusions and finally to expensive failures.*

# CHAPTER 2

# CHARACTERIZATION OF KARST AQUIFERS

## 2.1 KARSTIFICATION

*Karstification is a complex process that generally consists of two mutually connected processes: first, dissolution of carbonate rocks to create conduits or other voids, then mechanical collapse and erosion as a consequence of surface and groundwater kinetic energy.* It means karstification is a coupled process of chemical dissolution – corrosion and physical destruction – erosion.

The recorded history of interest in karst began long ago. A brief summary is presented here. The first known concepts on the nature of karst were formulated in the time of the Assyrians, Greeks and Romans, approximately 1000 BC (Herodotus, Olinius Secundus, Ptolomeus Claudius). Karst investigations were mentioned in Chinese documents from the Tsin period (316–265 BC). Based on detailed field investigations, the first example of a systematic scientific approach was the work of the Chinese administrator and caver, Xu Xiake (1586–1641, Fig. 2.1) who investigated about 300 caves in central and southern China.

Detailed description of karst features in Slovenia was given by Valvasor (1689). During the past 100+ years, many karstologists have focused predominantly on dissolution as the key or defining karstification process. At end of the 19th and beginning of the 20th century interest in karst increased tremendously. Among the many scientists involved in the development of modern karst science, Cvijić, Grund, Katzer and Ballif are prominent. Austrian engineer Ballif focused on engineering problems in karst, including construction of the first dam in the Balkan region – Klinje Dam in Herzegovina in 1896. He organized the first water gauging stations at karst springs, ponors and estavelles. In the second part of the 20th century dissolution as the key karst process was studied in further detail by Sweeting, 1972; Bögli, 1960; Sokolov, 1962; Ford, 1968; Lu, 1972; Jennings, 1971; Gams, 1974; White, 1977; Song, 1979; Yuan, 1981; Dreybrodt, 1988; Ford and Williams, 1989; Mijatović, 1990; and Palmer, 1991.

In many cases corrosion is the sole initial process, under predominantly laminar flow conditions. In the example shown in Fig. 2.2(A) the initial karst process was

**Figure 2.1.** Chinese karstologist Xu Xiake. Monument in front of the International Research Center on Karst, Guilin, Guangxi, China.

linear dissolution along a discontinuity such as a bedding plane or joint fracture. When a karst conduit is enlarged enough (about one cm in many cases), laminar flow will be replaced by turbulent flow where the hydraulic gradient is sufficient. Mechanical erosion processes may then be coupled with dissolution and the karst conduit size increased rapidly (Fig. 2.2B). In many cases tectonic activity, together with corrosion and the kinetic energy of turbulent groundwater flow, create huge caverns (Fig. 2.2C). The final stage of karstification occurs when the energy of the groundwater decreases to a minimum, usually as a result of lowering of the water table. As a consequence, plugging of conduits in any seasonally or temporarily saturated zone is a frequent process (Fig. 2.2D). Due to the changing groundwater dynamics introduced by

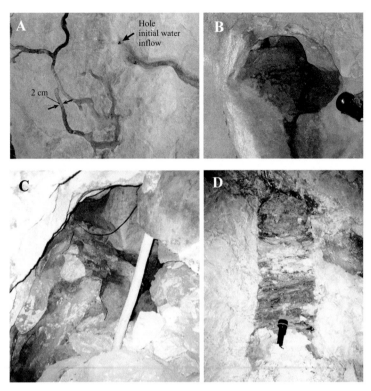

**Figure 2.2.** Genesis of karstic features. (A) Initial karst corrosion conduits along a discontinuity plane; (B) Karst conduit enlarged by the mutual energy of corrosion and mechanical erosion. (C) A large cavern (intercepted by a grout curtain) that was created by tectonic, corrosion and erosion processes. (D) An example of a karst conduit plugged during the final stages of karstifications.

construction of dams and reservoirs, such conduits can be reactivated, causing a large number of engineering problems.

## 2.2  SPECIFIC STRUCTURAL FEATURES OF KARST POROSITY

Soluble rocks (carbonates and evaporates) are distributed all over the world, at all altitudes and latitudes. Due to the brittle, fractured and karstified nature of carbonate rocks, their main physical properties from an engineering karstology view point can be defined as: discontinuity, heterogeneity and anisotropy. As a consequence of the lithologic, tectonic and karst evolutionary conditions, water filtering through a karst aquifer may display properties of a ***double porosity, triple porosity or multi-porosity hydrogeological system,*** with flow anisotropy expressed in both the vertical and horizontal directions. Commonly, karst porosity consists of three main types: in conduits, in fractures and through sediments deposited in the large cracks and cavities (Fig. 2.3).

    *A **triple porosity*** hydrogeological system, the most common type occurring in karst aquifers, is clearly expressed in the shape of the spring discharge graph.

Figure 2.4 shows a typical discharge graph for a deep and high developed karst aquifer. The shape of the recession section of the hydrograph is a direct consequence of the structure of the karst porosity.

Section **1** in Fig. 2.4 characterizes the outflow from caverns and conduits where turbulent flow prevails. Conduit systems, particularly master flow channels, are often under hydrostatic pressure. This type of porosity is rapidly depleted. Curve **2** is a consequence of the porosity comprised of well connected and partially karstified joints. Flows may be locally turbulent or laminar. Base flows are, mostly, under pressure. Section **3** characterizes the release of seepage water from narrow joints and cavities filled with different unconsolidated sediments. Base flow is in free surface conditions.

It is well known that hydrograph curves tend to plot as straight lines on a semi-logarithmic scale. The slope of each of these lines usually marked as $\alpha$ ($\alpha_1$, $\alpha_2$, $\alpha_3$) to

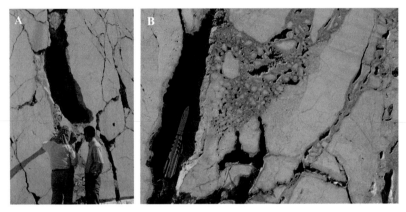

**Figure 2.3.** Karst porosity. (A) Open cavern, metric size; (B) Karstification developed along the joints and caverns filled with breccias with secondary karstification developed within those breccias.

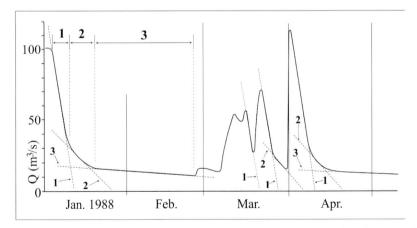

**Figure 2.4.** Spring discharge graph. Three different slopes in the recession curves show the progressive drainage of the different types of porosity. (1) karst conduits and caverns; (2) joints locally enlarged by karstification; and, (3) narrow joints and cavernous spaces locally filled with unconsolidated sediments.

represent the capability of the differing components of the karst to release water, i.e., they are consequences of the different types of porosity (Fig. 2.5).

The coefficient $\alpha$, well known in hydrology from Maillet's exponential function, has been analyzed by many hydrogeologists: Schoeller, 1962; Forkasiewicz and Paloc, 1965; Castany, 1963; Mijatović, 1968, 1990; Mangin, 1975; Drogue, 1980.

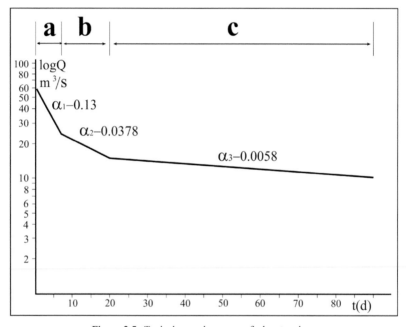

**Figure 2.5.** Typical recession curve of a karst spring.

Coefficient $\alpha_1$ characterizes the outflow from karst porosity that consists of large aperture karst conduits and caverns with turbulent flow and high pressure; coefficient $\alpha_2$ characterizes the emptying of porosity that consists of mechanically opened joints or joints slightly enlarged by solution, with mixed turbulent and laminar flows; and coefficient $\alpha_3$ characterizes outflow of 'dripping' water (seepage) from constricted joints, from intergranular porosity in the host limestone, or from sediments deposited in karst caverns and enlarged joints.

In some cases this multi-porosity can have consequences in the differing pumping capacity between closely spaced wells. It is not unusual for a well with discharge of 1 or 2 L/s and a large drawdown to be only a few meters away from a well with a capacity 100 L/s (1:100) and with negligible drawdown.

According to its origin the karst porosity can also be either ***primary*** or ***secondary***. There are four broad categories: epigenic (created by meteoric water); syngenetic (vuggy porosity and travertine, created during deposition and diagenesis); hypogenic (created by water rising from great depth, frequently mixed close to the surface with meteoric water); and cavernosity, a consequence of ore fluid intrusion into the soluble rock.

***Vuggy porosity*** represents a specific type of syngenetic porosity. Common sizes of vugs range between a few millimeters and a few centimeters, usually between 2 and

30 mm (Fig. 2.6). In some cases there are much larger vugs with ellipsoidal shapes and lengths of 100 to 300 mm in the direction of the longer axis.

In many cases this type of porosity is developed within single limestone beds or other layered deposits (Fig. 2.7A), or in massive limestone structures such as reefs (Fig. 2.7B). The vugs may be partially connected, partially isolated or fully unconnected with neighbors. In sections with isolated vugs, in spite of high porosity (sometimes 5–15%) water pressure tests may indicate a very low to impermeable rock mass (Lu < 1). In these sections grout mix consumption is very low also. However, if vuggy rock is tectonized and the vugs enlarged by dissolution to attain interconnected (sponge-like structure) permeability, the grout consumption will be high to very high.

Vuggy types of porosity are present and frequently encountered at a number of the dam sites in the Asmari limestone in Iran. Ten vuggy zones with different thicknesses

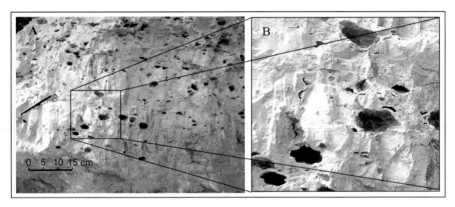

**Figure 2.6.** Different size of vugs, between 0.2 and 30 mm in diameter.

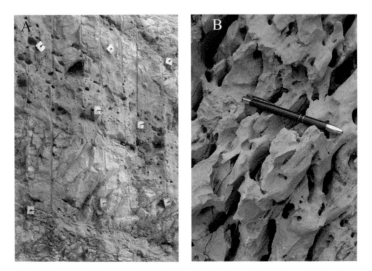

**Figure 2.7.** Limestone layers with vuggy (A) and spongy like (B) porosity.

(from 1 to 10 m) were registered in investigation boreholes at the Seymareh Dam site, for example. The permeability of this syngenetic type of karst may vary over a wide range. In solid limestone vugs are closed and in spite of very high porosity (5–10 times higher than average karst porosity) permeability is very low to impermeable (Lu < 1). The weakest parts of the rock mass are found in the contact zones between vuggy and crystalline limestones. These zones can be heavily corroded and the possibility of significant flow along them is very high. Particularly high permeability and grout mix consumption occurred in the fractured vuggy sections at Seymareh.

***Discontinuities*** are the main prerequisite for development of the secondary porosity. The carbonate rock mass is brittle, composed of different lithological varieties of limestone and dolomite with different thicknesses of beds. Commonly this rock mass is dissected by different discontinuities into a number of blocks. On the base of regional tectonic forces (particularly, new tectonic movements) orientation of regional structures generally follows the direction of megastructures. In general, karstification is a structurally controlled process. However, locally karstification tends to be developed along lesser secondary or tertiary discontinuities and drain toward the nearest local erosion base level, not necessarily trending in the direction of the statistically primary system.

Secondary porosity consists of bedding planes, joints and faults enlarged by karstification, caverns and karst channels. In thick-bedded pure limestone, karst features are developed mostly along joint discontinuities cross-cutting the carbonate beds (Fig. 2.8A) and along the bedding planes (Fig. 2.8B). The most common karst systems are a consequence of solution along both of these discontinuities (normal joints that pass through beds in different directions, and bedding plane joints) with mutual close connections.

**Figure 2.8.** Common types of karstification. (A) along joint discontinuities perpendicular to the bedding and (B) along the bedding planes.

In many cases, due to renewed tectonic movement, karstification proceeds in several successive phases. After dissolution concludes in a given phase a proportion of the opened joints can become plugged with sediments, breccia fragments or precipitated carbonates (Fig. 2.9).

The resistance of breccias depends on their matrix properties. If the matrix consists of precipitated minerals or well lithified clay, the breccia becomes resistant to rapid and turbulent flow under high pressure. However, if clay deposits or breccia matrix in joints and caverns below the dam or at a grout curtain route are unconsolidated and plastic, intensive erosion will begin on their becoming exposed to high water pressures.

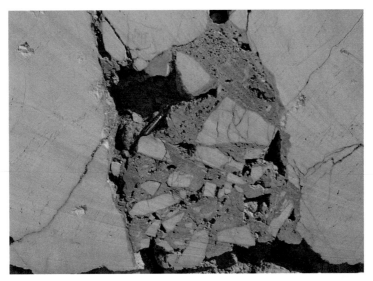

**Figure 2.9.**   Cavern filled with fragments of tectonic breccia in a partially-lithified carbonate and residual clay matrix.

## 2.3  ROLE OF EROSION AND EVORSION IN KARST FEATURES DEVELOPMENT

Karstification is a result of water penetrating into permeable and soluble rock masses. The process starts by a chemical action, i.e., dissolution of the soluble rock along narrow discontinuities. Development of initial dissolution channels along interbedding joints is presented in Fig. 2.2.

The chemistry of this process is explained in detail by many authors and is not the subject of this publication. However, processes of ***erosion and evorsion*** come into action as cavities enlarge and must be analyzed as important factors in the enlargement of karst channels and caverns. According to 'A Lexicon of Cave and Karst Terminology with Special reference to Environmental Karst Hydrology' APA, 1999, **evorsion (attrition)** is defined as "mechanical erosion by rotating or swirling water carrying sand, gravel, cobbles, or boulders in suspension or as bedload". As this process can have an important role in engineering karstology, a short explanation is presented.

With enlargement of fractures and other discontinuities by corrosion, apertures are increased, karst channels (conduits) are formed and the minimum size requirements for turbulent flow are met and then exceeded. The fast flowing water entrains and carries breakdown blocks, bedload and lesser clasts which widen the channel walls by abrasion. Due to swirling current effects, large volumes of rock fragments become trapped in erosional bowls or siphon sectors of conduits, are crushed, spun and rounded by milling action (Fig. 2.10A). After 30 to 50 hours of rapid rotation, instead of blocks and pieces of irregular shape and with sharp edges, rounded pebbles remain (Fig. 2.10B).

This process was monitored during the excavation of a long investigation adit to the site of the underground power plant Dabar at Herzegovina. After blasting, the cavern was opened at the adit invert (adit floor). Blasted limestone blocks and lesser fragments with sharp edges were deposited in the cavern. Following heavy rain there were large and turbulent (3–5 m³/s) discharges of water from it. The rocks trapped there were subject to strong swirling flow and milling action for between 40 and 50 hours. During this period all pieces were rounded, transformed into pebbles or cobbles (Fig. 2.11). In toto, about 25 to 30% of the volume of each limestone was eroded

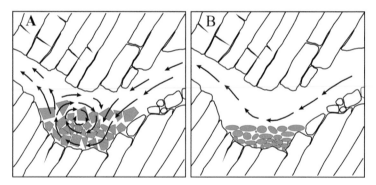

**Figure 2.10.** Rock clasts and blocks in a strong swirling current (A); and remaining pebbles deposited after flow velocity decreases or stops (B).

**Figure 2.11.** PP Dabar, investigation adit. Crushed limestone pieces transformed into sub-rounded to well-rounded pebbles by turbulent flow and swirling currents in a cavity—the evorsion (attrition) process.

and transported away as finer particles. It is clear that mechanical erosion coupled with dissolution have a major influence in formation of karst channels and caverns.

Understanding the processes of erosion and milling is important when plugging caverns that have turbulent water flow through the grout curtain routes and beneath dam foundations. The usual approach is first to fill the cavern with crushed rock. After it is fully filled, immediate injection of grout is mandatory to provide a compacted concrete plug and to prevent evorsion. Without grouting, the fill material can be subject to milling (evorsion) and the efficiency of the crushed rock fill will rapidly decrease, with possibly destructive erosional effects on the conduit walls and in the already grouted rock mass.

## 2.4  DEPTH OF KARSTIFICATION

The depth of karstification depends on many different conditions but, first of all, on the thickness of the soluble rocks, their structural features and the history of the evolution of the karst aquifer. On the basis of thousands of borehole logs, groundwater measurements, geoelectrical soundings and direct observations in underground excavations, a simplified zoning of highly developed karst is presented in Fig. 2.12. The approximate boundary between the epikarst and the vadose zone (base of epikarst) and between vadose zone and solid rocks (base of karstification) generally are not changeable. However, the 'boundary' between the saturated (phreatic) area of the karstified rock and water-free rock (vadose) zone above it can fluctuate from a few meters up to 100 or more meters. Geoelectrical data are based on five year measurements of 300 points in the same area; a generalized graph of the electrical soundings is presented in the figure. One important specific property is that the shape of electrical sounding graphs can differ at same location at different times due to the different groundwater levels (Arandjelović, 1976).

In the case of thick carbonate deposits karstification tends to decrease exponentially with depth. The results of permeability investigations from 146 deep boreholes in eastern Herzegovina showed that karstification decreased with the depth according to the exponential law:

$$\varepsilon = 23.9697\ e^{-0.012\,H}$$

where $\varepsilon$ is the index of karstification; H is the depth in meters; e is the base of natural logarithms; a and b are coefficients.

Graph 2 in Fig. 2.12, and the above equation are based on results associated with registered solution porosity along the boreholes. Five meter increments were tested along the 165 boreholes. Results are subdivided into the four categories: (1) Empty caverns identified by drilling and logging; (2) Increments with undefined high permeability (Lu > 100); (3) Increments with great permeability (planed pressure of 10 atmospheres could not be achieved; and (4) At this group are increments with permeability greater than 30 Lu. Values less than 30 Lu have not been considered because they could be the result of joint system permeability, not karstified or negligibly karstified, Milanović (1981).

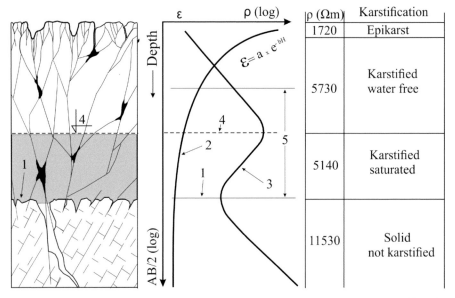

**Figure 2.12.** Generalized relationship between the intensity of karstification (ε) and the depth (Milanović, 1981), and the graph of electrical sounding in highly developed deep karst (Arandjelović, 1976). (1) Base of karstification; (2) Graph of karstification intensity by depth; (3) Graph of electrical sounding; (4) Groundwater level; (5) Fluctuation range of groundwater level and electrical resistivity.

According to this law the index of karstification will normally approach a minimum value at depths between 250 m and 300 m, roughly. In his analysis of karst aquifers in the Norwegian marble, Lauritzen (1988) confirmed this exponential law (karstification coefficient).

However, karst features can be detected locally much deeper than shown in the above diagram. Usually deep karstification follows deep discontinuities. The minimum level of groundwater measured in some piezometers, about 350 m below the surface, indicates presence of karstification at greater depths. Large caverns have been discovered in deep mines at depths between 500 and 1000 m or more. A large number of vadose shaft cave systems deeper than 1000 and 1500 m have been discovered and explored all over the world.

A giant water-filled cavern in marble under gneisses in the Rodope Mountain was discovered in depth of about 1200 m. Water temperature ranges between 86°C at the top, 130°C at the bottom, Sebev, 1970. In an area of Montenegro, the presence of caverns was established in boreholes at a depth of –2,236 m, or 1,600 m below modern mean sea level.

According Yuan (1981) in the Sichuan Basin, China, a 45 m high cavern with artesian water was discovered by borehole at a depth of 2,900 m (2,400 m below the sea level). In Hebei Province a borehole entered a cavern with stalactites at a depth of 1,000 m below sea level. According to the latest information the deepest karst features now detected in China are at depths of about 8000 m (pers. comm., Yuan, 2016).

## 2.5 HYDROGEOLOGICAL PROPERTIES OF KARST AQUIFERS

### 2.5.1 General Hydrogeological Properties

Karstified rock masses are extremely complex entities and may produce a great variety of geomorphological, hydrogeological, hydrological, hydraulic and engineering geological conditions.

The basic concept of a representative *Elemental Hydrogeological Volume (EHV)* for an aquifer does not exist in karst. The most important hydrogeological parameters (hydraulic coefficient, transmissivity) cannot be defined precisely either by field measurements or in the laboratory. In general the EHV, known as the 'effect of scale', usually must be taken to include the entire aquifer karst. The principal hydrogeological property of karst can be defined as 'hydrogeological heterogeneity'. The regional porosity of karstified rocks is low, in general between 0.5% and (rarely) 3% or more. Boreholes with permeability only 1–3 Lu, situated close to karst channels, are very common.

Underground water flows in karstified rocks are turbulent and (mostly) occur through large conduits. The average *flow velocity* varies within a range from 0.002 to 60 cm/s. However, in some cases (under a pressure of 12 bars) the measured velocity in karst outlet channels is more than 16 m/s.

The length of underground flow paths is sometimes more than 30 km. Infiltration of up to 80% of precipitation is common. Concentrated infiltration through large ponors (dolines or swallow holes) is one of most important characteristics of mature karst drainage. The sink capacity of single ponors can vary from a few liters per second up to 120 m³/s. Fluctuation of the discharge reported in the sample springs varies from 0 to 200 m³/s (Ljuta, Montenegro), or from 3 m³/s to 380 m³/s (Buna, Herzrgovina) and 6 to 515 m³/s (Frio, Mexico), respectively. As a consequence of the nature of karst, thousands of collapses have occurred in urban areas and at the bottoms of reservoirs, large numbers of tunnels have been flooded, leakage from reservoirs is frequent and, in some cases, very expensive dam structures have been abandoned or operated at only part of their designed capacities. In karst areas, site specific investigation, designs, technical construction and remediation approaches all have to be applied for successful construction and operation of large engineering structures at the surface or underground.

From the engineering karstology point of view the most common destructive influence of groundwater is a consequence of: the large solutional capacity of the rock (particularly in evaporites); the rapid erosion of any unconsolidated filling deposits in caverns and joints and erosion of carbonate conduit walls; - the great kinetic energy of underground flows; and the enormous hydraulic pressures created in periods of full aquifer saturation (piston-flow effects), including water-hammer and air-hammer effects that can arise due to rapid increases in groundwater levels. As a consequence of very fast water table rise, the measured velocity of air flow in some piezometers has been as great as 15 m/s. Seismicity triggered by rapid flooding in karst aquifers has been recorded in different karst regions of the world.

## 2.5.2  The Base of Karstification

In dam geology establishing the ***base of karstification*** is very important (Fig. 2.12). For any dam projects in karst terrain it is one of the most frequently required pieces of information. Physically, the base of karstification is an underground zone separating upper, highly karstified rock from the lesser- or non-karstified rock below. The base of karstification can be conceived as a specific geological structure, limited to karstic rocks only.

The base of karstification is usually not a simple surface between karstified and non-karstified (solid) carbonate or evaporite rock. Rather, it is transition zone below which the rock mass is not karstified or is considerably less karstified. This does not mean that locally, along deep faults or fault zones, steeply dipping bedding planes, etc., karstification cannot be deeper. To define an effective base of karstification is not an easy task. Borehole rock core quality, groundwater levels, results of water pressure (Lugeon) tests and geophysical methods (geoelectrical sounding) are measures frequently applied to determine it.

Generally speaking, the depth of karstification can vary widely, from a few tens to a few hundred meters. However, locally the rock mass can be karstified much deeper. In dam design procedures and protocols the base of karstification usually represents the basal contour line of the proposed grout curtain. In reservoir banks, the topography of the base of karstification is one very important indicator for detecting possible seepage zones.

For determining how deep to make a suspended grout curtain (its lower contour) or deciding where the ground water divide between reservoirs and lower erosion base levels such as neighboring valleys, lies in less karstified or underlying impervious rock, establishing the base of effective karstification has a crucial role to play.

Karst aquifer base flows are situated in the deepest zone of effective karstification.

## 2.5.3  The Evolution of Karst Aquifers – Epigene and Hypogene Karstification

The hydrogeological properties of karst aquifers clearly distinguish them from aquifers found in other rocks. However, due to the wide range of different basic hydrogeological properties and complexity of the evolution in karst aquifers there is no universal karst hydrogeological concept or model like that of Darcy flow for sand aquifers. One of the well known, original conceptions of the karst aquifer is that of orogenic deep karst, as known in the classical Dinaric karst. Figures 2.13 and 2.14 show simplified cross-sections *along* the principal direction of base flow and *perpendicular* to it respectively.

One of the key requirements to solve practical problems in engineering karstology, particularly in the construction of reservoirs, is to understand the evolution of the host karst aquifer. This depends on a variety of geological conditions and groundwater processes mutually interacting in space and time: the lithological composition (presence of soluble rocks); tectonic activities (particularly new tectonic movements); quantity of water; topography (distribution of erosion base levels); and the proportional relationship between fluvial and corrosion processes in the region. Generally,

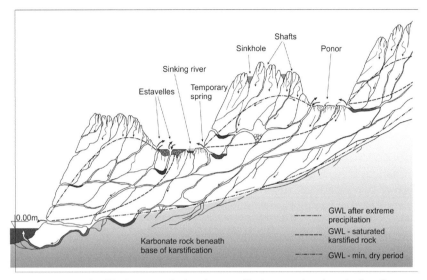

**Figure 2.13.** Simplified concept of a deep karst aquifer. Long section along the flow lines.

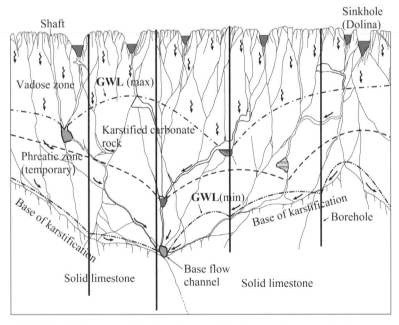

**Figure 2.14.** Simplified concept of the deep karst aquifer. Cross-section perpendicular to the general direction of underground flow.

infiltration of meteoric water from the surface into the soluble rock plays the principal role in dissolution (**epigene karstification**): however, deep rising water can be also a source of intensive karstification in some localities (**hypogene karstification**).

Before karstification began in the Dinarides karst networks of river valleys were well developed at the surface. Karstification transformed the regional hydrogeological and hydrological system created by the pre-existing topography in two steps:

- Disorganization of pre-existing river networks to create separate catchments and independent karst aquifers in the underground, and
- Integration of many initially independent small karst aquifers into larger single aquifers.

As a consequence of differential movements in separate tectonic blocks succeeded by intensive karstification, the surface drainage network thus becomes disorganized or completely destroyed. Groundwater flow replaces permanent surface flow, and only temporary flow in dry valleys, karst poljes and uvalas remains at the surface. The numbers of temporary springs, ponors and estavelles are created in closed depressions. Original catchment areas become divided into a few separate independent catchments. New watersheds are not in accordance with the surface topographic divides. In some cases the groundwater flow follows directions not in accordance with the trends of the ancient system of surface drainage.

As consequences of tectonic movement and rapid fluvial erosion on non-karst rocks the regional erosion base level becomes deeper and deeper. The trend of the groundwater flow is directed first to the nearest erosion base level outlet (local base level). As soon as this is reached, the process continues with adjustment to another, lower local base level until the sea coast itself is reached, sea level being the final base level of erosion. The processes of karstification reach into the deeper parts of the rock mass as the flow system adapts itself to the lowest discharge zone. Karstification can connect them together into a very complex hydrogeological entity provided that the process intensity increases from higher levels to the lower erosion bases.

One of the best examples of such karst aquifer evolution was caused by the Messinian salinity crisis in the Mediterranean Basin (Bakalowicz et al., 2007; Fleury et al., 2007, 2008, 2013; Bakalowicz, 2014). A lowering of the Mediterranean Sea level (base level for surface and underground water) caused lowering of the spring outlets, compelling adaptation of the underground karst systems to new lowest points of discharge. The maximum fall of sea level was approximately 1500 m. A simplified presentation of the consequent karst evolution along the Mediterranean Sea coast is presented by Flury (2013) in Fig. 2.15.

In reverse, subsequent rising of the marine base level resulted in the flooding of at least the lower parts of the conduit systems and the overflow of the aquifers at higher elevation, potentially rejuvenating abandoned karst structures (Bakalowicz, 2014). Due to deposits of thick marine blue clays carried by the large rivers some lower spring discharge points were completely blocked.

One consequence of the Messinian salinity crisis is the occurrence of a large number of deep coastal springs and submarine springs along the Mediterranean coast (e.g., Estramar and Vaucluse, near the French coast; Almirous, Heraclion (Crete); Chekka (Lebanon); and numerous submarine springs along the coast of Turkey, Greece, Syria, Israel, Italy, Spain and Libya. More than 50 submarine springs have now been cataloged along the Adriatic coast in the Dinaric karst region.

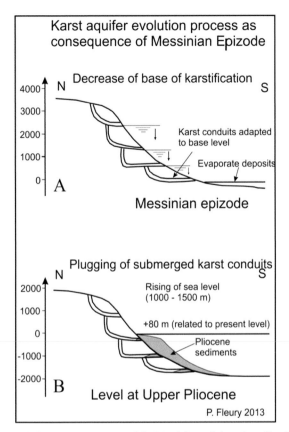

**Figure 2.15.** Karst evolution as a consequence of the Messinian salinity crises. Simplified presentation. Flury, 2013.

In some specific cases, due to new tectonic movements, thick impervious rocks became a barrier between the erosion base level and the carbonate rocks. In the case of the Ombla aquifer (Croatia) deep siphonal conduits were created (Fig. 2.16). Conduits created in the earlier phases of evolution are presently part of the vadose zone. In this zone only vertical groundwater flows are usually active, but in some cases after extreme rains, the former horizontal flows become temporarily reactivated.

From the engineering karstology point of view, particularly the water tightness of reservoirs, the nature of the karst evolution at higher elevations and far from the sea coast plays an important, sometimes crucial, role. Because Neotectonic movement and water flow through the soluble rocks are both continuous processes over geological time the ***evolution of the karst aquifer is thus itself a permanent process***. The intensity of this process is variable in space and time.

Figure 2.17 presents two selected stages in the evolution process. There are strong evidences for downstream shifting of karst springs, i.e., of karstification shifting into the deeper zone. The exact understanding of the evolution of a given karst aquifer thus is important in engineering karstology, particularly the definition of the base of karstification below a dam foundation and below a reservoir bottom.

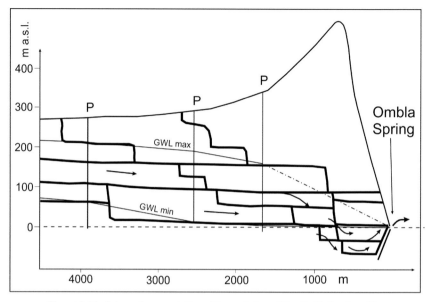

**Figure 2.16.** Schematic presentation of the evolution of the Ombla karst aquifer.

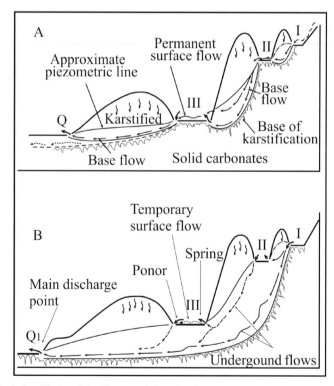

**Figure 2.17.** A simplified model of karst evolution showing the transformation of three separate karst aquifers into a single one.

The evolution of the karst aquifer in the Haditha Dam area (Iraq) was analyzed by Kondratyev (1979). He presented a picture of several different steps in a complex karst evolution, which includes fluvial erosion and karstification from Middle Pleistocene until the Holocene at three different sections across the Euphrates River (Fig. 2.18). A particularly important role in this process was played by a less permeable, brecciated bedrock layer. However, in the final stage of karst canyon entrenchment the brecciated layer was locally disintegrated, allowing for deep ground water penetration.

An analysis of local karst evolution was undertaken to construct a tailings pond at 4000 m a.s.l. in the Andes of Peru (Fig. 2.19). The approximate base of karstification was established from a number of piezometers and analysis of groundwater level fluctuations.

***Hypogene karstification.*** The origin of a number of well-known caves cannot be explained by solution of meteoric water and flow by gravity. It is established that many investigated and analyzed examples of karstification close to the surface were formed by water of deep origin, often mixing with meteoric waters. By analysis of the cave passage patterns, morphology, water chemistry and sediments the term ***hypogene caves*** has come to be applied as a relatively new term in karstology. The proposition that hypogene karstification is a consequence of ascending water (cold or thermal) rich with $CO_2$ or $H_2S$ has been proven and accepted after detailed analysis of many caves around the world (Palmer, 1991; Palmer and Palmer, 2009; Ford and Williams, 2007; Ford, 2009; Klimchouk, 2000; Audra et al. 2009; Dublyansky and Dublyanski, 2000; Derybrodt, 1988; and many others).

Caverns with no relation to surface topography, known as caverns of hydrothermal origin, were known centuries ago. A number of Zn/Pb and gold ores are located within karstified carbonate rocks by processes of hypogene (hydrothermal) origin. In many mines, large open and sometimes water-filled hypogene caverns have been

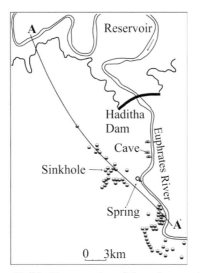

**Figure 2.18.** Euphrates River, Haditha Dam area, Iraq. Schematic presentation of the karst evolution (Kondratyev, 1979), from Lykoshin, A.G., Molokov, L.A., and Parabutchev, I.A., 'Gidroproekt' Moskva, 1992.

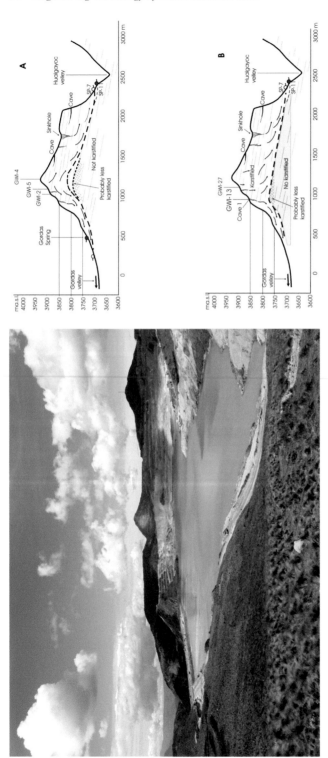

**Figure 2.19.** Cerro Corona tailings pond, Peru. Cross-sections A and B show the approximate boundary between high karstified and not karstified rocks.

discovered by interception: e.g., Pine Point (Canada), Yatela (Mali), Olkusz and Pomorzany (Poland), Berg Auks (Namibia), Daguanshav iron mine (China) and others. In some instances the ore was deposited into pre-existing caverns. In many other cases, the cavern genesis was syngenetic with the ore deposition. In Trepča Mine (Serbia), at depth of 900 m, a huge cavern with large gypsum crystals is formed at the contact between a Pb/Zn ore body and carbonate rocks. In the Berg Auks mine (Namibia), Zn/Pb/V hydrothermal karstification and ore deposition was controlled by vertical fractures in dolomite during several periods of karstification, starting probably around 600 Ma (Misiewicz, 1988); the largest cavity had dimensions of 200 m x 100 m x 120 m in height (Fig. 2.20). The base of the deepest cavern is 600 m below the surface.

There are a number of well known hydrothermal caves in Hungary, particularly inside the Buda Hills (Budapest, the right bank of Danube River). Jewel Cave, Wind Cave, Lechuguilla and Carlsbad Caverns are famous examples in the United States, and El Zacaton in Mexico.

At many dam sites on soluble rocks (limestone, dolomite and gypsum) the problem of karstification is (partly) the consequence of hypogene karstification (Fig. 2.21). The presence of thermal water has been reported at many dam sites: Salman Farsi and Karun IV (Iran), Francisco Morozan (Honduras), Višegrad (Bosnia), Haditha (Iraq), Hammam Grouz (Algeria), Salanfe (Switzerland), Samanalawewa (Sri Lanka), Pengshoui (China), Pueblo Vijeo (Guatemala) and Chicik Dam (Uzbekistan). In some cases, hazardous deep karstification was discovered although no discharges of hot waters were detected at the dam site itself, but were known to occur nearby: Berke Dam (Turkey) and Špilje Dam (FYUR Macedonia).

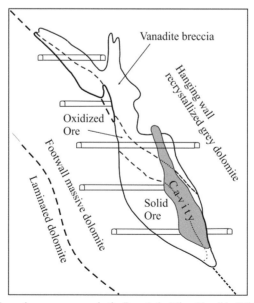

**Figure 2.20.** Large hypogene cavern in the Berg Auks Mine, Namibia (Misiewicz, 1988).

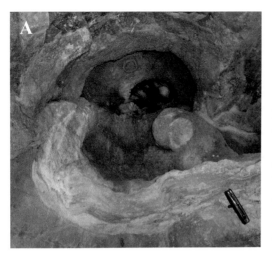

**Figure 2.21.** Karst channel (A), and cavern (B), created by hydrothermal water (hypogene karst features).

Usually, the quantities of ascending hypogene water at the dam sites are not large, but their influence may be considerable. To apply successful measures against leakage at the dam sites and from reservoirs the correct diagnosis of origin of the cavern is essential for the correct design of the grout curtain (alignment, depth, inclination, and properties of the grout mix).

In particular the existence of karstic features of hypogene/epigene origin at dam sites makes construction of watertight barriers much more complicated. From the practical dam engineering view point two general questions arise in cases of hypogene karstification at dam sites:

1. How to determine proper alignment of grout curtains (length, inclination and, particularly, their depth); and,
2. How to select the proper technology for grouting and cavern plugging above and beneath the water table.

For selecting the proper depth and alignment of the grout curtain it is essential to determine the three-dimensional spatial locations of the karst channel zones with principal ascending $CO_2$-rich water or hydrothermal $H_2S$-rich flows. To create watertight reservoirs all of these flow routes should be blocked by the grout curtain. Regarding the chemistry of these flows, the grout mix has to be resistant to pressure and the destructive action of $H_2S$ and its products. Plugging weakly consolidated cave sediments below the water table is an important and technically complicated task. Cases where there is high pressure and a rapidly rising flow of hot mineralized water are particularly sensitive. In the case of the Salman Farsi dam site, Iran (Fig. 2.21) two deep rising waters are present: thermal and cold. Sometimes in very closely spaced grouting boreholes, some yield hot water and others cold water.

If upward flows are present at dam sites, the design and construction of effective grout curtains requires considerable flexibility. In practice, the ability to modify and adapt grouting technology on the basis of *in situ* geological findings in the underground is essential. The verticality of rising flows and the (mostly) unknown depths and

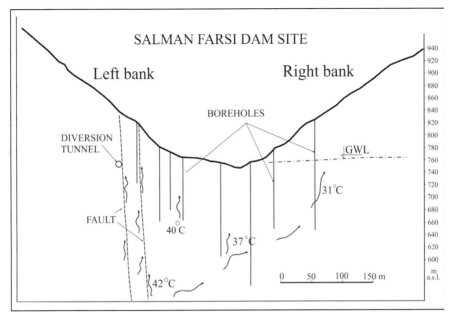

**Figure 2.22.** Upward groundwater flow detected at the Salman Farsi dam site, Iran.

locations of their sources are very unfavorable conditions because the normal purpose of grout curtains is to prevent mainly sub-horizontal groundwater filtration below the dam. Grout curtains should be deep enough to block rising flows. In the Salman Farsi case, the different stages of hypogene karstification are partially connected with the local history of fluvial erosion.

High temperature and presence of $H_2S$ in galleries and caverns is dangerous for people. For example, temperatures of 42°C (Salman Farsi Dam) or 55°C (Pueblo Viejo Dam) are indicators of deep hypogene karst activity.

### 2.5.4 The Hydrogeological Hierarchy of Groundwater Flows in Karst

The hydrogeological properties of karst aquifer are distinctive: a hierarchic system of underground flows (vertical and horizontal), the existence of solution caverns and conduits, the presence of vadose flow (underground rivers), but also phreatic flow (hydraulic systems under pressure, creating piston flow effects), siphoning lakes in the vadose zone, the huge kinetic energy of underground flows, the rapid filling (saturation) and emptying of the aquifer.

*The hydrogeological hierarchy* of karst conduits (hydrogeological singularities) is created and governed, first of all, by the multiple porosity system. In the hierarchies, the top ranking conduits are open karst channels and caverns where turbulent water flows prevail. Typically these channels are found at different levels. In upper parts (the aeration zone) they may not be active or only temporarily activated. All base (principal) flows, known as underground rivers, belong to the top ranking conduits in the hydrogeological hierarchy. In China alone, more than 3,358 well developed karst groundwater river systems with an aggregate discharge in the dry season of 420 x 10[8] m³ have been cataloged (Lu et al., 2006).

One of the best examples of a well-established hierarchic underground channel flow system in the world is the Disu System in Guangxi Province, China. The total length of the Disu underground flow system is at least 241.1 km (Chen, 1988). The length of the principal channel flow line is 57.2 km (Fig. 2.23). Twelve lower hierarchy (secondary and tertiary) underground flows merge into it. The recorded discharge of the underground river at Disu Spring varies from 4.03 m³/s minimum to a measured maximum of 544.9 m³/s.

The Trebišnjica aquifer, Herzegovina, similarly consists of a number of connected channel flows of different sizes. The length of the principal flow line (master channel, No. 1 in the hierarchy) is more than 35 kilometers. Together with secondary and tertiary flows the total length of this system of underground conducting channels is more than 150 km, with a maximum capacity of about 300 m³/s. The recorded rates of discharge have varied between 2 m³/s and 300 m³/s, with an average flow of 80 m³/s. In dry periods of the year, the secondary and tertiary flows in the hierarchy are not active. At times of full aquifer saturation all underground channels become active and the master channel comes under pressure flow. Due to insufficient conduit capacity the system is then not able to accept all water and part of it (that exceeding 300 m³/s) discharges into karst poljes to be temporarily stored there as flood water.

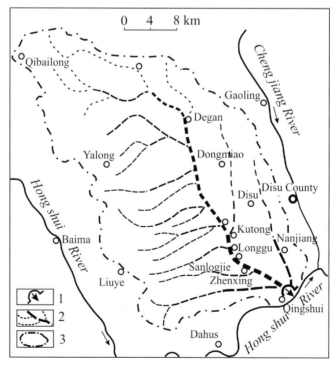

**Figure 2.23.** Sketch map of the Disu Subterranean System (Chen, 1988). (1) Disu Spring; (2) Underground river net; and (3) Border of catchment area.

The underground course of the intensively studied Reka River between Škocjanske Jame (Slovenia) and Timavo Spring (Italy) is about 40 km long with an average rate of flow of 8 m³/s studied by Martel, 1894 and many other karstologists. Figure 2.24 is a graphical presentation of longitudinal cross-section (Peric et al., 2006).

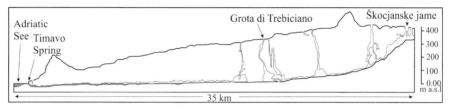

**Figure 2.24.** Reka River, longitudinal cross-section Škocjanske jame – Timavo Spring (Peric et al. 2006).

Another well studied underground case is the Wamme/Lomme karst system in Belgium. This system, 10 km long, recently celebrated one century of investigations (Willems and Ek, 2011). Average discharge of the system (Epave Resurgence) is $Q_{av} = 0.8$ m³/s.

Usually the master flow and secondary channels represent the base level for water transfers throughout the entire karst aquifer. During periods of saturation, such channels are usually flowing under pressure. Next in the hierarchy are fractures and partially karstified fractures, with more or less laminar flow and partially retarding properties. In many highly developed karst aquifers a significant number of the caverns and large conduits are substantially filled with different deposits (clay, clayey sand, sand, pebbles, local fallen rocks and, sometimes, huge blocks) which have intergranular porosity. In periods when the karst aquifer is fully saturated, so are these deposits. In dry periods when there are base flow conditions only in the channels, large amounts of ground water are still retained in these cave sediments.

In many deep and highly developed karst aquifers, the role of some of the conduits has changed during the evolution. After the aquifer discharge points (the springs) have descended to lower erosion base levels, the upper channels lose any master flow and stay permanently above the saturated zone. These channels become cave systems with temporary flow (flood overflows) and, sometimes, are partially or completely plugged with cave sediments and other deposits.

The recession curves of the spring discharge can be used to simplify the understanding of the karst aquifer hydrogeological hierarchy. The recession curve is a complex function which represents the discharge regime of the aquifer, i.e., the hierarchy of the active karst conduits. Different slopes in a recession curve correspond to different discharge coefficients, i.e., to the different hierarchic levels of karst water flow, as noted. In general: the large karst conduits represent the top of the flow hierarchy (master flows); joints enlarged by dissolution represent the second hierarchic level; and fractures and caverns filled with cave sediments represent a third hierarchic level. All of them are mutually connected into one well-organized hydrogeological flow system.

**Cross flows.** One of the distinctive properties of karst groundwater is that flows circulating in two or more directions may cross each other. An example is presented in Fig. 2.25. Tracer tests from two closely spaced boreholes B-1 and B-2, at the right bank

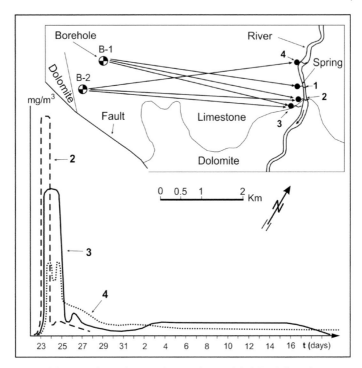

**Figure 2.25.** Crossing underground connections and their breakthrough curves.

of the Trebišnjica River show some transverse circulation through hydrogeologically independent drainage zones. Tracer from borehole B-1 appeared at Springs 1, 2 and 3 but not at the closest spring, 4. Tracer injected in B-2 appeared at all monitored springs, including Spring 4, meaning that some of the underground conduit flows crossed each other. Breakthrough curves show some rapid and concentrated underground flows, particularly between B-1 and Spring 2. The long duration of the dye outflow 'tail' at Springs 3 and 4 is due in part to the dye only slowly passing through the perforations in the piezometer pipe.

## 2.5.5 Hydrogeological Zoning of Karst Aquifers

Due to the great variety of karstified rocks it is practically impossible to formulate a unique hydrogeological model for karst aquifers. One general approach based on a large number of measured data is presented in Fig. 2.12.

There are a number of parameters that can be used as general descriptors for karst aquifers. From an engineering point of view, the hydrogeological and hydrodynamic properties have a crucial role. Based on these properties the most simple and generalized model can be defined as a hydrogeological zoning of the karst aquifer by depth: (1) Aeration zone (uppermost), with mainly an infiltration role (with or without shallow epikarst zone); (2) Zone of seasonal fluctuation of ground water level (temporarily aerated or temporarily saturated); (3) Permanently saturated zone, including deep siphon sectors and 4. In special cases, zones with upward underground flows (mostly hydrothermal water).

*The aeration (vadose, unsaturated) zone, or zone of 'vertical' water filtration* is the zone between the land surface and the water table. In highly developed karst, the depth of the aeration zone can be more than 1000 m. Thousands of caves, including almost all tourist caves, are situated in this zone.

At many dams and reservoirs in karst, properties of the aeration zone play an important role. If the aeration zone is always permanently deep beneath the surface, the effectiveness of any underground sealing technology will usually be problematic. In such cases surface prevention measures have a much better chance of success. However, if during rainy periods the water table rises to the surface (i.e., to the floor of a reservoir) then standard preventive measures at the surface are not always successful alone. To protect them, additional structures may be necessary.

A number of dams and reservoirs in karst areas are located in karst valleys or karst poljes where the water table lies deep beneath their bottoms. This means that below the dam sites and reservoirs there exists an unsaturated zone with highly variable properties. As examples, the known water table depths below the surface at some dam sites and reservoirs are: Wujiangdu Dam (China), 210–220 m; Lar (Iran), more than 200 m; Hutovo (Herzegovina), about 100 m; Akköpru (Turkey), 100–116 m; Abolabas (Iran), 55 m; Tangab Reservoir and Havasan Dam (Iran), 30 m; Ourkiss dam site (Algeria) 50 m, Vrtac (Montenegro), 50 m below the reservoir bottom and Boqaata Reservoir (Lebanon) more than 20 m. Due to the existence of such aeration zones beneath dams and reservoirs, some have been abandoned (for instance, the Havasan Dam) and new locations selected.

Caves and karst channels with strong natural air flow are common in the aeration zone. Air currents from deep boreholes are also common phenomena in karst. This indicates that there are good connections between karst channels in the zone, including connections with the open land surface. Movement of air and other gases in the aeration zone is a consequence of:

- Difference of air temperature between the underground and the surface (strong air currents are found in a large number of caves and from many boreholes in karstified rocks); Change of atmospheric pressure. Rise of water tables, expelling the air from flooding caverns and conduits; and
- Artificially, by injection of air or gas into the conduit system.

Due to saturation by water, large volumes of air may be compressed or forced out of caverns and channels. Aeration of karst porosity is registered in some piezometric boreholes in the form of rapid air currents. In some piezometers and aeration pipes measured air-stream velocities are up to 15 m/s.

*The zone of seasonal fluctuation of ground water level can be defined as a temporary aeration zone.* It is hydrogeologically the most dynamic part of karst aquifer, particularly during rainy seasons. Within this zone, there are karst channels, caverns, free surface flows (underground rivers), phreatic flow under pressure, and siphoning lakes.

The lower limit of the zone is the upper level of the permanently saturated zone as determined by minimum ground water levels in piezometric boreholes. The upper limit is indicated by maximum water levels in the piezometers. Both limits may have extremely irregular forms, with great differences occurring over short distances.

Due to the overall low porosity of karst aquifers, the transformation of the aeration zone into a seasonal aquifer and its emptying afterwards is very rapid, and during a rainy season may change permanently in space and time.

***The permanently saturated zone.*** Depending on the evolution of a karst, the depth and volume of its basal, permanently saturated, zone can vary from negligible to huge. In some cases the master channels are limited to this zone. Permanently saturated siphon zones behind large springs include Vaucluse and Estramar in France, Ombla in Croatia, Orahovačka Ljuta in Montenegro, Buna in Bosnia and Herzegovina and many others.

## 2.5.6 Epikarst

The uppermost zone of karstified rocks is called the ***epikarst*** or *subcutaneous zone.* These terms are relatively new and without precise definition. In the Glossary and multilingual equivalents of karst terms (UNESCO/FAO, 1972), they are not mentioned. The epikarst zone consists of different features and fissures created by solution and weathering in the uppermost few rock beds, without or with any cover such as residual soil, terra-rossa, bauxite, sand or thick masses of alluvium. Ford and Williams (2007) define epikarst as "a weathered zone of limestone that lies immediately beneath the soil, and it gradually gives way to the main body of the vadose zone that comprises largely unweathered bedrock". According Palmer A. and Palmer M. (2009): "A network of solution fissures and pores tends to develop in upper parts of soluble rock that is exposed at the surface, with or without a soil cover. This zone is called the epikarst".

In many cases the epikarst is exposed at the surface without any soil cover (bare karst). Approximate thickness (depth) ranges between 10–20 m, only locally deeper. Most openings (karst channels, dolines, sinkholes, karren and fissures) drain the water immediately down to the saturated zone. A negligible part of the water can be perched or retarded in the unconsolidated sediments deposited in the karren feature or bottoms of dolines (sinkholes). Thickness of epikarst cover can be a few meters, up to hundreds of meters. Epikarst is mostly a groundwater-free zone. However, it can be temporarily saturated (e.g., in temporarily flooded karst poljes) or can be permanently saturated. For instance, in the Dzershinsk area (Russia) the epikarst is covered by 50–70 m of sandy-clayey sediments. In this area the groundwater level is permanently a few meters, or a few ten of meters above the (buried) epikarst zone. Consequently, in high developed exposed karst the epikarst has no particular hydrogeological meaning, or particular hydrogeological properties.

For the epikarst zone sometimes other terms are used: pedogenetic weathering, pseudo-endo karst, subcutaneous zone, ghost-rock karstification, isovolumetric weathering zone, epi-fissure-karst-zone, instant epikarst, faraday cage, dirty karst, superficial karst, and some more. However, the meaning of term epikarst is simple: ***The epikarst is the upper, most densely karstified zone of the rock mass, with or without soil cover, permanently or temporarily saturated or permanently water free.***

In some cases ***paleokarst*** relief is also detected, at different levels depending on the number of emergences. Paleo-epikarst or fossil epikarst is best preserved beneath bauxite deposits. In the case of the Dinaric karst, the bauxite lies unconformably over karstified Jurassic limestone and beneath karstified Cretaceous carbonates. In some cases this unconformity is detected by boreholes at depths of a few hundred meters.

Locally, where there is thick residual soil on the surface and deposited into the karren and karstified joints, much water may be retained in storage as a perched epikarstic aquifer (temporarily or permanently).

From the dam and reservoir point of view, epikarst means particularly weak rock and requires different sealing approaches. In the case of dam foundations, the epikarst needs to be completely removed (excavated), and sometimes to be consolidated. In the case of reservoirs, epikarst contains a great variety of uncertainties that are in many cases hidden beneath the top soil or thick deposits of other sediments. Usually the epikarst zone beneath the reservoir floor and along its banks has a great erosion and drainage capacity and, locally, is potentially unstable. A number of reservoir failures have occurred during first filling or operation because karst conduits (sometimes large) are hidden below the reservoir bottom, in the banks or in the vicinity (Keban, Logan Martin, Mosul, Hammam Grouz, Mavrovo, Lar, Samanalawewa, Bileća, Hutovo and many others).

## 2.5.7 Dynamics of Air Flow in the Aeration Zone

In karstified rocks, caverns and conduits above the water table contain air at atmospheric pressure and other gases, chiefly carbon dioxide. Under natural conditions, strong air currents have been recorded in many caves. One of the best examples is Vjetrenica (Windy) Cave in Herzegovina (Fig. 2.26). Velocities of air currents in this cave range between 5.1 and 9.8 m/s (Pers. information, Lučić, 2009). In the cave under the Nullarbor Plain, Australia, velocity of ~ 40 km/hr has been recorded at narrowing points of some big and long conduits.

In many different karst regions of the world strong air currents have been noted in many piezometric boreholes. Directions of air currents are often reversible, depending on the temperature and pressure differences from conditions prevailing at the surface. Obviously, the presence of air currents is one of the important properties of the unsaturated zone in karst.

**Figure 2.26.**   Vjetrenica Cave. Air current measurements.

One of the well-known characteristics of the karst aquifer is its rapid saturation. During the wet season, the reaction of an aquifer to heavy rain may be quickly expressed. The aquifer reacts rapidly, sometimes as fast as 10 to 15 hours and in some cases even faster, 4 to 6 hours after rain. As a consequence, filling of karst channels and caverns can be extremely fast, as well as increasing of general groundwater levels, sometimes as much as 90 m in 24 hours. The rising water forces out the air and 'occupies' the space in caverns (Fig. 2.27).

Enormous amounts of air have to be evacuated from karst channels and caverns in the vadose zone. Due to abrupt rise of the water level, the air currents out of some piezometric pipes have reached a velocity of 15 m/s, as noted, and water pressure in conduit flow increased up to 15 bars.

When the water level in a karst aquifer rises, pulsating air currents have been recorded in some piezometers. According to laboratory tests and measurements done at the Hutovo Reservoir by Kovačina et al. (1978), over a cycle of 30 minutes air first rushes out from the piezometers (exhales) and then rushes back in (inhales). This means that changes of pressure and vacuum in karst channels were occurring frequently (Fig. 2.28). Laboratory tests confirmed two different kinds of forced air flow: air squeezed from turbulent water flow in the form of bubbles, and air forced out by the rising water table.

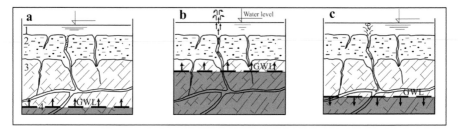

**Figure 2.27.** Destructive role of air under pressure in the reservoir bottom.

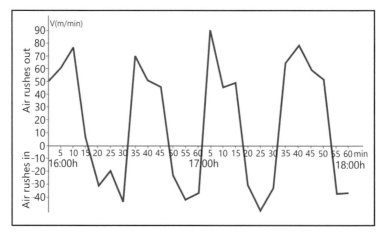

**Figure 2.28.** Diagram of cyclic changing of air current direction from piezometric boreholes due to rapid water table rising. Hutovo Reservoir.

The relationship between the volumes of air expelled from the cavernous spaces at Hutovo Reservoir and rates of water level rise are presented in Fig. 2.29. Test results show there is a much stronger relationship between the air volumes expelled by rising water and rate of groundwater rise than between air bubble expulsion and rate of groundwater rise (dh/dt) in the borehole.

However, during rapid increases in turbulent water flow significant volumes of air become entrained in bubbles. Release of this air from the water towards the aeration zone provokes the micro-pulsations of pressure recorded in the piezometers (Fig. 2.30).

Usually there is a strong functional relation between the volume of air forced out of the karst voids and the rate of increase in the ground water flow. The frequency and intensity of air velocity changes proportionally to the depth of water in piezometric pipes.

In solving the problem of water tightness of reservoirs where the bottoms are covered with alluvial sediments, the greatest difficulty appears to be dealing with the air captured in caverns and conduits in the underlying karstified bedrock.

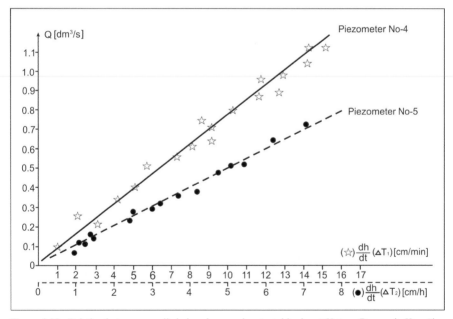

**Figure 2.29.** Relation between expelled air volume and water table rise at Hutovo Reservoir, Kovačina et al., 1978.

## 2.5.8 Hydraulic Properties of Karst Water Flow

The complexity of hydrogeological parameters within karst aquifers requires special approaches. Because the majority of karst flows are turbulent flows under pressure through rough-walled conduits, with some local losses, the groundwater movement can be expressed by a square law of resistance. According to a large number of analyzed examples, in the case of hydraulically well-connected karst conduit systems the flow is proportional to the square root of the piezometric difference between two selected points.

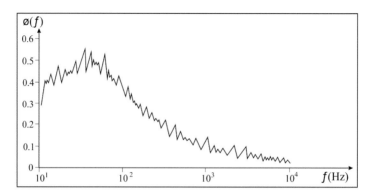

**Figure 2.30.** Micro-pulsation of water level in piezometer, Skopljak and Kovačina, 1978.

Recent knowledge concerning the hydraulic properties of karst aquifers has been gained by application of spring discharge data, groundwater fluctuation data in standpipe piezometers, pumping from karst siphons, and by injection of huge amounts of water into the aquifers.

***Laminar flows become turbulent*** (along rough conduits) when the critical Reynolds Number increases to 2300. Many authors have analyzed the properties of underground flow through karst conduits and their relation with Reynolds Number (Lehman, 1932; Krasnopoljsky, 1947; Mangin, 1975; Mijatović, 1990). Because the conduits are geometrically extremely irregular and rough, their shape and diameter frequently changes over short distances, and both horizontal and vertical changes of direction are frequent also, the eddying type of torrential turbulent flow is common in karst channels. This type of flow is much more destructive than laminar or turbulent flow in man-made pipes.

Depending on karst aquifer hydrogeological properties relations between the amount of turbulent and laminar flows can be different. Based on tracing tests Atkinson (1977) concluded that active storage in diffuse component (in fine fractures) is 30 times greater than in phreatic conduits (turbulent component, 'quick flow'). However at some well developed aquifers turbulent component is much greater, which means retardation capacity is poor and the aquifer empties very fast (temporary springs).

Experiments have shown that the critical Reynolds number which indicates the end of laminar flow and the start of turbulent flow is 2300, as noted. However, some investigations have shown that if there is great roughness in the conduits, the transition may occur at Reynolds numbers much smaller than 2300. The minimum diameter of a conduit for laminar water flow to become turbulent depends of many factors, particularly on velocity of flow. According to many authors, the minimum ranges approximately between 5–15 mm. However, in the case of low flow velocities, laminar flow can be maintained even in conduits apertures of a few hundred millimeters.

Swirling and eddying turbulent flow is dangerous during any underground works in karst, particularly during grouting and plugging large karst conduits and caverns. Turbidity increasing at springs is also one of consequences of turbulent flows.

Mijatović (1990) analyzed the role of friction losses on the water flow through karst conduits (Fig. 2.31).

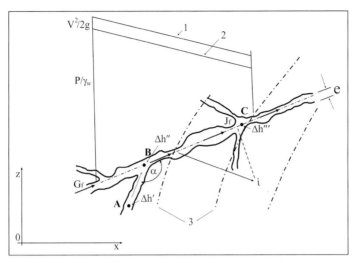

**Figure 2.31.** Schematic presentation of water movement along a karst conduit. (1) Energy Grade Line; (2) Piezometric Line; (3) Equipotential Lines; **i** – Gradient, Mijatović, 1990.

The energy of the karst water flow is first influenced by friction losses between points A and B. At point B the sum of losses increases due to a change in the direction of the flow. Around point C local losses appear as consequences of changes of karst conduit diameter. These losses appear at different points in karst systems and are variable. Accordingly, the hydraulic gradient in a complex system of individual conduits can be defined only if friction losses and the other losses of different origin are considered simultaneously. ***Piston effect (propagation of hydraulic pressure)*** is one of the key properties in karst aquifers. During rapid filling the major conduits, particularly inclined or sub-vertical sections, become saturated in a short time. Because a large proportion of the main karst conduits then come under hydraulic pressure, the piston effect changes the ground water regime over large distances. To investigate the hydraulic properties of a deep karst aquifer, large scale tests were undertaken by injecting a very large amount of water into the Ombla karst aquifer during the dry period of the year (Fig. 2.32).

Approximately 3.3 million cubic meters of water were injected over 24 hours into a large ponor zone along one section of a dry river bed. The distance between this zone and the aquifer outlet (Ombla Spring) is 12–15 km. Before injection, spring discharge was 10–12 m³/s. About 35 hours elapsed between the start of water injection (point A in Fig. 2.32c) and the first response of the spring (point B). Tracer tests showed that to travel the same distance the labeled wave required 137 hours. The piston effect was registered in piezometric boreholes hydraulically connected with base flow (P1 and P3). The correlation obtained between spring discharge and water level fluctuation in the two piezometers is very high: r = 0.979. Clearly, pressure propagation (the piston flow effect) along karst channels under pressure is confirmed as one of the important properties of the karst aquifer.

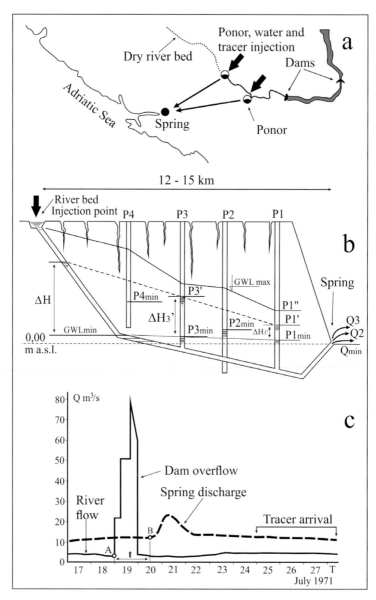

**Figure 2.32.** Confirmation of piston flow effects by large scale experiment. (a) Layout with position of injections (ponors) and the spring; (b) Schematic cross-section between the injection and discharge points; and (c) Simultaneous graphs of water injection (dam overflow) and the spring discharge.

## 2.5.9  Complexity of Karst Aquifers

As a consequence of the varying thickness and deformation of soluble rock formations, the geologic structure and movement of tectonic blocks, and the history of the aquifer evolution, a great variety of karst aquifers have developed, ranging from small and hydrogeologically simple up to huge and complex. To construct dams and reservoirs in such a complicated environment is a complex task. To find good storage space and a watertight dam site requires lengthy investigations.

The example in Fig. 2.33 presents a part of a fully developed karst aquifer. This very complicated karst aquifer was created in a complex geologic structural setting and has adapted to the lowest erosion base level. In general, any well developed karst aquifer can be represented as a system of reservoirs (surface and underground) and

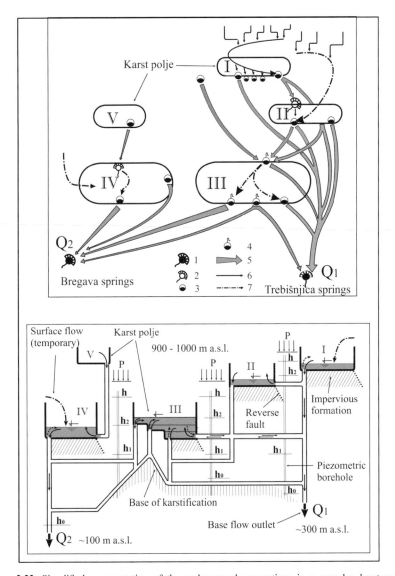

**Figure 2.33.** Simplified representation of the underground connections in a complex karst aquifer. Upper—schematic plan of underground connections: (1) Large karst spring (end of regional master flow); (2) Karst spring (end of local master flow); (3) Ponor; (4) Estavelle; (5) Approximate route of key underground flows; (6) Permanent surface channel flow; (7) Temporary surface channel flow. Lower—schematic vertical projection of the hydrogeological connections between the karst poljes.

conduits. In many cases the complex karst system will be composed of a few distinct hydrogeological subsystems. The flow regime in the system and subsystems depends on the groundwater level (**h**) and flow (**Q**). Both parameters are variable in time. Points for their measurement should be carefully selected. The location of piezometers is particularly important. The quality of the data depends on the distance between the array of piezometers and base flow (master channel). If a piezometer has a direct hydraulic contact to this flow zone the quality of data will be high.

One of the most complex karst aquifers is schematically presented in Fig. 2.33 in plan-form and as a vertical projection. Six karst poljes (labeled I–VI) are located in an area of approximately 1550 km², with differences of elevation between 900 m and 450 m. This hydrogeologically complicated aquifer discharges through two concentrated spring outlets at approximate elevations of 100 and 330 m above sea level. The combined average annual discharge from both outlets is approximately 100 m³/s (20 + 80 m³/s).

This eastern Herzegovina karst system consists of six large karst poljes that are connected together by deep underground conduit systems only. There are no surface channel connections between them. About 90% of the water flows through the deep karst conduits during dry periods of year. During rainy periods the entire karst aquifer becomes fully saturated. When its transmission capacity is exceeded large volumes of water are discharged into the karst poljes. In those periods the underground and surface waters become one, hydraulically well connected, 'body'. Depending on the intensity and distribution of precipitation (P in Fig. 2.33) the hydrogeological properties of this 'body' change abruptly in space and time. The piezometric line fluctuates between **h** and **h1**. The capacity and hydrogeological function of the estavelles change also.

Based on a large number of measurements, the hydrogeological relationship between waters sinking in the upper karst polje (I) and recharge in the lower polje (III) is defined by the graphical correlation presented in (Fig. 2.34, Avdagić, 1987).

This correlation was crucial for management of optimal power production at a hydro-electric power plant located about 50 km downstream from the sinking zone in karst polje (I).

From the above it is evident that crucial data for karst aquifer analysis must include aquifer saturation, ground water level fluctuation and spring discharge. Two different sources of karst aquifer saturation are:

**Surface infiltration** as consequences of rain (i.e., vertical percolation over broad catchment areas), measured by installing precipitation stations.

**Concentrated infiltration** through ponors and estavelles, i.e., through karst solution channels in the form of compact masses of flowing water. This regime is measured by water gauging stations installed in front of the ponor opening.

Flow behavior in the case of concentrated infiltration is the consequence of the relation between water level above the karst channel opening at surface (H1) and the saturation of downstream parts of the aquifer, recorded by the water level in piezometer (H2). $Q = f(H1, H_2)$. Four possible (most frequently encountered) regimes are schematically presented in Fig. 2.35.

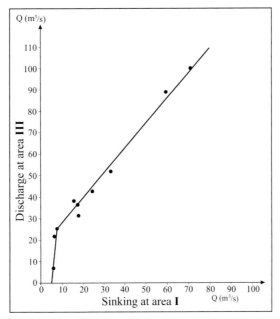

**Figure 2.34.** Correlation between volumes of water sinking in the upper karst polje (I) and discharging in the lower karst polje (III), Avdagić, 1987.

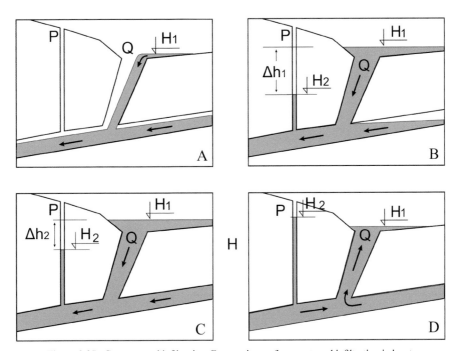

**Figure 2.35.** Concentrated infiltration. Four regimes of concentrated infiltration in karst.

A – Sinking regime in periods of unsaturated or partially saturated karst aquifer – free surface flow underground.

B – Flow under pressure with limited influence from downstream parts of the aquifer. Downstream part of aquifer is partially saturated (H1 > H2).

C – Entire system is under pressure. However, the piezometric level in the downstream aquifer is lower than the water level in sinking zone. Level in piezometer (H2) still is lower than H1 but difference Δh is not as high as in case B. The ponor still 'swallows' water, but its capacity is considerably decreased. If downstream discharge capacity is larger than saturation capacity H2 never equalizes with H1.

D – In this case the pressurized flow is strongly influenced by back up of groundwater in downstream parts of the aquifer because its storage capacity is higher than the discharge capacity of the springs. When downstream channels are completely 'blocked' by water under pressure, the water level in the piezometer increases above the water level in the sinking area (H2 > H1). Because of this there is upward water flow into the sinking zone ponor (estavelle regime). In such cases, occurrence of rapid surface infiltration (extreme rains) usually plays a crucial role.

Well developed karst aquifers usually discharge through single spring outlets or small clusters of springs around the downstream end of the master channel flow. Analysis of the hydraulic properties in the discharge zone can be adapted to the hydraulic law of system discharge under pressure. The same law can be applied in the case of flow along a karst channel. The most frequent relationships between two arbitrary points along the karst channel (or between the piezometric level and spring discharge) are presented in Fig. 2.36.

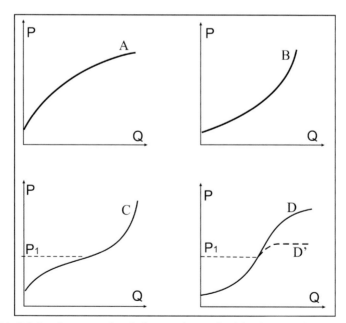

**Figure 2.36.** Relations between spring discharge and water level in piezometer located at base flow, Jaćimović et al., 2015.

The curve in graph A indicates simple overflow. The outlet channel is not totally filled with water.

The shape of curve B indicates flow under pressure. The capacity of the outflow channel is not large enough for all of the water. As a consequence there is rapid increase of the piezometric level but spring discharge increases slowly.

Curve C shows a complex relationship—the transformation from an overflow regime to flow under pressure.

In the lower part of curve D, behavior up to the GWL elevation P1, is a consequence of flow under pressure until water rises to P1; above this elevation additional conduit(s) become incorporated into the system (free flow conduits). In some cases the incorporated conduits (large caverns and channels) have huge storage capacity. In such cases the upper part of the curve becomes horizontal in spite of the tremendous increase in the spring discharge.

Spring discharge depends on the square roots of the difference between the piezometric levels measured at two arbitrary points (spring and piezometer) as presented as curve 1 in Fig. 2.37. During aquifer saturation and ground water level rising, upper conduits become included in the system. In that case curve $2^b$ change direction to $2^d$. The relation Q = Q (P2) does not maintain the trend $2^c$ because the new (upper) karst channels become part of the active conduit system.

The relation between spring discharge and the water level in a piezometer (if the piezometer is in the master flow or very close) is proportional to the square root of the difference in water level between the two, (Hajdin and Ivetić, 1976; Bonacci, 1987), expressed as curve 1 (the so-called 'square rule'):

$$Q = \alpha\sqrt{2g(P1-P0)}$$

Where: α is the approximate aperture of the karst channel,

$g$ – gravitational potential
P1 – GWL in piezometer
P0 – spring discharge

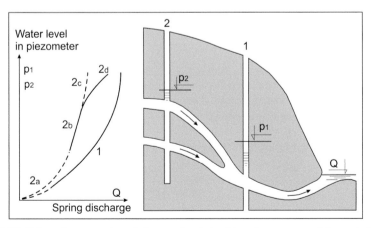

**Figure 2.37.** Schematic presentation of piezometric relationships when upper conduits are included in the karst aquifer system.

If rainfall intensity is so high that the karst channel becomes submerged and flow comes under strong outside influences, the applicability of the 'square rule' becomes problematic.

## 2.5.10  The Origin of Deep Karst Aquifers as Consequences of Coupled Cold/Thermal Flows

Among the crucial questions in karstology are determination of the residence time, depth and genesis of water being discharged at the springs. These are particularly complex where discharge waters have different temperatures. In such cases the standard hydraulic properties and rules have only a limited capacity to explain what is happening in the deep part(s) of the karst aquifer. Since it is difficult to determine the residence time and depth of origin of the water with sufficient accuracy by using only chemical methods (for instance, the Mg/Ca ratio), a comprehensive approach to investigation is necessary: detailed geological analysis, measuring properties of the recharge zones, properties and geometry of accessible underground karst features, hydrogeological monitoring and, particularly important, the temperature regime and isotope composition of water.

A multidisciplinary approach to deep karst aquifer analysis has been applied in the karst of the Kučaj-Beljanica carbonate massif in eastern Serbia (part of the Carpatian-Balkan mountain arch). The massif ($1076$ km$^2$) is completely impounded by surrounding impervious rocks, with the springs developed along the contact. It is assumed that the depth of karstification in the massif is more than 600 m. For analysis, the massif was divided into 15 sub-aquifers and each of them carefully analyzed. For presentation here Krupaja Spring, the most complex example, has been selected (Fig. 2.41). The main Krupaja Spring outlet is a deep siphon channel developed at the contact between the karstified carbonate rocks and impervious Permian sandstones. Spring discharge varies between 0.25 m$^3$/s and 8.7 m$^3$/s, T $= 10°$C–$11.4°$C. Extreme $Q_{max}$ recorded once only is 31 m$^3$/s. The channel has been explored by cave divers down to $-130$ m and is known to extend deeper. Close to the main outlet is a small thermal spring ($Q_{const} = 2$ l/s, T $= 26°$C), and a 400 m deep borehole with a permanent thermal outflow at 5 l/s and T $= 18.5°$C.

The basic data are more than 500 analyses of the $^{18}$O and $^{2}$H, $^{13}$C, $^{3}$H, $^{3}$H/$^{3}$He and $^{14}$C isotope ratios, the ionic and physic-chemical composition of the waters. Samples were taken from all three sources: main outlet, borehole and thermal spring. Analyses were undertaken at laboratories in Serbia, China and Hungary. On the basis of these results, considering all available data, the Krupaja aquifer could be divided into four sections with different flow properties (Vasić, 2017). In Fig. 2.38 the four flow sections are presented schematically in the vertical plane as: (1) A fast flow aquifer (t1); (2) Section with retarded flows (t2); (3) Section with deep and slow filtration (t3); and (4) Very deep and slow filtration (t4). A volcanic rock (dacite) is, most probably, the heating source.

The deepest point of the explored main channel is approximately 150 m lower than the modern spring outlet. It is apparent that this highly developed karst includes at least one phreatic loop 150 m or more in depth. The base flow channels in Section t1 are large, with huge flow capacity. The lag between rain and a response at the

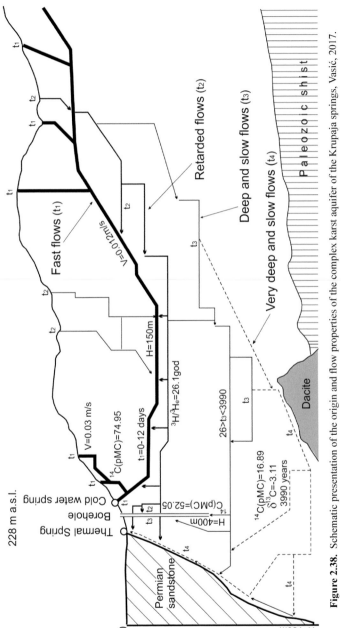

**Figure 2.38.** Schematic presentation of the origin and flow properties of the complex karst aquifer of the Krupaja springs, Vasić, 2017.

main spring outlet vary between 16 hours up to 12 days. The second flow type (t2) is ascribed to less developed karst porosity, plus water 'trapped' in cavities filled with cave sediments (mostly clay and sand). The estimated mean residence time of about 26 years indicates slow filtration, including long periods of groundwater immobility and partial mixing with deep (thermal) water. Flow type t3 lies beneath the cold (i.e., largely meteoric) water, at depths of more than 400 m and is strongly influenced by admixture of thermal water (up to 25°C). According $^{14}C$ isotope analysis, the residence time is up to 3.990 radiocarbon years in the deepest parts of Section t3, and beneath it Section t4 residence times are greater than 3.990 years. The transition zones between these deepest sections are irregular and sometimes mutually coupled without the possibility of being separated.

From view point of dam, reservoir and tunnel construction the transition zone between Sections $t_1$ and $t_2$ is of primary importance. The most problematic rock permeability conditions are usually encountered in this zone. However, if thermal water or any other deep upward flow is present at the site, hypogene karstification is also likely to be important and a possible source of problems.

## 2.6  KARST FEATURES RELEVANT TO DAM GEOLOGY

From the engineering karstology viewpoint the most important karst features are: springs, ponors, caverns, karst channels, sinkholes, estavelles, shafts, caves, sinking flows. In many dam and reservoir projects at least one of these features is present and can be (or has been) the source of failure if not carefully investigated and analyzed.

### 2.6.1  Springs

During construction of many dams and reservoir, particularly underground dams, the hydrogeological properties of karst springs have an important role. In some cases, depending on their characteristics, an entire project can be questionable. A number of the properties of karst springs are quite different from springs in non-karstified rock masses:

- The range between minimum and maximum discharge can be huge, in extreme cases 1:150 or greater.
- Maximum discharge from single outlets can reach 500 m³/s.
- Many karst springs are not permanent, their discharge varying between 0 and 200 m³/s or more.
- Many springs discharge from deep siphon channels.
- Submarine springs are common along limestone coasts and are mostly temporary, but can be permanent, also.
- 'Ebbing and flowing' springs with discharge that increases and then decreases to the starting value over a steady time cycle (mostly during low flow periods) are rare, but are found in the karstified rocks only.

The world's largest permanent karst springs, with minimum and maximum discharges are: Tisu (China) 4 m³–545 m³/s; Frio (Mexico) 6–515 m³/s; Chingshui (China) 4–390 m³/s; Buna (Herzegovina) 2.9–380 m³/s; Trebišnjica (Herzegovina)

2–300 m³/s; Matali (Papua New Guinea) 20 –> 240 m³/s; Vaucluse (France) 4.5–200 m³/s; Coy (Mexico) 13–200 m³/s; Dumanli (Turkey) 25 –> 100 m³/s; Ombla (Croatia) 3 –>130 m³/s; Bunica (Herzegovina) 0.7–207 m³/s and many others with Q$_{max}$ between 10 and 150 m³/s.

Photos of a few large springs are presented in Fig. 2.39.

Discharge of all karst springs is directly dependant on precipitation and rapid saturation of the karst aquifer. The hydrograph of Buna Spring is a good example to show this dependence (Fig. 2.40).

Huge springs with almost constant discharge are rare in karst. However, one of the world's largest springs—Ras-el-Ain in Syria, is a good example. This spring has an average annual discharge of Q = 36.667 m³/s and ranges only between Qmin = 36.00 m³/s and Qmax = 41.894 m³/s. The average annual precipitation in the catchment area of Ras-el-Ain is only 280 mm (Burdon and Safadi, 1963).

**Figure 2.39.** Karst springs: (A) Buna Spring, Herzegovina; (B) Sekerpinari Spring, Turkey; (C) Vaucluse Spring, France, in a dry period; (D) Sopot Spring, Montenegro, an overflow active in rainy periods only.

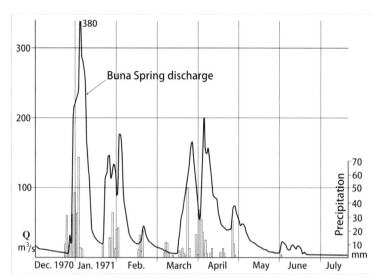

**Figure 2.40.** Buna Spring (Herzegovina). Relationship between precipitation and spring discharge.

Many karst springs dry up during the dry season. These temporary springs belong to the highly karstified type of aquifer with negligible retardation capacity and limited catchment area. For instance, discharge of the springs Sopot and Ljuta in Kotor Bay (Montenegro), range from 0.0 to > 150 m³/s. Discharge of Obod Spring in Fatničko Polje (a potential reservoir) ranges 0.0 to 60 m³/s.

When springs are submerged by reservoir water prediction of their behavior is particularly difficult. More detail about such springs is presented in Chapter 11.

## 2.6.2  Ponors (Swallow Holes)

Ponors are known to cause the most frequent problems for reservoir integrity in karstified rocks. A significant number of reservoir failures have occurred due to them, some projects have been frozen and others abandoned. Ponors can display a great variety of shapes and sizes in reservoir bottoms covered with alluvium and other weakly consolidated sediments, in the bedrock itself around the perimeter of reservoir bottom, or at different levels in the reservoir banks. Their effective apertures can be from a few centimeters (large cracks enlarged by karstification) up to openings a few meters in diameter. Ponors in alluvial bottoms of reservoirs are consequences of karstified bedrock at depths of 1 to 40 or more meters below.

Detection of these features beneath weakly consolidated sediments is extremely difficult even when the most sophisticated methods of investigation are applied. Many ponors that were not active during the preceding natural conditions, become reactivated after reservoir filling. Some of them are only reactivated after many years of operation.

If thick alluvial sediments have low permeability and the ponor zone beneath has only a small sinking capacity, there is not likely to be any surface evidence such as collapses (e.g., the Elati Basin, Aliakmon Reservoir, Greece).

Some ponors with well developed karst channels and a high gradient between the inlet and outlet function as permanent sinks. The transmission capacity of the underlying channels is usually much larger than that of the ponor openings. If such ponors are situated in the reservoir area they are generally protected by special structures (dams and water gates) to prevent leakage, or allowed to function as controls of reservoir overflow: examples include Slivlje Ponor (Montenegro, Nikšićko Polje) with a swallow capacity of ~ 120 m³/s; Biograd Ponor (Herzegovina, Nevesinjsko polje) ~ 110 m³/s; Doljašnica Ponor (Herzegovina, Popovo polje) ~ 60 m³/s; Kovači Ponor (Bosnia, Duvanjsko Polje) ~ 60 m³/s; Vranjača (Herzegovina, Gatačko Polje) > 50 m³/s; Ključki Ponor (Herzegovina, Cerničko Polje) 25 m³/s (Fig. 2.41).

All of these particular ponors are situated at the borders of karst poljes. All are the inlets into large diameter karst channels with huge flow capacities, and therefore never exhibit an estavelle regime.

**Figure 2.41.** Herzegovina, Cerničko Polje. Water sinks into the Ključki Ponor, Qmax ~ 25 m³/s.

## 2.6.3 Estavelles

The function of an estavelle and its regime are well known. Due to its dual function, an estavelle is more dangerous if situated inside the reservoir area. Estavelles can exist only where the discharge capacity of the karst aquifer downstream is not high enough to handle all inflowing water, meaning that the aquifer becomes fully saturated. Depending on the location and groundwater regime, there are two different types of estavelles in general: estavelles that regularly change their function (outflow to inflow and *vice versa*) depending on wet and dry season changes, and estavelles where outflow is much less frequent than inflow (sinking). The latter are active in the discharge regime (as spring) a few days per year, only after extreme rains. In some cases discharge from such estavelles occurs only once every few years.

Disastrous consequences of estavelles have been reported during the operation of many reservoirs. To protect reservoirs against seepage losses, different barrier structures have been constructed at estavelle outlets. In some cases, due to extreme water table rise in the estavelle area, all remediation structures attempted were not efficient and the reservoir was abandoned (Vrtac Reservoir, Montenegro).

## 2.7  TURBIDITY

In natural conditions, after heavy rains the water discharged from karst springs almost always contains suspended solids that consist mostly of particles of varying sizes. The majority are particles of high plasticity such as clay. During periods of slow (laminar) flow some of the particles are large enough and heavy enough to settle on the bottom in lower parts of karst channels. Highly developed karst aquifers incorporate a number of such depressions in which smaller particles, sometimes colloidal size, settle slowly. Consequences of heavy rain are: rapid saturation of the karst aquifer and groundwater level increase; flows become turbulent, and turbidity in spring water increase almost proportionally to these events. If rain occurs after a long dry period when slow (laminar) flow conditions prevailed, the intensity of the turbidity is higher than in the case of more steady high spring discharge. In karst springs with steady discharge ($Q_{max}$:$Q_{min}$ = 5:1 or less) turbidity is less intensive then at springs with high discharge fluctuations. The distribution of rain showers over the catchment area is also very important. If rains fall on areas with shafts and ponors directly connected to master karst flow, turbidity increases rapidly.

In some cases, the turbidity is so extreme that the water becomes non-potable for short time periods.

A few examples are presented here. One instructive case is the increase of turbidity at the deep siphonal Krupaja Spring in Serbia. The vertical channel of the Krupaja outlet has been investigated by divers to a depth of 105 m (Milanović, S. 2007). Its bottom is not known. After one sudden and extremely heavy rain the turbidity increased to 1000 NTU (Fig. 2.42). This rain storm was large in amount and focused over a part of the catchment directly connected with Krupaja Spring.

**Figure 2.42.** Krupaja karst spring, Serbia. (A) Spring water in dry period; (B) Spring water after extremely high rain (Photo Milanović S.).

Figure 2.43 illustrates the case of the Sekerpinari (Sugar) Spring in Turkey. Turbidity changes as a consequence of spring discharge; however its magnitude is not precisely correlated.

The turbidity in some large springs a short distance downstream of the Karun 1 Dam in Iran was studied before impounding the reservoir. This turbidity correlates directly with rainfall. Fluctuations of the water level in the reservoir correlate well with heavy rain and with spring turbidity. In some cases without rain, in spite of a rise of water level in the reservoir of 25 m over a short period, no increase in turbidity was reported. This implies that the spring turbidity does not directly correlate with the reservoir fluctuations. The relationship between turbidity and the discharge of the Big Spring here is presented in Fig. 2.44. Note that, in 1989, there was no increase of turbidity during the period of high spring discharge.

In the case of Ombla Spring (Dubrovnik, Croatia) the turbidity occurrence takes place at the beginning of the arrival of the flood wave, practically at the same time as the abrupt increase of discharge. Figure 2.45 shows the comparative graph of turbidity, expressed in mg/l of silicate soil, and discharge of the spring in $m^3$/s. It is clear that rainfall of 35–40 mm/24 hours causes an increase of turbidity above 10 mg/l of silicate soil. As soon as the discharge begins to decrease, the turbidity is also rapidly reduced.

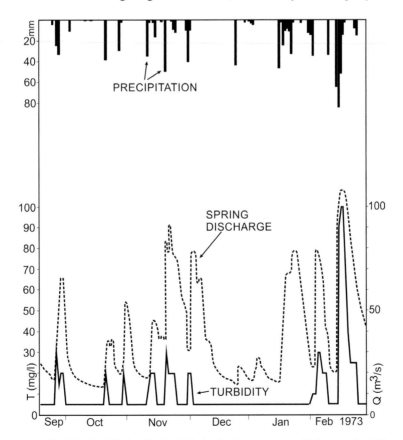

**Figure 2.43.** Sekerpinari Spring, Turkey. Relationship between spring discharge and turbidity.

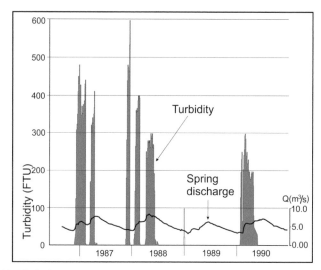

**Figure 2.44.**  Big Spring, Karun I, Iran. Relationship between turbidity and Big Spring discharge.

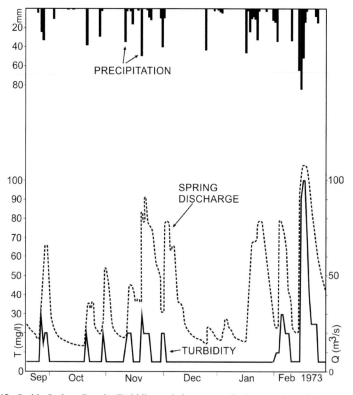

**Figure 2.45.**  Ombla Spring, Croatia. Turbidity variation versus discharge and precipitation, Milanović, 1977.

According to Bonacci (2016) average annual turbidity ($NTU_{av}$) of Ombla Spring does not correlate with average annual spring discharge ($Q_{av}$, $m^3/s$) and average annual precipitation. For instance, for the dry year 2008, when P = 872 mm and $Q_{av}$ = 20.6 $m^3/s$, the $NTU_{av}$ was 4.70. In the extremely wet year 2010, when P = 1722 mm and $Q_{av}$ = 42.9 $m^3/s$, turbidity was $NTU_{av}$ = 0.607. This situation is explained as consequence of mutually connected natural and anthropogenic events.

Sudden and extensive increases of turbidity can occur after earthquakes when, due to the seismic waves, huge amounts of clay are squeezed from the opened karstic joints.

Increasing turbidity can be a consequence of anthropogenic events in the catchment area, and also a consequence of dam and grout curtain construction. At many dams, due to increasing the water pressure beneath the dam foundation or in the area of the grout curtain, there was intensive washout of clay. Increasing turbidity downstream from a dam (at springs or in rivers) also indicate that there is flushing of clay from joints or caverns, i.e., that there is the possibility for structural failures to occur.

# CHAPTER 3

# THE REGIONAL APPROACH TO SELECT PROPER SITES FOR DAMS AND RESERVOIRS

## 3.1 THE KARST TYPES – ENGINEERING APPROACH

In the Introduction it was stated that "for engineers the term 'karst' suggests likelihood of failures". In a discussion at the Tenth Congress on Large Dams, Montreal, 1970, the General Reporter in Section Q.37, Londe P., emphasized: "No doubt, for engineers, karst has a bad reputation". Due to this opinion the concept *"keep away from karst"* was for many years a prevalent premise for many dam engineers. However, there are a large number of potential dam sites with excellent morphology and perfect geotechnical properties in carbonate rock masses prone to karstification. Increasing demands for drinking water, irrigation and hydropower, particularly in countries with huge potential water supplies in their karst regions, has created strong pressures for dam construction. In many karst regions of the world, an optimal strategy for water resources development has been a key requirement for regional socio/economic development. Because of that the concept *"be familiar with karst to adapt them for human needs"* became the prevailing premise. So, a substantial number of dams, even as high as 200 m or more, have been constructed in spite of the potential risks. By increasing the role of geologists and hydrogeologists in the building of large structures in karst, the number of failures can be decreased considerably. One of the crucial findings has been that successful construction of dams and reservoirs in karst is possible, while acknowledging that in any given instance some risk cannot be avoided absolutely.

The potential risk for off-site seepage from a reservoir is one of the key questions in engineering karstology. To minimize future risks, a sound geological concept is the key for the selection of proper dam sites or reservoir areas. The selection of the right location for building hydraulic structures such as dams is of primary importance for the construction and optimal operation. The initial step in selecting a suitable location is to build a good model of the site geology, for which an accurate large-scale geological map is essential. Such a map will guide the program of investigations that provide the basis for preparation of comprehensive hydrogeological maps and elaboration of relevant hydrogeological models.

The basic investigation procedure consists of:

- detailed geological and hydrogeological mapping of the catchment area that includes inventory of all specific karst feature such as ponors, estavelles, sinkholes, caves, shafts, etc., and tracer test results,
- geomorphological analysis,
- application of a range of the appropriate geophysical methods,
- detailed structural analysis,
- drilling of structural and piezometric boreholes,
- geological analysis of the borehole cores
- simultaneous hydrological measurements of surface flows;
- measurements of all permanent and temporary springs to determine their periods of minimum and maximum discharge (so-called 'zero data'),
- monitoring of groundwater level fluctuations in piezometric boreholes,
- excavation of investigation adits, and
- speleological investigations, including geological mapping of caverns and channels.

Particularly important is determining the properties and role of ponors (including zones liable to develop suffosion sinkholes), estavelles and caves. Each of these listed features has to be carefully investigated and analyzed. Ponors, estavelles and springs inside a reservoir area are particularly sensitive and risky. However, their presence does not automatically exclude the possibility of constructing a watertight reservoir.

However the significance of the karst conditions differs from project to project and site to site because each karst region must be presumed to have unique features. The complexity of different lithological units, tectonic activities, geological structures, type of karstification, evolution of the karst aquifer and many other influences make it difficult to accept one single, entirely valid model that will include all of the parameters necessary for dam site selection. But some common characteristics make it possible to divide karst areas into a few general types.

From the view point of the integrity of reservoirs and dams, karstified formations can be divided into five broad types: (1) Karst with hydrogeologically isolated base levels of erosion; (2) Karst with hydrogeologically open base levels of erosion; (3) Dammed (impounded or barred) karst; (4) Hypogene karst; (5) Evaporite karst. Each of these types displays a number of variations, and combinations of two or more of them can occur.

***Karst with hydrogeologically isolated base levels of erosion*** are direct consequences of several simultaneous processes—new tectonic movement, fluvial entrenchment and karstification. This type of karst usually develops during the creation of deep canyons by fluvial erosion, or as a consequence of intensive folding at a regional scale.

The canyon bottom is the local base level of erosion. The karst aquifers in both banks adapt their discharge to the river bed (canyon bottom). Karstification penetrates deeper into those banks in pace with entrenchment of the canyon. Figure 3.1 presents a

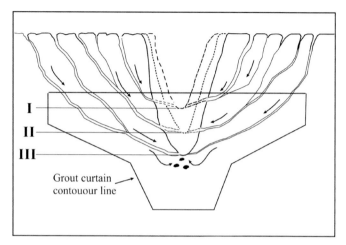

**Figure 3.1.** Schematic presentation of the three stages of karst evolution in a deep river canyon.

simplified schematic model showing three consecutive stages of the aquifer evolution. Caves and springs in cliff faces and along the river bed are evidence of this process.

In some cases karstification occurs locally below the river bed. Shallow underground flows follow the direction of the river. Usually this is a consequence of shallow longitudinal karstification coupled with siphon flow.

In this type of karst, only by-pass leakage at the dam site, including filtration beneath the foundations, is possible. The likelihood of leakage into adjacent catchment areas is limited or does not exist. In this setting, the water tightness of the entire reservoir can be ensured by a properly designed grout curtain at the dam alone. Lengths of lateral parts of the grout curtain depend on how far into both banks the rock mass is karstified. The depth of the curtain below the foundation depends on the depth of karstification.

However, in some cases the karstification process is not able to follow fast erosion and decreasing of erosion base level (river bottom). Because of geological reasons the process of karstification is limited or is obstructed. Figure 3.2 presents

**Figure 3.2.** Depth of karstification limited above the erosion base level.

two examples with obstructed karstification due to presence geological formations not prone to karstification. Particularly instructive is example (B) with thick shaly formation. The karstification process is stopped high above the river bottom. At the presented cases, the possibility of leakage through the reservoir bank practically does not exist. If reservoir level is higher than discharge outlets the expected and required remedial works are shallow.

In the case of Inguri Dam (Georgia) the river canyon was created very rapidly. Development of karst in the canyon cliffs was not able to keep pace with the fluvial process, i.e., to reach the local base level of erosion (Fig. 3.3). Because of that rapidity, the intensity of karstification in both banks, and particularly below the river bed, is limited (Djigaury et al., 1980). Due to these facts the grout curtain required for such a high concrete arch dam (271.5 m) was relatively shallow.

***Karst with hydrogeologically open base levels of erosion*** is an extremely complicated environment for the construction of dams and reservoirs. Presence of wide valleys, karst poljes, rivers with sharp bends, and intrusion of volcanic rocks into the soluble formations, are all hazardous factors for reservoir construction. Dry valleys, periodic rivers, deep shafts, large ponors, estavelles, high amplitude groundwater level fluctuations, a deep base of karstification, and large permanent and temporary springs are common and important features where groundwater is open to flow in different directions to different local base levels of erosion in adjoining drainage basins. The water in reservoirs is hydrogeologically 'open' to redirect flow to one or more of the adjoining erosion base level(s) or to downstream sections of the same river. The possibility of leakage from reservoirs in such settings is real and must be assumed during the feasibility investigations.

***Dammed (barré) karst*** is found where the karstified rock masses are in direct contact with steep or sub-vertical impermeable formations (hydrogeological barriers). In such cases the impervious formations play the role of natural 'dams' for groundwater flow within the karst aquifer. Very often large siphon (phreatic) springs

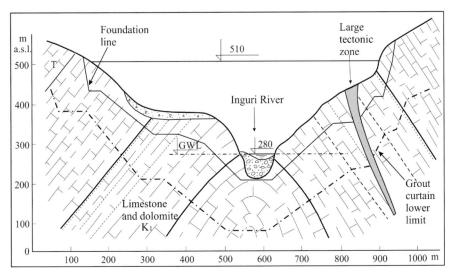

**Figure 3.3.** Inguri dam site, Georgia. Simplified cross-section, Djigaury et al., 1980.

are developed along the contact. If such a formation is present at a dam site its role usually is hydrogeologically positive. Based on this structure, the layout of the grout curtain should be designed to be tied into the impervious formation (positive cut-off).

*Hypogene karst* is a consequence of upward groundwater flows, mostly hydrothermal flows. If present at the dam site, water tightness problems become complicated, sometimes extremely complicated, because the origin of the sources and depths of upward flows usually are not detectable. Karst features are developed deep below the dam foundation (e.g., more than 160 m below the Višegrad dam site, and 100 m below the Chichik Dam site). Prevention of upward flow is a complicated task geotechnically. More about hypogene karstification is presented in Chapter 2.

*Evaporite karst* is the most risky geological environment for construction of dams and reservoirs. General properties of evaporite rocks and some experience with dam construction in them are presented in Chapter 7.

## 3.2 GEOMORPHOLOGICAL PROPERTIES

One of the most important characteristics of karst is the great variety of morphological forms that may be produced by solution. The close genetic relationship between geology and geomorphology has consequences on the dynamic evolution of karst aquifers. Many specific landforms make karst distinctively different from terrains on insoluble rocks. Narrow gorges, dry valleys, karst poljes, sinkholes (dolines), swallow holes, caves and shafts are the most important morphological features from the perspective of dam and reservoir construction. All of these geomorphological features are closely connected with underground morphological forms, particularly with karst conduits (cave passages or channels) and caverns. Very narrow gorges are typical in carbonate rocks and offer suitable morphological conditions for dam construction. But, during the creation of gorges by intensive fluvial entrenchment, carbonate rock masses in the gorge area and close surroundings will respond to these stimulating increases in the hydraulic gradient in influencing with intensive karstification. Depending on which process is faster (fluvial erosion or corrosion) the natural properties of the gorge can vary from acceptable to hazardous. It is particularly risky if the karstification penetrates deep beneath the floor of the gorge (Fig. 3.1). Because of this factor, the apparently ideal surface shape for installing a dam does not mean that a given gorge is a good and safe dam site. *A number of dam failures occurred due to the site being selected on the basis of its suitable exterior geometry and the favorable geotechnical (strength) properties of the rock mass only.*

River valleys, dry valleys, karst poljes and sinkholes (dolines, uvalas) are significant karst geomorphological forms, particularly if they are within the proposed reservoir area or its vicinity. Dry valleys or valleys with only seasonal or briefer periods of flow are risky places for man-made reservoirs. One of the key concerns in these sites is the depth of the groundwater level. If the dry valley is situated in the same hydrogeological basin as the reservoir, but at lower elevation, it will be the local base level of erosion. In that case, it is possible that underground connections from the reservoir will develop and drain it to the dry valley.

Sinkholes are frequent, indeed 'the diagnostic' (Cvijic, 1893), landforms in karst regions. Their shapes and locations are mostly consequences of the tectonic fracture patterns. In terrain covered with soil, the faults and joints are not visible on the surface.

However, a series of aligned sinkholes can indicate the direction of faults or major joints beneath the soil. Their long axes are usually in the direction of the main fracture or fractured zone.

Together with a geological map, geomorphological analysis is part of the first step that provides information necessary to determine whether it is possible to construct a dam and reservoir in a particular karst area. The importance of geomorphological analysis is confirmed by many karstologists from Cvijić (1893), Jennings (1971), Roglić (1972), Sweeting (1972), Marković (1973), Gams (1974), Zhi Ping (1979), Ford (1980), Gvozdecki (1981), Yaoru Lu (1985), Ford and Williams (1989, 2007) and a number of later contributors.

An inventory of all relief forms at the surface was the key procedure for many years. By application of new technologies (remote sensing methods), analysis of geomorphological features has become a much more reliable method in the proper selection of reservoir sites than topographic maps alone. Morphometry is frequently used for quantitative geomorphological analysis of karst forms. By application of the new techniques, for instance LiDAR imagery, the location, shape and size of sinkholes and dry valleys can be precisely determined even where the karstified surface is covered by dense vegetation (jungle). Quantitative geomorphological analysis becomes improved by a qualitative approach.

## 3.3  STORAGE PROPERTIES OF VALLEYS AND KARST DEPRESSIONS - POLJES

For preliminary selection of the right location for dams and reservoirs in karstified rock masses large-scale geomorphic features play an important role. Based on a number of case studies, the storage properties of deep valleys and large karstic depressions (poljes) are significantly different in general.

In karstified rocks, river valleys and deep gorges usually are less risky for dam and reservoir construction than karst poljes. The creation of deep valleys and narrow canyons is the consequence of rapid fluvial entrenchment. The valley bottom is normally the erosion base level for surface water and also for any underground water that passes through the rock in the valley sides. The intensity of karstification in valley sides generally will be much slower and unable to compete with the fluvial entrenchment, at least until the river enters its meandering, floodplain-widening, stage. Penetration of karst groundwater circulation below the valley bottom is usually limited (not deep), although it may exceed 100 m where strata are dipping steeply. Laterally, the intensity of karstification in valley sides decreases with distance from the thalweg (valley low center line). The base of karstification (rock of lower permeability) at a depth beneath the dam foundation is easily reached by a simple and efficient grout curtain. Restricted permeability at the abutments in the valley sides, to eliminate any by-pass leakage (seepage), can be achieved by the construction of short lateral sections of grout curtain. However, this generalization has to be understood as a very broad one, not an absolute rule. In the case of deep faults, hypogene karstification or buried paleokarst, deep karstification can develop even where the river entrenchment is narrow and appears to be young and rapid. In such cases, defining the nature and extent of watertight cut-off structures beneath dam foundations is a complex task.

Experience with reservoirs in karst poljes is quite different. A karst polje is a closed geomorphological structure with mostly underground inflow and discharge. Poljes are local erosion base levels, and the largest areas of concentrated infiltration at higher levels in larger catchments. By definition, groundwater levels in dry periods are below the polje level, sometimes very deep – more than 100 m. In the rainy period of the year, when the surrounding karst aquifer is fully saturated, many poljes get flooded. Ponors (sinkhole drains) and ponor zones are located along the perimeters of poljes, generally closer to any adjoining lower erosion base level; i.e., in the downstream area with regards to the direction of groundwater flow. The largest capacity ponor is usually located at the lowest point in the polje floor. Estavelles are another important feature characteristic of poljes. The sinking zone along the perimeter of a polje can be a few kilometers long. The bedrock floors of many poljes are covered with alluvial sediments and evolved top soils. The paleo-topography beneath these sediments is often typical karst topography with sinkholes, karren fields and shafts. In natural conditions, many of these features have become plugged and are no longer active drains into the underground. Some of them remain active, however, and will be manifested at the surface in the form of alluvial ponors. Because so many of these features are hidden, the possibility of detecting them precisely with the techniques currently available is limited. Treatment of these features, even when they are detected, is a very complex task. Treatment of estavelles with their reversible functions (ebbing and flowing, with destructive uplift forces when discharging upwards and out into the polje) is a particularly challenging problem. To prevent seepage from the surface water storage in a karst polje is a long-term task. Usually, the preventive measures undertaken during construction will be only partially successful. Corrective measures after the experimental filling and during the reservoir operation should be expected. Underground treatment alone (grouting, plugging) is not an efficient approach. To achieve required water tightness of reservoirs in karst poljes, a combination of surface treatment with local underground measures provides the best results.

## 3.4  IMPORTANCE OF LITHOLOGICAL AND GEOLOGICAL STRUCTURES

The most important geological characteristics in an area are: the lithological characteristics, the discontinuities (faults and joints) and folds. Structural relationships determine the direction of karstification, i.e., development of karst conduits. Discontinuities, folded structures (particularly anticlines), the depth of the base of karstification, and river bends have important roles in dam site selection.

*Discontinuities (joints, faults and bedding planes)* are of fundamental importance in the development of karstification. A large number of authors have confirmed that there are substantial increases of karstification and permeability along fractured zones: Cvijić (1893), Rižikov (1954), Jennings (1971), LeGrand and Lamoreaux (1975), Parizek (1976), Milanović (1981), Dreybrodt (1988), Ford and Williams (1989, 2007), Palmer and Palmer (2009) and many others.

Discontinuities, together with bedding planes (inter-bedding joints), play a decisive role in forming the net of karst conduits (Fig. 3.4). At a regional scale, fault traces at

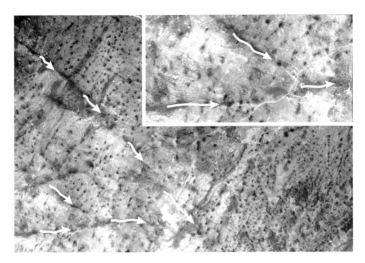

**Figure 3.4.** Aerial photograph—a limestone mass dissected by faults and with a large number of sinkholes. Arrows indicate preferred zones for possible conduits.

the surface indicate possible and general groundwater flow directions. Many of them will be indicated at the surface by lines of sinkholes (dolines).

Joints and faults are of fundamental importance in dam geology, particularly in dam foundations and the execution of a grout curtain. Every joint, even if tight and almost invisible, if intersected by a similar joint or not, can be karstified. Figure 3.5A shows karst channels along a tight and practically impervious joint. This joint was not detectable in the borehole core rock sample, despite the presence of open solution channels along it on both sides of the core and less than 0.5 m from it. During a water pressure test conducted in this particular section the registered permeability was very low. During the grouting, however, due to the higher pressures being applied and to hydro-jacking effects, the consumption of grout mix became enormously high.

Thin bedded limestone, with interbedded shaly layers, usually is not prone to karstification and has a role of hydrogeological barrier. However, if the joint cross this formation water filtration is possible along the discontinuity (Fig. 3.5B). If formation consists of thick bedded limestone this sequence can be karstified.

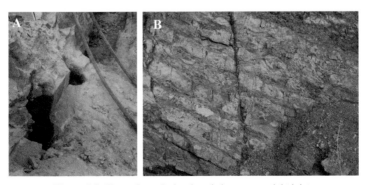

**Figure 3.5.** Karst channels developed along a very tight joint.

68    *Engineering Karstology of Dams and Reservoirs*

In contrast, wide fault zones (from a few to the few ten of meters) are usually filled with mylonitic fine-grained and clayey material so they may not be prone to karstification. If such structures cross man-made reservoirs, the likelihood of leakage can be negligible—but this should not be assumed.

***Structurally controlled karstification*** is important in a regional analysis of reservoir impermeability. Depending on its lithological composition, particularly on the composition of its core, presence of a fold in the bedrock can reduce the development of karstification perpendicular to it and increase it along its trend (along the strike). If the core of an antiform structure was subjected to horizontal stresses then the core will be in a compressive state of stress. If the core of a geological formation is resistant to karstification, groundwater flow will be concentrated in the limbs of its folds, with a tendency to go around the anticline core which serves as a hydrogeological barrier. The role of fold structures in development of karstification was analyzed in the case of the Tennessee Valley Project (1949); and also by Sikošek (1954), Jennings (1971), Lu (1972) and Milanović and Aghili (1990). A few examples are presented below to illustrate the role of anticlinal structures in the selection of reservoir sites.

Figure 3.6 presents a cross-section perpendicular to an anticline structure with a low permeability core and systems of faults on the both limbs. The large spring at the bottom of anticline limb indicates that the karst base flow developed along the limb (parallel with the strike of the core).

The Karun 3 dam site (Iran) is situated upstream of a large river bend (Fig. 3.7). This section of the river channel is developed around the southwest limb of an anticline. After the reservoir was filled, a small amount of seepage water began to flow parallel with the core along the northeast limb of the anticline, including an adjoining synclinal section. Seepage occurred along the bedding planes. The length of the seepage route from upstream to the downstream discharge area is shorter than the river channel length between same points; i.e., the hydraulic gradient of the underground flow is higher than the gradient of the river between same points.

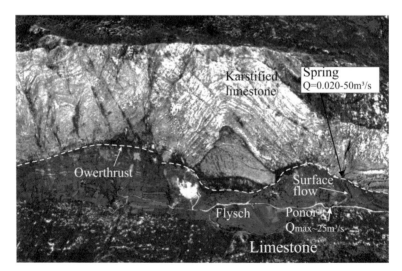

**Figure 3.6.** Cernica Polje, eastern Herzegovina. An anticline with a compressed, less permeable core.

**Figure 3.7.** Karun 3, Iran. Leakage through bedding planes along the strike in the limb of an anticline.

One of the best examples is the hydrogeological role of the regional 'Lastva Anticline' in eastern Herzegovina. The viability of the large Bileća Reservoir is based on the assumption that this anticlinal structure is impermeable (Fig. 3.8).

The Bileća Reservoir is located entirely in highly karstified limestones and dolomites. The reservoir volume is 1.28 billion $m^3$ and depth of water is between 75 and 100 m. Between the reservoir and the nearest base level of erosion (the Trebišnjica River in Popovo Polje), an asymmetrical anticline with a core of Triassic dolomite provides an absolutely impermeable hydrogeological barrier and prevents any leakage from the reservoir. Although it displays much grus, this dolomite has been subject to heavy compression, preventing any hydrogeological connection across it to provide direct groundwater between the two levels. More than 10 ponors were detected inside the reservoir area, with tracer tests being performed at five of them. It was established that the reservoir area is a hydrogeologically closed basin, with its only exit being dammed by a concrete arch structure on the surface and a grout curtain underground. In spite of the substantial number of ponors in the reservoir floor, there has been no leakage during the reservoir's 50 years of operation. It is an excellent example of a hydrogeologically isolated area. All inflow, surface and underground, into it (approximately $Q_{av/a} = 81$ $m^3$/s) can only flow out through the existing dam site.

The Hutovo Reservoir (Herzegovina) is located along the north-west limb of an anticline in Cretaceous dolomite. In spite of the marine base level of erosion being close by to the south, karstification has developed in a north-west direction parallel with the anticline limb. A number of tracer tests using the radioactive isotope B-82, and some deep boreholes, established the existence of a large underground flow zone

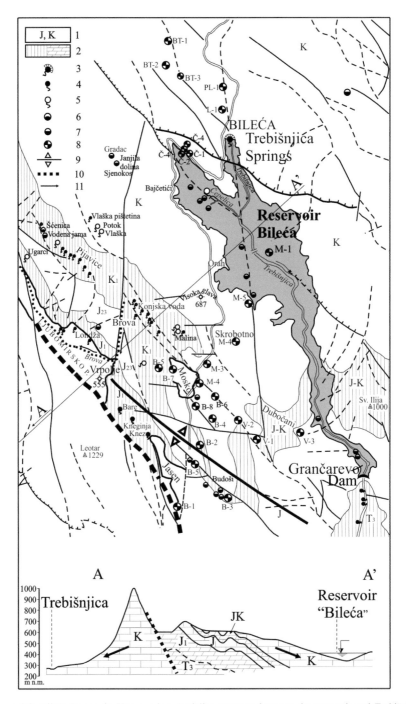

**Figure 3.8.** Bileća Reservoir, Herzegovina. Anticline structure between the reservoir and Trebišnjica River (Popovo Polje). (1) Limestone; (2) Dolomite; (3) Large spring (Q = 2–300 m³/s); (4) Small spring; (5) Temporary spring; (6) Ponor; (7) Karst shaft; (8) Borehole; (9) Anticline axis; (10) Large fault; (11) Fault.

along the north-eastern anticline limb (Fig. 3.9). The role of the anticline with its compressed dolomitic core as a hydrogeological barrier, was confirmed.

One branch of the Karun 4 Reservoir (Iran) is located in the deep Bazuft River canyon. Downstream, the main Karun River (site of the present Karun 3 Reservoir) is much lower than the Bazuft River (Fig. 3.10). Between the reservoir and the Karun River valley however, there is the long and continuous Safid-Kuh Anticline, with the

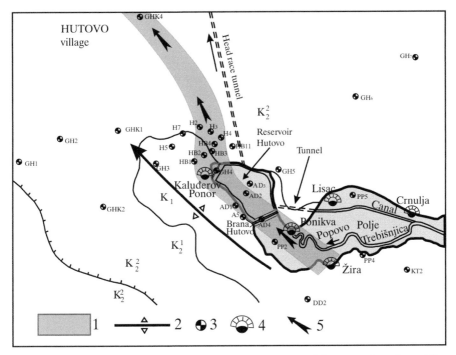

**Figure 3.9.** Hutovo Reservoir, Herzegovina, showing the underground flow zone along the anticline limb. (1) Route of underground flow; (2) Anticline axis; (3) Borehole; (4) Large ponor; (5) Direction of underground flow.

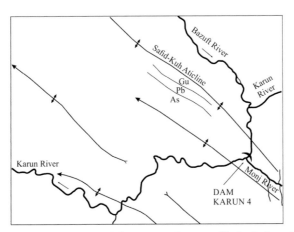

**Figure 3.10.** Karun 4 Dam, Iran. An anticline structure forms an effective hydrogeological barrier.

Pabdeh Formation (Pb - marl, marly limestone, shale) and Gurpi Formation (Gu - marl, marly limestone and shale) in its core. The upper Pabdeh strata reach elevations of more than 2,000 m asl, which is about 1000 m higher than the level of the Karun 3 reservoir. The lithological composition and location of the anticline core combine to form a good hydrogeological barrier between the branch reservoir in Bazuft Canyon and the much lower erosion base level of the Karun River.

Figure 3.11 shows photographs of some the caves situated at the river level in Bazuft Canyon. They are now at the bottom of the reservoir, submerged by 200 m of water in Karun 4 Reservoir. No inter-basin seepage losses from this reservoir have

**Figure 3.11.** Karun IV Reservoir, Iran. Caves in cliff sides along the Bazuft River gorge.

been recorded. All karst features in both sides of the canyon (banks of the reservoir) were left in a natural condition, without any waterproofing works.

## 3.5 THE HYDROGEOLOGICAL ROLE OF THE BASE OF KARSTIFICATION

In engineering karstology, the ***base of karstification*** may be defined as a specific geological 'underground surface or narrow zone' where the permeability of the rock mass is considerably decreased due to major reduction in effective karstification there. This basal condition has been confirmed in many dam projects: Geheyan Reservoir (China); Bileća Reservoir (Herzegovina); Piva Reservoir (Montenegro).

If fluvial erosion is too rapid and the distance between neighboring base levels of erosion (usually two valleys) is great, the karst aquifer develops discharge points separately for each base level. As a consequence, the middle of the rock mass in between (the interfluve) may be left less karstified or non-karstified (Fig. 3.12). The location of the base of karstification in the Bileca example was indicated by geophysical investigations (geoelectrical methods) and confirmed by boreholes, water level measurements and tracer tests (Na-fluorescein and Lycopodium spores). At Bileca the high base of karstification played the role of hydrogeological barrier, i.e., prevented seepage from the reservoir into the Bregava valley which is almost 300 m lower than the reservoir (Fig. 3.11).

In the case of the Geheyan Project (China), a less karstified rock mass (high base of karstification) between the reservoir area and the downstream section of the Qingjiang River prevented leakage from the reservoir (Fig. 3.13; Ruichun and Fuzhang, 2004). The occurrence of two base levels, with limestone ridge in between, led to karstification being directed in direction of each of them. Because of that, the interfluve rock was not karstified down to these base levels of erosion. This less pervious rock mass acts as hydrogeological barrier.

Based on experience, if dam sites and reservoirs are located upstream of major river bends there are potential risks and massive preventive works must be anticipated: e.g., Dokan and Haditha (Iraq), Salakovac and Rilja (Herzegovina), Seymareh and Karun 3 (Iran) and Keban Dam (Turkey).

During fluvial entrenchment the direction taken by the surface channel flow depends on many factors. As a consequence, rivers tend to create a number of bends. Particularly dangerous are those bends where large differences between the elevations

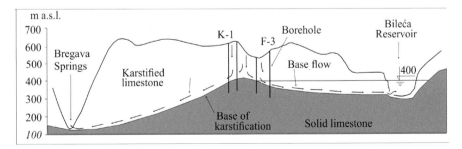

**Figure 3.12.** Illustrating a higher base of karstification between a reservoir and adjoining, lower valley.

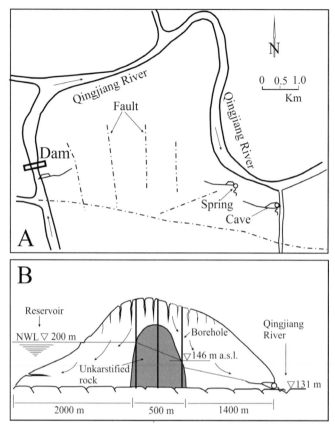

**Figure 3.13.** Geheyan Dam (China). A high base of karstification under the interfluves prevents leakage from the reservoir (Ruichun and Fuzhang, 2004).

of the upstream and downstream river sections can develop. If there are no lithological or structural obstacles, the possibility for karst drainage (a cut-off cave) developing between the two sections depends on the hydraulic gradient between them. In all cases, this gradient will be greater than that measured along the river bed between these sections. Due to this higher potential water energy, the intensity of karstification becomes stronger, i.e., underground flow between the two sections increases and the rock between them is karstified down to the elevation of the lower river section. Thousands of cut-off caves of this kind are known in river valleys and canyons, particularly across the necks of full meanders.

As an example, Fig. 3.14 shows the leakage through the left bank of the Salakovac Dam. The water by-passes this bank towards downstream sections of a river bend. As a result, the left bank has become intensively karstified since impounding of the reservoir, with leakage increasing up to15 m³/s.

For the same reasons, a dam proposed in the Zalomka River valley in Herzegovina was abandoned after detailed investigations. A large ponor zone in the reservoir area and underground flow between upstream and downstream sections of the same river was detected. The difference in elevation between the sinking and discharge areas was

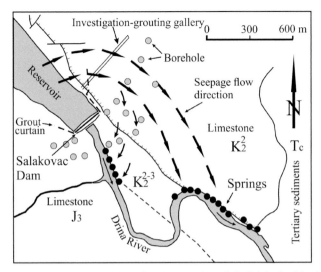

**Figure 3.14.** Salakovac Dam, Herzegovina. Seepage through the left bank of the dam site.

35–40 m and the river channel length between them was 15–20 km. The straight-line course of any underground flow(s) was only 4–5 km, implying that the underground gradient for flow is three to four times larger than the gradient along the surface (Fig. 3.15). The proposed dam site was abandoned and a new location shifted seven km upstream. The new site is on Triassic dolomite with much better hydrogeological properties for the dam and the reservoir behind it.

The Seymareh dam site (Iran) and nearby parts of the reservoir are very complex from the engineering karstology perspective. Karstification over a wide area around the site has been a multiphase and a very complex process before, during and after an extremely large landslide that occurred 15,700 years ago. Due to very rapid filling of the narrow river valley behind the slide by lacustrine sediments, the intensity of karstification also changed rapidly but the underground karst flow connections

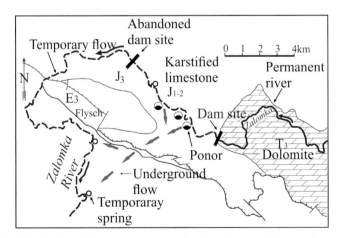

**Figure 3.15.** Zalomka River, Herzegovina. An example of a underground 'short cut flow' or 'cut-off'.

developed during the different stages were all retained. There was intensive groundwater communication between areas upstream and downstream of the dam site. An underground network of karst conduits and caverns is well developed, permanently submerged, and rich in underground fauna. The conduit system is developed along faults parallel to the main anticline structure, plus joint systems: however, the principal groundwater flow is more or less perpendicular to the orientation of the anticline (Fig. 3.16). A number of different processes and structural features (low dip on the anticline limbs) have as a consequence determined the modern direction of underground flow. More details about Seymareh Dam are presented in Chapter 13.

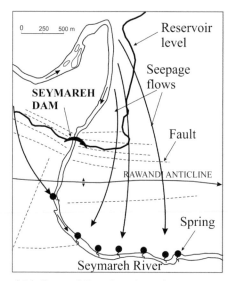

**Figure 3.16.** Seymareh Dam, Iran. Approximate seepage traces.

The banks of abandoned river channels cut deeply into limestone are usually intensively karstified along the relaxation (pressure release) joints that develop in them. Any dam site selected along a bend in the modern river is a risky place, particularly where a prior, now abandoned river bed has been infilled and masked with alluvial deposits. Figure 3.17 shows the example of the Ufa River in Russia. According to detailed analysis by Lykoshin (1968) a karstified and permeable zone has developed along the system of discontinuities. If, for instance, A – B is the possible axis of the dam, massive grouting works will be needed to achieve a watertight left bank.

To select the least risky dam site of eight possible locations along the Da (Black) River in Vietnam, crucial attention was focused on possible seepage from the potential reservoir toward the adjacent Mo River valley. The watershed between the Da River (Hoa Binh Reservoir) and Mo River consists of karstified Devonian, Permian and Triassic carbonate rocks. There were very detailed investigations to define the hydrogeological role of this carbonate massife: geomorphic analysis, geological mapping, geophysical investigations, boreholes, permeability tests and ground water level measurements. 79 caves and 95 springs were cataloged. Two of the exploration

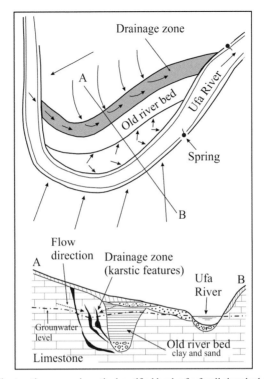

**Figure 3.17.** 'Short-cut' seepage along the karstified bank of a fossil river bed. Lykoshin, 1968.

boreholes were as deep as 450 and 550 m. On the basis of these investigations, it was concluded that there would not be significant seepage loss from the Hoa Binh Reservoir (Skiba et al., 1992).

# CHAPTER 4

# SPECIFIC METHODS OF INVESTIGATION IN KARST

## 4.1 KARST NATURE – NEEDS FOR SPECIFIC INVESTIGATIONS

The greatest impacts on engineering structures in karst terrains occur when there are caverns, karst solution channels (conduits) and lesser solution features in the rock. Precise detection of the spatial position of the caverns or conduits is a key requirement in engineering karstology. No single method of investigation can provide sufficient information to locate the precise position of a cavern nor can it provide an understanding of hydrogeological characteristics of the surrounding karst aquifer. In addition to conventional geological mapping, structural and geomorphological analysis, the most common investigation methods in engineering karstology are: tracer tests; a variety of geophysical methods; hydrological measurements and monitoring; exploration and/or piezometric boreholes; water pressure tests; and investigation galleries large enough for human entry. A hydrogeological karst map should contain data related to the underground flow connections and a catalog of all specific karst features: ponors, estavelles, springs, shafts, sinkholes – dolines, caves, sinking streams, dry valleys and karst poljes.

The unavoidable starting point of the investigation is a good geological/hydrogeological map and detailed tectonic structural analysis. An accurate interpretation of regional and local geological properties is an essential prerequisite for a rational design and successful construction. The key diagnostic tools in karst hydrogeology are piezometers, rainfall recorders, spring discharge recorders and tracer tests. Speleological investigations and investigation adits are the only methods which allow direct observation and detailed investigation of karst cavities. Speleology is limited to caverns and conduits of accessible size and dry. Diving speleology (cave diving) is extremely dangerous and is not recommended. Adits are time consuming and expensive but are the best way to find and investigate many underground karst features.

However, no investigation methods from the surface have been developed yet to confidently detect the spatial locations of karst cavities and turbulent water flow underground, especially when the karst is at considerable depth. Methods used for other geological formations have been adapted for use in karst: geoelectrical and seismic methods, downhole TV cameras (video-endoscopy), gravimetry, radioactive tracers, geothermal methods, radar techniques, LiDAR topography, echo-sounding and WLF. An important feature has been the development of completely new methods and equipment generally unknown in investigations of non-karst terrain (geo-bombs and different kinds of tracers). The lessons learned by applying some of these methods are presented briefly in this chapter. The basic steps (geological and hydrogeological mapping, geomorphological and regional structural analysis) are presented in Chapter 3.

The most important karst-related techniques are those investigating the flow of water through the karstified rock mass. Determination of the following properties of a karstified rock mass is crucial for successful dam projects: (1) Analysis of the fractures and other discontinuities in the bedrock; (2) Groundwater regime monitoring (piezometry); (3) Geophysical properties; (4) Groundwater flow directions (groundwater tracing); (5) Speleology (investigation of underground morphology); and (6) Hydrological characteristics (measurements and monitoring).

## 4.2  PROPERTIES OF DISCONTINUITIES

In natural conditions the hydrogeological role of discontinuities such as joints and bedding planes is to channel seeping water or greater flows. During dam construction a different kind of fluid, the grout mix, should penetrate the discontinuities to plug them. Successful plugging depends on the aperture, roughness, and nature of any natural fillings of the discontinuities, the nature and extent of their interconnection, and the grouting procedures themselves: the first four important conditions here determine the properties of a given discontinuity. Grouting procedures then add a number of further parameters: among them, properties of the grout mix and the grouting pressure are of primary importance.

The collection of deep-seated structural geological information on the properties of all types of discontinuities and karst features (caverns) from boreholes is a difficult enough problem. However, it must be anticipated that other caverns may remain undiscovered even close to boreholes. When the borehole is finished it should be investigated by applying all available methods of investigation: the different kinds of logging (particularly radioactive logging and thermal logging) and application of downhole video television.

Due to their role in hydrogeology and karstification, the mapping and analysis of joints at the surface and underground is one of the basic investigation methods. Because the purpose of grouting is to plug every water-conducting void (joint or other cavity) all properties of these fractures need to be determined: fracture genesis, elongation, aperture, any type of filling, the connections with other fractures and bedding planes as noted, plus their geometry at regional and local scales, and the intensity of karstification along the fractures. Hence, serious consideration should be given to the core logging procedures. Although borehole cores represent only a very small proportion of the rock mass, the core pieces are usually the first geological data taken from underground. The probability of discovering hazardous caverns with boreholes drilled in standard exploration patterns is negligible; however, the borehole cores contain valuable information, particularly about the frequency and properties of joints and other fractures or discontinuities.

The shapes and variations of joints in karstified rocks ranges across a wide spectrum, from very simple to extremely complicated. Some of them are clean and straight, crossing each other simply, or only slightly undulating (Fig. 4.1A and B). Joints with wide, open and clean walls are less frequent. Many open joints will be filled with clayey deposits, coated with crystals of calcite or filled with crushed fragments of rock (Fig. 4.1B and C).

The aperture and roughness of the joint walls crucially affects the natural flow of water and penetration by the grout mix. Usually the aperture of open joints is between 0.2 mm and 10 mm. Joints in Fig. 4.1(A) are filled with weakly consolidated clayey material. The interface between these sediments and the limestone is weak. The aperture of the joint in Fig. 4.1(B) was wide enough for precipitation of calcite crystals along both walls, substantially increasing the friction. There is still enough space along this joint for water flow by gravity (as a Newtonian Fluid): however, in the case of a grout mix (which is a Bingham Fluid), even in very thin mixes this roughness will play a negative role in the grouting process. The discontinuity shown in Fig. 4.1(C) is extremely rough and will present serious problems for water percolation and grout mix penetration. Increasing friction between the slurry and the rough joint walls will require increasing the grouting pressure or changing the consistency of the slurry.

The apertures of joints can be considerably enlarged, up to a few decimeters, as consequences of karstification (Fig. 4.2). Frequently such joints tend to be filled not with groutable sandy clays but by other sediment mixtures with different geotechnical properties.

Many joints filled with secondary precipitated pure calcite have tight interfaces between the calcite veins and the limestone (Fig. 4.3). In some cases, due to action of flowing water, such veins become preferentially karstified.

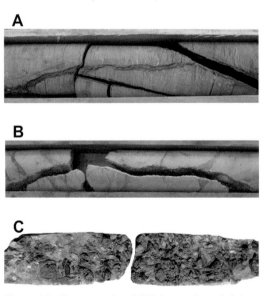

**Figure 4.1.** Three examples of differing roughness in joints.

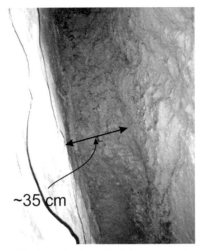

**Figure 4.2.** A joint enlarged by karstification and filled with more than 30 cm of sandy clay.

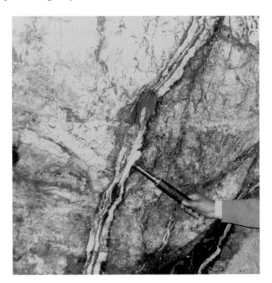

**Figure 4.3.** Joint in limestone filled with secondary deposited calcite

Some examples of borehole cores with frequent joints or differing karst porosity are presented in Fig. 4.4.

Figure 4.4(A) is an example of karst with a high density of small cavities—mostly vuggy porosity. Ease of water filtering through this type of porosity depends on the interconnection between vugs. Where the individual vugs are isolated from each other a water pressure test (Lugeon test) will indicate a low or very low permeable rock mass. Grout mix consumption is negligible.

The cavernous space in the core shown in Fig. 4.4(B) is filled with redeposited old and lithified clayey-sandstone. The geotechnical properties of the cavern infilling become almost the same as those of the surrounding carbonate rock. As a consequence

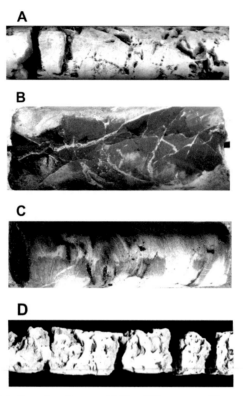

**Figure 4.4.** Examples of cores exhibiting different types of karst porosity.

of neotectonic movement, the sandstone filling then disintegrated and newly created discontinuities became completely filled with precipitated calcite, forming an entirely sound rock mass.

Figure 4.4(C) gives an example where limestone vugs are filled with oil. In practice this situation is very rare. As a result, the efficiency of grout mix intrusion into the oil-filled cavities still is not known well enough for general recommendations.

Spongy like porosity shown in Fig. 4.4(D) is an example of karst with a high density of interconnected cavities. Because permeability in spongy porosity is high or very high the grout mix penetration is fast and consumption is very high, also.

Statistical analysis of joints is an important geological procedure for selecting the optimal spacing and orientation of grouting holes. The use of spherical projections for presentation and analysis is routinely applied in dam geology as an important basis for deciding the directions of grouting boreholes.

In addition to interbedding discontinuities (bedding planes), a number of different joints are likely to be present in each rock mass. At each dam site the majority of these joints can be placed in just two or three joint systems (Fig. 4.5). Solution features can be developed along the each of these systems. After detailed analysis of all available data, particularly after geological mapping of adits and solution features if they are detected, the inclination of the grouting boreholes should be selected. Grouting boreholes should be designed to cross the maximum number of potentially karstified joints.

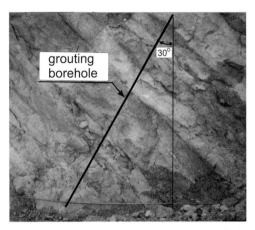

**Figure 4.5.** Illustrating the orientation of grouting boreholes to intersect bedding planes, and a wing diagram with the main joint systems.

## 4.3 PIEZOMETERS AND THE GROUNDWATER REGIME

*The groundwater level* and its regime at and around reservoir areas and dam sites is the most representative and reliable hydrogeological parameter to establish during the first investigation step in karst. In engineering practice the term *piezometric level* is commonly used for the water level measured in a borehole, replacing the term *potentiometric level*. In karstified rock this will be the present groundwater level in saturated surrounding rock or the water level in the nearest karst conduit. It is assumed that this level represents the boundary between the phreatic (saturated) and the vadose (unsaturated) zone in a limited area around the borehole. Large differences in groundwater levels in closely spaced boreholes is common in highly developed karst.

The w*ater table* in karst is discontinuous and variable in space and time. Great GWL differences in neighboring piezometers are consequences of erratically distributed karst porosity and the position of the base of karstification. To represent the hydrogeological properties of the rock mass, piezometric boreholes should be in hydraulic contact with the active conduit zone. The ground water level in a given piezometer depends on its spatial location, the capacity and direction of master flows (locally) and base flows (regionally). If a piezometer is located in a compact rock mass without any hydraulic connection to a nearby karstified (flow) zone, its fluctuation regime may be considerably different. To plot a piezometric surface (water table) in such cases is problematic. Particularly questionable can be the piezometric surfaces plotted by standard computer programs. The concept of a water table in a karst area has been a hot topic of discussion between karstologists for more than 100 years. Theories that oppose each other began in the time of Cvijić (1893, 1918), Grund (1903) and Katzer (1909) and continue down to modern times.

The occurrence of a base flow groundwater level deep below a river bed or reservoir bottom can be of concern for the entire dam project. In many cases it has been the crucial parameter determining that a potential dam site must be redesigned or even abandoned. However, there have been many dams and reservoirs constructed in areas with deep groundwater levels. A list of some examples with groundwater levels deep below the dam foundation and reservoir bottoms is presented in Chapter 2.

Depending on the waterproofing methods and technologies selected, some of these noted projects operate successfully but others experience significant seepage problems. Some of the dam sites are now abandoned and replaced by alternative sites with better hydrogeological properties (e.g., the Havasan Dam site).

***The piezometric borehole*** is one of the best diagnostic devices in hydrogeology, particularly in karst hydrogeology. However, due to the hydrogeological complexity of karstified rocks the piezometric data needs careful analysis.

A stand pipe piezometer is a borehole equipped with a perforated pipe wide enough for ground water level measurements, manually or electronically.

Because every karst conduit represents the local base level for drainage and erosion in the surrounding parts of the aquifer, the groundwater level recorded in a piezometric borehole depends on the hydrogeological conditions in the nearest conduit. Due to the fact that each borehole acts as a drain for the immediately surrounding rock mass, even in the vadose zone, departure of piezometer levels from true aquifer levels is one of the frequent problems encountered in karst. A piezometer can serve its purpose only if it is located and installed in such way that it provides a direct contact with the circulation in an aquifer.

Standpipe piezometers have absolute priority in karst aquifer investigations. Multiple (sleeved) piezometers should be suggested only in cases where the hydrogeological properties require this kind of monitoring structure, i.e., if geological formations with differing hydrogeological properties are encountered down the borehole, primarily differences in porosity (e.g., volcanic or low permeable sedimentary formations). If a borehole is located entirely in a karstified rock mass and equipped with multiple piezometers (two, three or four separate piezometer pipes in one hole) a common water level may be found in two or more of them. It is well known that during the drilling procedure some water is almost always found at the bottom of a borehole in spite of the fact that the true water table is deep below and the rock belongs to a karstified formation. As stated above, this anomaly happens due to each borehole functioning as a drain for the surrounding local rock mass and does not necessarily represent the real groundwater water level. Frequently the water level in the upper piezometers(s) may be declared, by mistake, to be a perched water level. ***Because of this, systematic application of multiple piezometers may lead to erroneous conclusions and fatal mistakes.***

The rapid changes in water levels in karst do not permit an investigator to accept long time intervals between successive measurements. The best monitoring of water levels is by continuous recording. These specific characteristics of groundwater fluctuation in karst will be explained in some of the following paragraphs.

The role of piezometric boreholes in dam geology include:

* At dam sites and in underground man-made caverns, including tunnels, the key role of piezometers is to collect and analyze the structural and hydrogeological properties of the foundation rock mass. The intensity of the karstification is one of the important parameters to be determined also. For these reasons, the location, depth and inclination of each piezometer depends on the geometry of the dam

body and appurtenant structures. Rotary drilling with coring is mandatory. Before the borehole is equipped with piezometric tubes, comprehensive geophysical logging and use of borehole video cameras are strongly suggested. Piezometric boreholes are frequently used in tracer tests as injection or observation sites.

The second requirement is to analyze the hydrogeological properties of the reservoir area from view point of its water tightness, and to define the borders (watersheds) of the catchment area. To define hydrogeological properties of important geological structures, piezometric boreholes are frequently used for tracer tests. As has been emphasized above, the basic tool for selecting the best locations for piezometric boreholes is a good geological (hydrogeological) map of the entire catchment area and, particularly, a large scale map of the reservoir area. The location, depth and type of piezometer equipment depends on many properties of the rocks: depth and extension of the karstified formations; type of karst porosity; structural properties of the geological formations; occurrence and location of insoluble rock formations and their expected permeability. A piezometer can serve its purpose only if is located in such a way that it is in hydraulic contact with the karst aquifer, meaning that it should intercept at least one hydrogeologically active karst discontinuity, or to be close to a karst conduit. However, it should always be understood that the probability of intersecting caverns by boreholes is very low, even where boreholes are closely spaced.

In deep bare karst (holokarst), the systems of discontinuities (joints and bedding planes) are frequently seen at the surface. Areas where large discontinuities (faults) are merging or crossing at the surface indicate the possibility of karst channels in the rock mass beneath. In many cases piezometers constructed in these zones have entered the zone of karst base flows. Of course, the likelihood of penetrating directly into the karst channel is negligible but hydraulic influence of flow in the karst channel is propagated through the joint system in the surrounding rock mass. These piezometers are the sources of the best quality data for analysis of the hydrogeological and hydrodynamic properties of a karst aquifer.

It has been observed that some piezometers will show an accurate record of fluctuations of the water table when it is high however, provide incorrect data at its minimum because the borehole is not deep enough. The true GWL is much deeper than bottom of the hole but, due to a variety of reasons, water has remained trapped in that bottom and may be recorded incorrectly as the GWL. The following few examples illustrate the most frequent types of cases encountered.

Each borehole serves as a drain for the surrounding rock, particularly during rainy periods (Fig. 4.6A).

The volume of inflows (Qi) is greater than outflow (Qo) that percolates from the borehole into the aquifer because permeability is smaller in deeper parts of the rock. Due to temporary filling of the borehole the measured level does not correspond to the true water level in the aquifer.

Even when a borehole is at relatively great depth (200, 300 or more meters) but does not reach the minimum underground water level, it is rarely completely dry. It is common for the lowest couple of meters of a borehole to be filled with water. Two

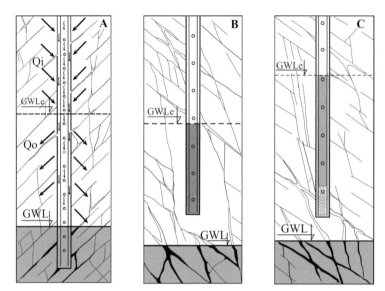

**Figure 4.6.** Erroneous piezometric levels as consequences of: (A) Lower permeability of the rock mass at greater depth; (B) Terminating borehole in a monolithic block; (C) Deposition of clayey material around the piezometer tube and at the bottom of borehole; GWL-Aquifer real water table; and GWLe-Erroneous level measured in piezometer.

most common cases are presented in Fig. 4.6B and C. The measured water levels in piezometers (GWLe) are not the true GWL (Fig. 4.6B and C). In Case B the erroneous GWLe is due to termination of the borehole in a monolithic (non-karstified) block (B), or by suspended deposits at the bottom and between the casing and the borehole walls (C).

In many cases data collected from one piezometer only cannot be accepted as a valid basis for conclusions. Sometimes at least three piezometric boreholes, placed close together, can provide enough valuable data for analysis of the groundwater regime in the surrounding parts of karst aquifer.

The use of angled (inclined; i.e., tilted) piezometers can create possible problems, particularly when crossing caverns. In such cases, attempts to install casing in the borehole can end in serious problems (Fig. 4.7).

***Borehole protection*** is an important requirement for every piezometric installation. Usually boreholes are located in remote areas where it is difficult to protect them properly. At many dam sites all over the world significant numbers of boreholes have been destroyed, mostly by plugging. A borehole is expensive and needs to be properly protected. The cost of redrilling a plugged borehole is 50% or more of the cost of an entirely new hole. The 'cost' of the data lost during the period when the piezometer is not operational may also be enormous. Consequences of lack of data can have an influence on project timing or can be sources of erroneous conclusions.

The cost of proper protection and monitoring is negligible compared to that of drilling a new borehole and the value of the lost data. Because of these considerations, a properly constructed piezometer collar must be resistant to destruction. One possible protective design with a double cap is shown in Fig. 4.8.

**Figure 4.7.** An inclined piezometer that has broken in an intercepted cavern.

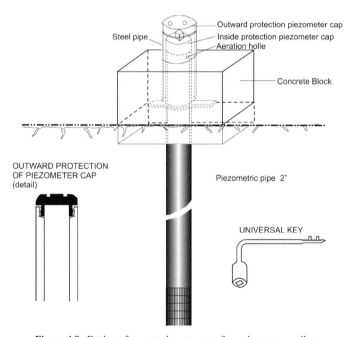

**Figure 4.8.** Design of a protective structure for a piezometer collar.

***Groundwater level fluctuation data (hydrographs)*** are among the most useful diagnostic data for assessing different hydrogeological conditions and the analysis of the hydraulic properties of a karst aquifer. In general, it can be accepted that the water level in the standpipe piezometer represents the approximate water level in the surrounding rock mass, mostly as a consequence of water pressure or its level in the closest active karst conduit.

Because of the special character of karst porosity (channels, caverns, cracks enlarged by solution, spongy porosity, combined with joint systems of the types common in hard rocks) water level fluctuations in the piezometric boreholes exhibit some distinctive features. Because every karst conduit represents the local base level of drainage for the surrounding parts of the aquifer, the groundwater level in the piezometric borehole depends on the hydrogeological conditions in this conduit. A piezometer can serve its purpose only if it was located and installed in such way that it provides a direct contact with an aquifer.

In bare karst formations (holokarst) after heavy rains the rates of **vertical percolation** down to the water table can be extremely fast, more than 100 minutes in 4–5 hours. The hydrograph data will show a very rapid response to such events, sometimes in less than one hour (Fig. 4.9). In some cases water percolates vertically more than 1000 m in less than 24 hours. As a consequence of such very rapid saturation, the **reaction of water level** in the piezometer is fast—sometimes less than 4 hours after precipitation begins. Rise of water level in piezometers can be up to 90 m/10 h. With heavy precipitation, spring discharges and the GWL in piezometers at distances of about 4–5 km from them may react almost simultaneously. In Fig. 4.9, the GWL in piezometers rose between 70 m and 175 m in a few hours and, simultaneously, the spring discharge increased from 20 m³/s to 85 m³/s (Ombla Spring, Croatia).

**Amplitudes of ground water level** fluctuation are high, in extreme cases during wet periods of the year being frequently more than 100 m. After precipitation stops, the fall of water level in piezometers is also fast, almost the same as the rate of rise

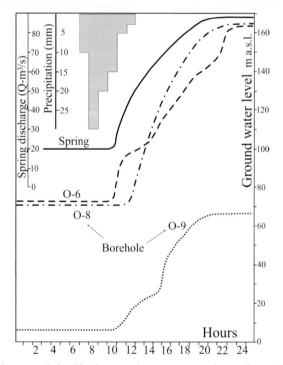

**Figure 4.9.** Simultaneous relationships between the occurrence and rate of a precipitation event, the GWL in piezometers and spring discharge.

during the rainfall. After 2–10 days, water levels in piezometers can decrease 10 to 50 m. This is consequence of rapid draining of the karst channels and cavities, as emphasized in Chapter 3.

In extreme cases the amplitude of ground water level fluctuation can be enormous. The largest fluctuations of water level measured in boreholes in Nevesinjsko Polje (Herzegovina) are more than 300 m (Fig. 4.10).

Another important characteristic of the groundwater regime in karst is the *difference of minimum GWL* even where piezometers are closely spaced. The drastic differences of minimal GWL seen in Fig. 4.10 are consequences of different stages of karstification in this aquifer. Groundwater minima and amplitude differences between the relatively close boreholes in Nevesinjsko Polje (between Z5 and Z3) is more than 200 m.

During dry periods of the year this karst aquifer continues to have a fully saturated phreatic zone. In piezometers along the base flow route (Z4–Z6–Z3), the maximum ground water levels are at almost the same elevation. In dry periods, the minimum water levels are stabilized at three different levels in particular highly karstified

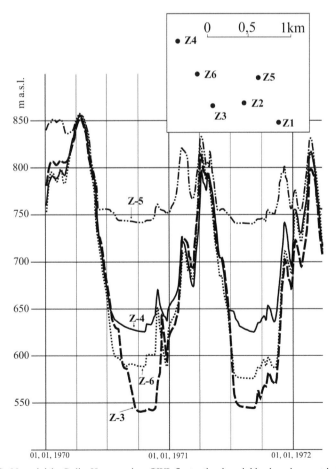

**Figure 4.10.** Nevesinjsko Polje, Herzegovina. GWL fluctuation in neighboring piezometric boreholes.

horizons. Ground water overflows from the upper to the lower stepwise horizons via near-horizontal and siphon-like passages. The differences between these karstified horizons, as determined by the piezometers, is about 500 m.

The hydrogeological cross-section drawn perpendicular to a karstified river bottom in Fig. 4.11 is an example of GWL complexity in karst. The large differences between groundwater levels in neighboring piezometers were registered in periods of maximum groundwater levels. Even when the river flow is more than 100 m³/s, the groundwater level lies beneath the river bed. Differences between GWL in close piezometers are relatively large. The measured data are connected with dashed lines only because the continuity of the water table surface in this case is questionable. It is obvious that the groundwater level in each piezometer depends on the local distribution and intensity of karstification, and on the capacity of this zone to receive seepage water and, at the same time, to release it to continue to flow toward the base level of erosion.

Based on the two above cases plus the experience gained at a number of similar sites, it appears correct to question the standard practice of presenting and interpreting such water table data in the form of equipotential lines and surfaces. Particularly questionable is application of the many different computer programs designed to construct equipotential maps. Acceptance of such output data without careful analysis can lead to erroneous conclusions and wrong decisions.

Typically, the duration of ***hydrograph peaks*** in karst is brief. In many cases it will be less than 10 hours. These rapid changes in water levels do not permit investigators to adopt long intervals between their successive measurements. The usual regime for GWL measurements, one or two per week, is not sufficient in rainy or flood periods because some peaks will have occurred between two measurements. In dam geology GWL data in real time (i.e., continuous recording) or measurements on an hourly basis are required for the reliable detection of each GWL peak.

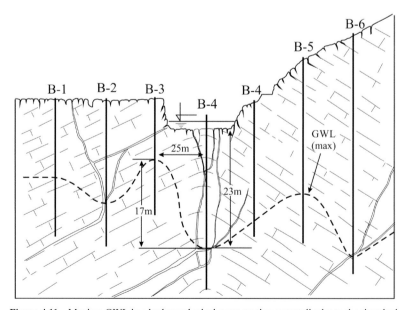

**Figure 4.11.** Maxima GWL in a hydrogeological cross-section perpendicular to the river bed.

When the piezometric borehole penetrates extremely karstified horizons, with caverns and conduit channels of great permeability, the maximum levels of underground waters often become steady. Due to the karstified zone locally having great permeability, rise of the water level above this zone is limited. The capacity of channels downstream of the piezometer is huge, much higher than the capacity of the inflow channels. In that case, the graphs of GWL fluctuation have the characteristics of overflow hydrographs. Figure 4.12 shows an example of such ***maximum steady level*** behavior shown in a hydrograph at an elevation of 194 m a.s.l., where the minimum water table is about 100 m a.s.l. (Milanović, 1981).

Only during extremely heavy precipitation it is possible for the water level to rise above the steady level for a short time in the piezometric borehole.

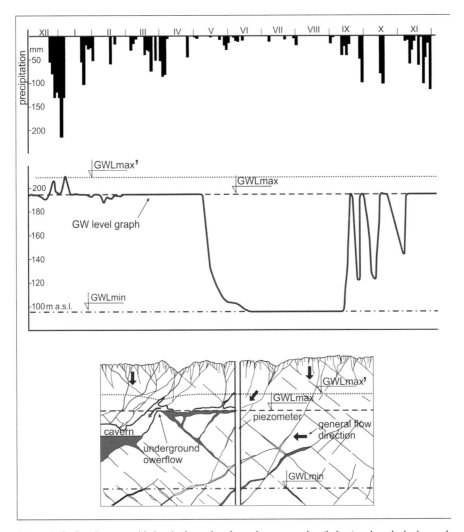

**Figure 4.12.** Steady water table levels shown in schematic cross-section (below) and on the hydrograph (above).

## 4.4  GEOPHYSICAL METHODS

Application of geophysical methods began in 1915 (gravimetry), and 1923 (seismic), mostly for oil field investigations. Intensive use of engineering geophysics in karst followed after World War II.

It is well known that ***geophysical methods*** are imperfect. However, if they are properly selected and coupled with other hydrogeological investigations, they can be very useful. The first book focused on the application of geophysical methods in karst was the monograph *Geophysics in the Karst* by Arandjelović in 1976. As distinct from boreholes, areal geophysical methods enable the investigation of a particular volume of rock mass, delineating zones with possible karst features. Applied geophysics creates a so-called 'reverse problem'; namely, results of geophysical measurements are always numerical physical values. Correct interpretation of these data enables their transformation into the relevant geological characteristics of the site. This means that *correct interpretation is possible only if geophysicist and geologist undertake this procedure together.*

Almost all available geophysical methods have been tested in karst environments, but with differing results. The rapid development of electronic and other technologies have had great influence on improving many of them. For instance, 40 years ago radar waves were not able to penetrate even one mm into a rock mass. Today, Ground-Penetrating Radar (GPR) technology is applied routinely. Instead of time-consuming interpretation of the measured data, a large number of different computer programs now provide high quality analysis in real time, i.e., during the field measurements.

The greatest impacts on dams and reservoirs occur when caverns are present. Signs of this risk include a deep base of karstification and, particularly, caverns or conduits in the foundation rock or along the grout curtain route. Detecting and characterizing them is very difficult despite the intensive application of geophysical methods. Caverns have been discovered hundreds of meters below the ground surface by drilling and tunneling, but the geophysical methods still have a hard time detecting the exact spatial position of caverns deeper than 20 m. A few geophysical methods that have been applied successfully in karst are listed below. Some of them were developed specifically for application to karstified rocks.

***Geo-electrical methods*** (sounding, profiling and charged body, mostly used together) have a long and successful tradition in karst investigations. By applying them, the base of karstification has been defined along the many tunnel routes in karst, within reservoir banks, and behind large springs to detect the base flow zones. Based on thousands of measurements it is evident that the electrical resistance of karstified rocks above the underground water table is greater than that of karstified limestone saturated with water, and smaller or equal to the electrical resistance of non-karstified limestone. This rule has been confirmed by measurement of the resistance curves in many boreholes in karst. The general resistivity graph for highly karstified carbonate rocks is presented in Fig. 2.12.

One example of the successful application of geoelectrical sounding and profiling is presented in Fig. 4.13. On the basis of measurements over five successive years at 300 points along the reservoir banks, the underground 'topography' of the base of karstification was identified, indicating the possible seepage zones, i.e., the zones with possible base flows.

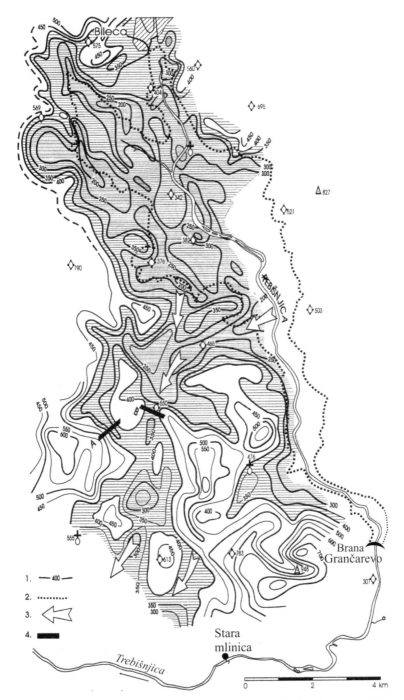

**Figure 4.13.** Base of karstification 'topography' in the right bank of Bileća Reservoir. (1) Contours of the base of karstification in meters; (2) Storage limits of the reservoir at an elevation of 400 m asl; (3) Zones where water losses (leakages) from the reservoir are possible; (4) Areas designated for future waterproofing measures (Arandjelović, 1976).

Geoelectrical investigations, were successfully applied during excavation of the tunnel for Pumping Power Plant Bajna Bašta, Serbia. After detailed geological mapping, as well as geoelectrical sounding and profiling, the proposed tunnel route was changed to avoid the cavernous rocks. Geophysical investigations have been done continuously during TBM operation, 2 to 4 km ahead of TBM head.

The ***charged body*** (*mise à la masse*) method can be appropriate for determining the routes of concentrated groundwater flow (streams) in karst conduits flowing away from a ponor (swallowhole) or in the hinterland of a spring. This method has been successfully applied in a few cases for tapping the flow behind large karst springs and also in detecting zones of concentrated seepage losses from reservoirs. Circular electrical sounding in a number of directions (azimuths) around one point has been applied to determine the apparent rock resistivity in various directions. Lower electrical resistivity is always measured when soundings are in the direction of intensive tectonic deformation or karstification.

Application of ***radar techniques*** in karst investigations has made great progress in the past 30 years. Ground Penetrating Radar (GPR) and borehole radar have been successfully applied in a number of projects. The limitations of standard GPR are that data are questionable below depths of more than 7–8 m. The new LOZA radar system (Russia) is based on same principles as GPR but is much more powerful. Penetration is up to 200 m but any clay present will have high absorption of the radar waves. According to some references, by applying this system a karst cavity has been detected at a depth of 38–40 m and confirmed by a borehole. These data are very optimistic and need confirmation.

The Swedish ABEM ***borehole radar*** system, RAMAC, was applied to detect the position of karst channels and underground flow (Q = 3–130 m$^3$/s) behind the Ombla karst spring in Croatia. At a distance of approximately 15–20 m from the borehole, the results indicated the presence of caverns and fractures; however, the exact azimuth of the reflection was questionable at that time (1990).

***LIDAR*** (Light Detection and Ranging) is a remote sensing method used to make high-resolution surface maps. Detection of the position and shape of karst polje borders, dry valleys and sinkholes provide useful data for geomorphological and structural analysis. Even where the surface is covered with tall and dense vegetation the images are still precise enough for good quality analysis.

***Geothermal methods*** are based on disturbances of the thermal field in the vicinity of piezometric boreholes. Results of thermal measurements show that they can be useful for karst conduit flow investigations. The thermal field in the surrounding rock mass will be disturbed as a consequence of water circulating through the karst channels. In the rock mass around a channel a low-temperature anomaly is usually formed. Such an anomaly can be measured in surrounding boreholes as an inverse thermal gradient. In the borehole section below the karst flow zone, a normal thermal gradient will then be re-established. Distribution of the temperature field depends on the geometry and thermal properties of the surrounding rock. On the basis of temperature measurements down boreholes and the construction of isotherms, the thermal field anomalies may clearly indicate the locations of conduit flow. Application of geothermal measurements to karstified rocks in order to detect active karst channels has been investigated and analyzed by many authors: Borić, 1980; Drogue, 1985; Bonacci, 1987; Chengjie, 1988; Ravnik and Rajver 1998; Riemer et al., 1997; Milanović, 2004.

This method was successfully applied in the case of the Buško Blato Reservoir (Bosnia and Herzegovina) to detect seepage zones through the grout curtain; at Maotiohe 4th Dam (China) to detect a large and active karst channel at a depth of 75–79 m below the storage level; at the Ataturk dam site to detect weak zones along the grout curtain; at Višegrad (Herzegovina) and Salman Farsi (Iran) to detect deep upward flows beneath dam foundations; at the Seymareh Dam site to investigate seepage through some sections of the grout curtain route. Groundwater isotherms for an envelope upstream of the curtain at Ataturk Dam indicated a cold spot in the right flank at an elevation of 300 m, showing the presence of an active karst channel (Riemer et al., 1995).

One example of the use of the inverse geothermal gradient method was the detection of a deep siphoning underground flow channel behind the large karstic Ombla Spring. More than 20 boreholes were drilled along an investigation gallery located about 200 m behind the spring outlet. The gallery was oriented almost perpendicular (normal) to the expected direction of concentrated underground flow(s). Measurements of temperatures in the boreholes were performed during four measurement periods, together with simultaneous measurements of the spring water temperature itself. The existence of two zones with large karst conduits at depths 50–70 and 130 to 150 meters below sea elevation were clearly indicated in this cross-section normal to the underground flow. The line in Fig. 4.14 that connects points of the boreholes where the geothermal gradient changes from negative to positive represents the 'zero temperature gradient line'. "It can be asserted that all significant temperature anomalies near the underground water channels are situated along this zero temperature gradient line" (Ravnik and Rajver, 1998).

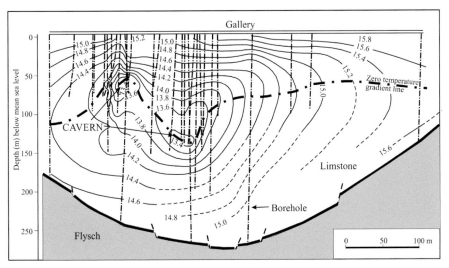

**Figure 4.14.** Vertical section of the temperature field between the investigation gallery and the impermeable flysch barrier at Ombla Spring (Ravnik and Rajver, 1998).

*Video-endoscopy* (downhole TV logging) of boreholes has become an important logging method for investigations of cavernosity close to the borehole. Small TV cameras are now available to make it possible to collect detailed observations even down in small diameter boreholes. Video-endoscopy was applied during investigation of a large cavern around the head race tunnel for the reversible Power Plant Čapljina (Herzegovina), for investigations of large empty caverns beneath the foundation of Ourkiss Dam (Algeria), for remedial works below the foundation of Višegrad Dam (Bosnia), and during remedial works along the grout curtain of the Logan Martin Dam (Alabama, U.S.A.).

Investigations of discontinuities and cavities encountered in boreholes are an important factor in karst investigations. The size and shape of cavities, including presence of cave sediments, are very important data if caverns are located beneath large structures (dams, reservoirs or along grout curtain and tunnel routes). This method makes direct observation possible for efficient control of grouting and cavern plugging operations. This method is particularly effective in the aeration zone. At the Ourkiss dam site (Algeria) a number of caverns were observed below the dam foundation and close to it (Milanović et al., 2015). Figure 4.15(2) is a photo of a cavern at depth

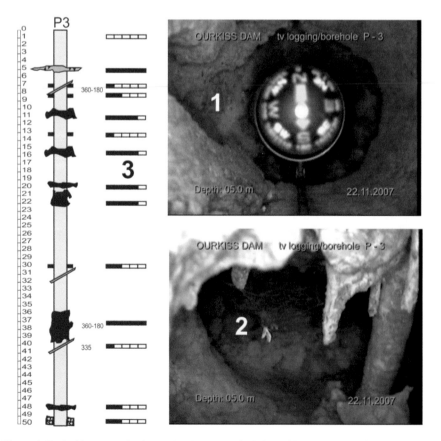

**Figure 4.15.** Ourkiss Dam, Algeria. TV logging. (1) Vertical view with compass; (2) Side-view; and (3) Caverns along the borehole observed by camera (Milanović et al., 2015).

of 5 m below the dam foundation. Two other cavernous zones below the foundation and above the water table are illustrated in Fig. 4.16. Based on these data, a standard waterproofing solution by grout curtain was considered to be too risky. Surface protection by a watertight blanket was adopted instead.

However, there are some limitations, particularly in the saturated zone. The presence of particles in the water (muddy water) is the most common disturbance of the image quality: it can be alleviated to some extent by letting the water settle for some days after the hole has been drilled. In the case of huge caverns, particularly beneath the water table, a very strong light is needed. At the present time, cameras still are not equipped with good enough lights. If a cavern wall is 10 or more meters from the borehole, the light will be too weak. In some cases two closely placed boreholes have been used, one for the camera and the other for a much stronger light.

***Echo-sounding*** is one of the successful methods for the investigation of air-filled or submerged caverns detected by boreholes. This method is based on propagation of ultrasonic waves and recorded echo travel times. Using the known speed of sound, the travel times are converted into distances from echo-sounder to a cavern wall. The shape of the cavity is determined by interpreting a large number of measurements: horizontal, vertical and inclined. On the basis of these data, the contours and volume of a cavern can be determined (Fig. 4.17).

This method was successfully applied in the case of the Ombla Spring underground dam project (SOCON Sonar Control Kavernenvernessung GmbH, Germany, 2005). From the investigation gallery (elevation of 3 m a.s.l.) behind the large karst spring, the contours of a large submerged cavern were investigated. The cavern was detected by boreholes but its shape was not known. Two boreholes were used to apply the echo-sounding method.

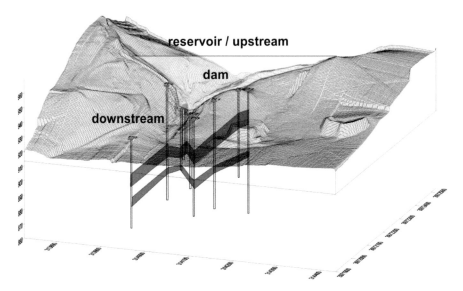

**Figure 4.16.** Ourkiss dam site, Algeria. Zone with caverns detected by TV camera (Milanović et al., 2015).

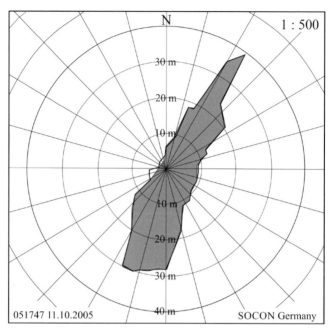

**Figure 4.17.** Contours of a cavern. A horizontal cross-section measured by echo-sounding (SOCON
Sonar Control Kavernenvernessung GmbH, 2005).

Borehole P-122.5 was scanned in 36 vertical and 46 horizontal sections. The highest point of the cavern was at depth of 112 m and lowest at 158 m. The cavern volume is $V = 2764$ m$^3$. Borehole PHD-2 was scanned with 36 vertical and 54 horizontal sections. Highest point of the cavern is at –37 m (35 m below groundwater level) and the lowest point at –85 m (83 m below ground water level). The cavern volume is $V = 574$ m$^3$.

The ***Geo-bomb*** method is a specific seismic method developed in Yugoslavia and also applied in China. The idea is to detect the spatial position of large, but inaccessible, karst channels that have rapid water flow. The Geo-bomb is a floating spherical time bomb, diameter of 10–11 cm, which contains enough explosive to produce a seismic wave easily detectable at the surface, using geophones set up in two perpendicular directions and connected to a standard seismometer. The activation time for each geo-bomb will be different. The point of each explosion then indicates the position of the karst channel at this location.

## 4.5  TRACER TECHNIQUES

The use of tracers to establish underground flow connection between sinks and springs has been dated back to the first years A.D. Josip Flavious, a Roman historian of Israeli descent, in his work 'The Judean War,' referred to use of the method in 60 A.D. to find the origin of the Jordan River. Chaff was used as the tracer.

Frequent use of tracers started in the later decades of the 19th century and beginning of the 20th century. During this period insoluble tracers were frequently used (bran, sawdust, chaff, yeast fungus). Water-soluble tracers used today include dyes, radioactive isotopes, salt, optical brighteners and many other chemicals. With the large number of new chemicals and monitoring techniques that have evolved during the middle and second half of the 20th century, application of soluble tracers has increased tremendously. During that period in Dinaric karst area alone, more than 600 tracer tests have been reported. More than 10,000 kg of Na-fluorescein and other tracers have been used so far. For investigation of properties of the aeration (vadose) zone gaseous (smoke) tracers have been tested and applied.

A substantial number of articles and books have presented different techniques and experiences with tracing technology, including detailed prescriptions for the execution of quantitative tracing tests. A few international Symposiums on Underground Water Tracing (SUWT) were organized during the second part of 20th century. Particularly important are the 6th International Symposium on Water Tracing (Karlsruhe, 1992) and the 7th International Symposium on Water Tracing (Portorož, 1997). The proceedings of Portorož Symposium have been published as 'Tracer Hydrology' 97 (Kranjc, 1997). Because of this, in the text below, tracer methods are presented at a summary level only.

## 4.5.1 Dye Tracers

Ground water tracing using different tracer dyes is a fundamental and efficient hydrogeological method for determining ground water flow directions, ground water travel rates, and hydrogeological connections between tracer injection places and discharge points. It is an easy and cheap method and is routinely applied at almost every dam project in karstified rocks.

One of first tracer tests was organized 1863 in England; 140 kg of ochre were used as tracer to investigate an 1,5 km long underground flow (Ford, 2015). The first application of sodium-fluorescein (Uranine) was on October 09, 1877. A geologist from Karlsruhe, A. Knop, injected Uranine into ponors in the Upper Danube, establishing some losses to the Rhine (Käss, 1976). Among the number of different kinds of dye tracers, sodium-fluorescein and rhodamine are most frequently used in engineering karstology. Two different kinds of tracer tests in karst can be distinguished according to their scale and purpose:

1. Large scale tracer tests that are usually organized to define the watersheds (i.e., catchment areas), and
2. Local tracer tests at dam sites and reservoir areas to establish and analyze the role of local karst features as a basis for the grout curtain design.

If seepage appears during the operation of a dam, particularly around the grout curtain, tracer tests are frequently applied to establish the seepage area, i.e., the area that needs regrouting.

In the Dinaric karst area (Croatia, Bosnia and Herzegovina and Montenegro) there have been more than 1000 tracer tests, using different tracers. More than 12 tons of different dyes (mostly Na-fluorescein) have been used. More than 60% of these tests were routinely organized as part of the hydrogeological investigations for different dam and reservoir projects.

## 4.5.2 Radioactive Isotopes

The use of ***radioactive isotopes*** as tracers is based on their easy detection in highly diluted solutions. Different lengths of half-time decay for various isotopes enable the selection of isotopes for each particular case that will contaminate the underground water for the shortest possible time. If the observation points are boreholes this type of tracer has several advantages over dyes and other types of tracers. A continuous measurement of radiation in the borehole, which intercepts the flow of the tracer, can provide a definite position for the underground flow zone. It is one of most important pieces of data ever in underground flow investigations. The quantity of isotopes in solution needed for an experiment vary in a range around 10 cc and the isotope carrier between 1–3 liters. It can be injected into the underground flow almost instantaneously. For the investigation of underground flow the best results have been with the following isotopes: Br-82 (half life of 35.9 hours); I-131 (half life, 8.1 days); Cr-51(half life 28 days) and Tritium ($^3H$, half life, 12.5 years).

## 4.5.3 Common Salt

Common salt (NaCl) is one of the frequently applied soluble tracers. Its solubility is 328 gr/L at 18°C. Usually salt is applied only for short distance investigations because for regional experiments the quantity needed will be very large and cannot be injected instantaneously into the flow, particularly if it has to be injected through a borehole. One of the largest tracer tests using NaCl tracer was organized to investigate the Pivka underground flow, Slovenia. There was used 1500 kg of the salt diluted in 8000 liters of water (Avdagić et al., 1976).

## 4.5.4 Post-activated Isotopes

Inorganic salts can be injected into the flow, collected at monitoring points and then activated for detection in a nuclear reactor. This **activation analysis** is one of the most sensitive analytical methods.

## 4.5.5 Optical Brighteners

This group of tracers includes synthetic dyes which are used in the textile and paper industries as brighteners. Mostly they have been applied in karst by American and British investigators. The use of optical brighteners started in 1969 (Drew and Smith) and Smart (1976), Quinlan (1976) and others. The detector for collection of samples is treated cotton. Though the toxicity of this tracer is negligible, Smart (1976) suggests that the maximum concentration in water should not exceed 50 gr per 0.1 liter of water.

## 4.5.6 Spores

*Lycopodium calvatum* spores belong to the class of tracers that are insoluble in water. The size of the spores varies between 30–40 microns. One gram of Lycopodium powder contains ~ 80 million spores. The preferred use for this kind of tracer is when several nearby sinking points need to be traced simultaneously. In that case spores of different colors can be simultaneously injected. The collecting points (springs, submarine springs or boreholes) must be equipped with plankton nets. Detection of the spores in water samples is possible by microscope only. It is a time-consuming method and prone to operator error.

## 4.5.7 Smoke and Air

Smoke and air have been used to investigate interconnections in inaccessible karst channels in any aeration (vadose) zone beneath dams, reservoirs and along the tunnel routes. The basic method is to inject smoke or air into a system of karst conduits with a strong artificial air current (a fan) and detecting its appearance at the surface or elsewhere through natural openings or boreholes. Air is about 1000 times lighter than water and so the forced air current requires about 30,000 times less energy in comparison with the quantity that would be necessary for corresponding water currents. Tracer diffusion in air is 1000 times greater than in water.

Movement of smoke and some gases through the aeration zone differs from groundwater movement through a karst aquifer in the following manner: Both water and smoke (or gas) circulate through the karst cavities toward their points of minimum potential ($h = Z + P/\gamma = min$). In the case of water, the potential minimum is at the point of lowest elevation, while in the case of smoke or gas circulation the point of minimum potential is on the open surface.

Different gases can be used as tracers. They must be lighter than air, easily detectable at the ground surface in very low concentrations (radioactive or smelling gases), adsorbed by water or water vapor in negligible quantities and, if possible, it should be visible (e.g., smoke of different colors).

In the first part of 20th century smoke (by burning crude oil) was applied in Macocha Cave (Czech Rep). Dry karst channels were investigated by smoke in China by burning chemicals or straw.

Smoke was applied for investigation of properties of the aeration zone beneath the Hutovo Reservoir (Herzegovina). In the dry season the ground water table is 80–100 m below the reservoir bottom. To detect points of possible strong upward air pressure, smoke was injected into the zone at two places: through an aeration pipe and through an artificial shaft connected to a large karst channel (Ponikva). The aeration pipe was connected to a karst channel beneath the alluvial cover (Fig. 4.18A). A strong air current was created artificially by a fan with a capacity 0.24 m³/s. About 8200 m³ of air was injected into the aeration zone through the pipe. The smoke was from common army smoke canisters (bombs). The duration of air injection was 9.5 hours. Smoke appeared the next day from one borehole at distance of 120 m from the injection point.

The next investigation was with smoke injected into the karst channel through the Ponikva ponor, located outside of the reservoir (Fig. 4.18B). About 350,000 m³ of air was injected into the channel by three powerful fans. Five smoke bombs were used to create a sufficient quantity of tracer. The emerging smoke was detected at the aeration pipe 1100 m away inside the reservoir, and at two other piezometers. The velocity of the smoke flowing out of the aeration pipe was 5.5 m/s. Further emanation of smoke, from a few karren cavities, was observed next day at a distance of 3.5 km from the injection point (Fig. 4.19).

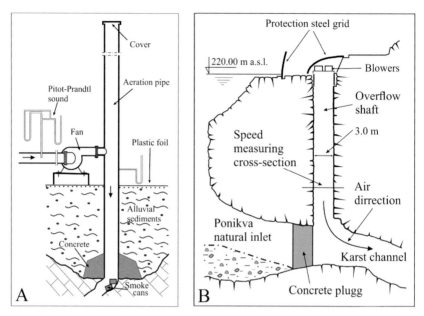

**Figure 4.18.** Investigations of karst conduit connections in the aeration zone by tracing with smoke. (A) Injection of smoke through aeration pipe. (B) Injection of smoke through a shaft connected with a karst channel.

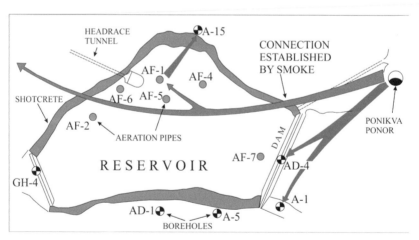

**Figure 4.19.** Hutovo Reservoir, Herzegovina. Connections in the aeration zone established by the smoke tracer. Arrows show connection between site of the smoke injection and points of smoke appearance.

Smoke tracing was also successfully applied in the case of heavily damaged concrete lining in the headrace tunnel. To investigate voids around the tunnel smoke bombs were lowered into them. The appearance of the smoke in boreholes showed that wash-out processes had created interconnected voids below, around and above the tunnel lining.

## 4.6 SPELEOLOGY

*Speleology* is a science but also a challenging and extreme sport, a paradise for biologists and, for many other people, fascinating underground landscapes. From the hydrogeological and geotechnical view point, however, speleology (including speleo-diving) is one of the very important methods of investigation in engineering karstology. Speleology is the only means that makes direct observations of underground karst features by humans possible. Based on the results of speleological investigations, a number of dam projects in karst have been redesigned, particularly the routes of grout curtains. Some large dams all over the world have serious problems because the investigations did not include speleology. Proper design of dams and reservoir, as well as remedial works after failures at dam sites, reservoirs and tunnels, is practically impossible without speleological investigations. In the case where it is proposed to install underground dams, speleology is one of the key investigation methods. Speleologists with a geological background are able to collect and interpret the geological information that is important from the engineering view point and for the building of different underground structures.

A speleological investigation program includes the following interactive activities:

- Mapping with a common co-ordinate system.
- The graphical presentation of results should include a layout plan, longitudinal sections and selected cross-sections on a scale 1:100.
- All important geological features (dip of faults, joints and bedding planes, characteristics of cavern deposits and infilling of faults and joints) have to be mapped and graphically presented.
- The air temperature must be measured in different levels of each cavern.
- If a cavern or karst channel is filled with clay or sand deposits, samples of this material have to be taken for analysis.
- If caverns are filled with water, their temperature should be measured and samples taken for comprehensive water chemical analysis.
- Any appearance of air currents in karst channels and caverns has to be noted.
- The main survey points along the cave channels have to be clearly marked with resistant paint.
- Submerged karst caverns and channels have to be investigated by speleo-diving (cave diving). However, cave diving is extremely risky and should be undertaken only if no other method can provide the required data.
- Selected karst channels should be used for geophysical investigations.
- Water samples should be collected during all tracer tests at a site.
- During the impounding of the reservoir, all known caverns downstream of the grout curtain have to be monitored.
- Detailed photo documentation has to be included in the final speleological report.

Speleology was successfully applied as part of the engineering investigations at a number of dam and reservoir projects: e.g., Canelles Dam, Spain; Sklope Dam, Croatia; Buško Blato Reservoir, Bosnia; Ombla underground dam, Croatia; Reversible Power Plant Čapljina, Herzegovina; Salman Farsi and Marun Dam, Iran; Wujiangdu Dam, Geheyan Dam and Wulichong underground dam, China; and Keban Dam, Turkey.

In some cases, based on results of speleological investigations, some dam projects were postponed (Bogovina Reservoir, eastern Serbia), or abandoned (Lower Gordon Dam, Tasmania).

During all phases of investigation, design, construction and operation of the multipurpose Trebišnjica Hydrosystem more than 50 caves, ponors and caverns along the head race and transmission tunnels were investigated speleologically.

Speleology was an important part of the investigations during the conception and development of the Ombla underground power plant. About 2500 m of karst channels was investigated in detail (Krašovec, 1989, unpublished - Fig. 4.20). Submerged karst channels were investigated down to 54 m below datum (sea level).

A particularly complex decision was the construction of a grout curtain in the right bank of the Salman Farsi Dam in Iran. During the curtain construction a large cavern (volume 150,000 m$^3$) was discovered (Fig. 4.21). This cavern consisted of a main chamber plus many channels, subsidiary rooms, and shafts partially filled with huge limestone blocks to an unknown depth. The essential data required for the analysis and selection of the final solution were collected during detailed speleological investigations. On the basis of these findings an upstream bypass was selected as the technically and economically most feasible location for the grout curtain (more details in Chapter 13).

In the case of Sklope Dam, Croatia, on the basis of speleological data also, the grout curtain route was shifted to bypass a huge cavern on the downstream side.

Speleo-diving (cave diving) has been applied chiefly for tapping into large karst springs and at a few hydro-power projects (Milanović, 2012). Diving speleology was successfully applied during the investigations for the Ombla underground dam project, for instance. However, it is too risky a method and sometime ends in tragedy. Based on current experience this method should be applied in very special cases only.

## 4.7  HYDROLOGICAL INVESTIGATIONS AND MONITORING

### 4.7.1  Hydrological Properties

In general, hydrology is defined as the science of the surface sector and hydrogeology as science of the subterranean sector of the water cycle. However, in many instances it is not possible to separate the hydrogeological regime in karst areas from the strictly surface regime of hydrology. Surface and underground water flows in karstified rocks are variable in space and time and strongly interconnected. The transformation of one discontinuous event (precipitation) into a continuous one (surface flow), followed by the water sinking into the underground (aquifer saturation), and finally its discharge through karst spring(s) is the cycle of the hydrologic/hydrogeologic dynamic karst water regime. To analyze hydrogeological properties of a karst aquifer without knowledge of the regional and local hydrological properties of the area is not possible. *Hydrogeological maps, precipitation stations, piezometers, water gauging stations and tracer tests all provide essential basic data for the proper analysis of karst aquifers.*

Among the number of publications and articles that refer to the hydrology in karstified rocks, the mostly cited and used as references are: 'Karst Hydrology and

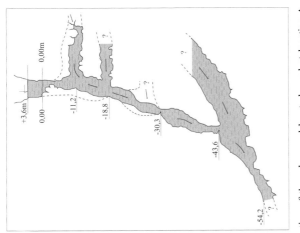

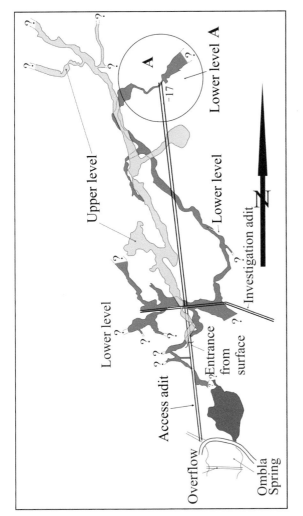

**Figure 4.20.** Speleological investigations for the underground dam at Ombla, Croatia.. Left–layout; Right–cross-section of the submerged karst channel at location A (Krašovec, 1989).

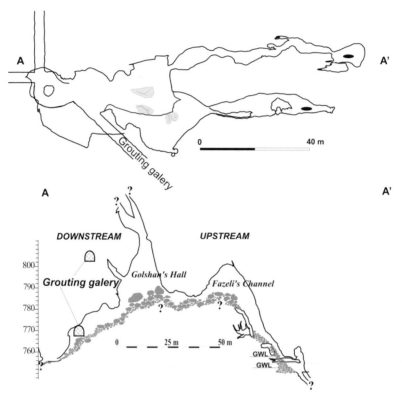

**Figure 4.21.** Salman Farsi Dam. Large cavern in the right bank. Layout and cross-section A–A'. Vučković and Milanović, 2001.

Water Resources' (1976, edited by Yevjevich; Water Resources Publication, Fort Collins, USA), and 'Karst Hydrogeology, with special references to the Dinaric Karst' (Bonacci, 1987).

Standard hydrologic methods are among the basic procedures in any dam project in karst. Many of them, when applied in karst areas, will be considerably more specific in their design than the same methods used in non-karst regions. Accurate assessment of infiltration is one of the crucial differences.

In considering infiltration into karst aquifers it is necessary to distinguish two different conditions; at the extremes there can be

- more or less uniform infiltration over all of the outcrop (the surface of the entire catchment area), or
- concentrated infiltration along linear infiltration zones (river beds) and/or through single openings such as ponors and estavelles.

The infiltration capacity depends on many factors: the depth to the base of karstification, properties of the epikarst zone, fillings of joints and fractures, properties of the in-filling deposits, thickness of any surface soil cover, slope of the ground, and type and density of vegetation cover. In extreme cases, the infiltration capacity of bare karstified outcrops can be more than 80% of the precipitation, with rates of vertical percolation downwards of 50 m per hour or more. In some cases water can percolate hundreds of meters into the underground within a few hours. In the case of Orjen Mountain (Montenegro), after heavy showers of more than 100 mm/24 h, the water penetrated to depths of 2000 or more meters in 15 to 24 hours. Average annual precipitation are more than 5000 mm. Maxima measured is more than 8000 mm (Fig. 4.22).

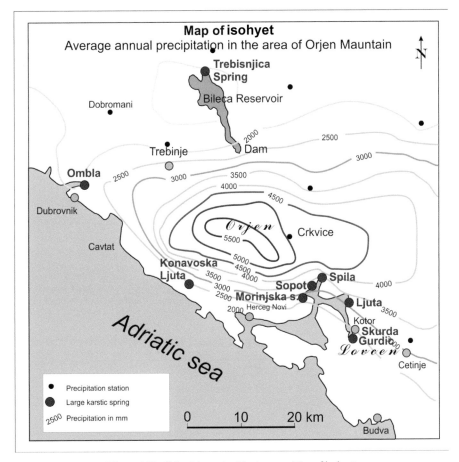

**Figure 4.22.** Orjen Mountain, Montenegro. Map of isohyet.

## 4.7.2 Simultaneous Hydrological Measurements

Simultaneous hydrological measurements are one of the first steps during the feasibility investigations. River flows can be permanent or temporary, depending on the intensity of karstification and the groundwater regime. Frequently, in spite of the surface flow in a river channel being permanent, there are some seepage losses along the bed. If the river flow is a few cubic meters per second and seepage only ranges up to a few hundred liters per second this difference cannot be registered by visual observation only. The method of simultaneous hydrological measurements is the best way to determine the permeability of a river bed, dam site or reservoir bottom. If there is a river channel at the bottom of a proposed future reservoir, the results of simultaneous hydrological measurements can confirm the feasibility of the project or, alternatively, end with the opposite conclusion and abandonment of the project.

Determination of the permeability along the riverbed upstream of a dam site is an important diagnostic method. Depending on the hydrogeological properties of the karstified rock mass, the crucial parameters of seepage loss may vary and, as consequence, seepage between two points can be different.

Bonacci (1987) distinguishes six combination of the ratio between discharges over different two sections of an open natural flow (river channel) in karst (Fig. 4.23).

Case A refers to surface flow with no losses, due to the groundwater levels being continuously higher than the water level in the surface channel. It can also occur if $\Delta Q$ is constantly equal to zero (B). Case C shows the situation where the losses are constantly increasing with increase of discharge between the upstream and downstream stations, due to groundwater levels always being below the river bed. The dashed line represents an alternative for the same case in which, with increase in the water level in the river, new ponors start to swallow water so that the losses along the analyzed section constantly increase. In Case D the groundwater levels vary. Graph E shows a situation where the ponors are located at some significant height above the riverbed. Due to clogging in sediments in the river bed and banks there are no losses until a critical water pressure is attained at discharge $= Q_p$. In this situation, the groundwater

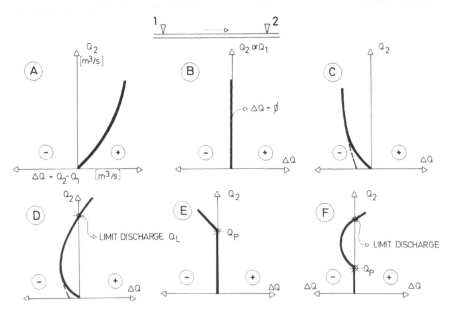

**Figure 4.23.** The possible combinations of discharge relationships between an upstream (2) and downstream (1) station in an karst open channel flow. (A) $Q_2 > Q_1$ no losses with increase in discharge; (B) $Q_2 = Q_1$ no losses, no recharge; (C) $Q_2 < Q_1$ losses constantly increasing with increase of the discharge; (D) $Q_2 < Q_1$, discharge (seepage) loss increases up to a limit, $Q_L$; (E) the situation where the ponors are located at an elevation above the bed of the river; (F) combination of cases (B) and (D) (Bonacci, 1987).

levels are constantly below the level of the river bed. Case F shows a combination of cases B and D. Until discharge $Q_p$ occurs, the river bed is clogged and there are no losses, although the groundwater levels are below the river bed. When the limiting discharge $Q_L$ is reached, the groundwater levels have increased so much that they begin to recharge the river. Finding that the behavior fits Cases A and B are positive and possibilities for establishing a watertight reservoir are high. All other situations require additional investigations. Case C is particularly risky.

The complexity of river flow in a highly developed karst can be represented by the example of the Trebišnjica River in Popovo Polje, East Herzegovina (Fig. 4.24). The existence of different relationships between the surface and groundwater

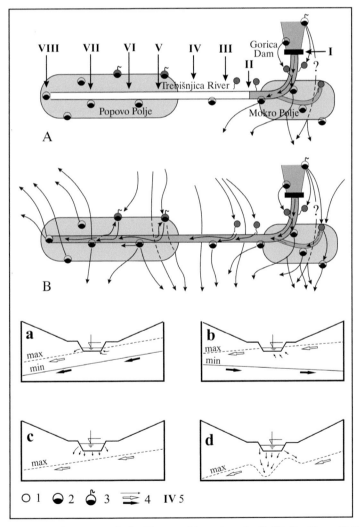

**Figure 4.24.** Herzegovina, Popovo Polje, Flow conditions along the Trebišnjica River. (1) Temporary spring; (2) Ponor; (3) Estavelle; (4) Direction of groundwater flows; (5) Number of hydrological gauge stations.

was established by analyzing the data from many surface flow measurements and piezometric boreholes during different stages of aquifer saturation. During dry periods there is river flow for a few km only (A in Fig. 4.24). The flow includes seepage from the reservoir and dam overflow. Considerable amounts of seepage water flows underground towards a lower base level of erosion. In wet conditions (B) the flow exists, along the length of 65 km, to the end (last ponor) in Popovo Polje.

Due to Popovo Polje being the regional base level, hydrogeologically open in direction of the absolute base level (the Adriatic Sea), there is a very complex hydrological and hydrogeological relationship that is changeable in both space and time. Surface and underground waters are mutually connected through more than 500 temporary springs, ponors and estavelles. This communication is particularly intensive in periods of significant rainfall. By generalizing these relationships, the four most frequent cases are shown in Fig. 4.23, *a, b, c,* and *d.* In dry periods the minimum groundwater level is deep beneath the river bed along the entire channel. Generally, the piezometric surface dips gently towards the absolute base level of erosion—the sea (*a*); however, it may be locally inclined in the opposite direction due to the influence of a large karst channel (*b*). After the karstified rock mass is fully saturated, the water table rises to the river level (*a*) and (*b*). A number of temporary springs become active but some ponors are activated also. However, in some sections of the river the water table never rises up to the surface (*c*). Water from the river percolates into the aquifer. This is a permanent leakage zone. The groundwater level is inclined towards the base of erosion. The piezometric surface is generally inclined toward the erosion base level also, but is locally very uneven (*d*). As in case (*b*), disturbances of the water level in piezometers are consequences of nearest local conduits that play role of local base levels of erosion.

Similar relationships between surface and groundwater have been found along the river bed. Part of the sinking water follows the direction of the river bed, and part flows away from it towards the sea coast. A lot of water from the upstream part of the catchment flows deep beneath the river bed directly toward the general base level of erosion (sea coast).

These relationships were based on the results of simultaneous measurements along the Trebišnjica River bed (Fig. 4.24A). The main aim of these measurements was to select effective waterproofing structures to be applied along the bed, and to establish whether any sections would potentially be endangered by hydraulic uplift pressure. Measurements were organized at eight hydrological gauge stations (from Gorica Dam, Station **I** to Station **VIII**) during three characteristic hydrologic situations: Q1-natural conditions, rainy season; Q2-dry season—from the upstream reservoir an artificial water pulse of 50 m$^3$/s was discharged as overflow at the dam; and Q3-dry season also but with an artificial water pulse of 150 m$^3$/s. Results of these measurements are presented in Table 4.1.

During measurements under natural conditions (Q1) a large proportion of the bedrock beneath the polje was saturated and the swallowing capacity of all ponors along the river bed was at a minimum. Experiment Q2 was performed during the dry season, when the water table was deep below the river bed. A constant discharge of 50 m$^3$/s was released from the reservoir. Water sank through the ponors and joints that are in the river bed only. The measured losses were 39.4 m$^3$/s. The third simultaneous

**Table 4.1** Seepage through the river bed between hydrology gauge stations.

| Section between gauge stations | I/II | II/III | III/IV | IV/V | V/VI | VI/VII | VII/VIII | ΣQ m³/s |
|---|---|---|---|---|---|---|---|---|
| Length of section (km) | 10.2 | 17.4 | 11.6 | 3.4 | 8.2 | 11.6 | 6.7 | |
| Seepage Q1 (m³/s) | 3.1 | 5.5 | 5.8 | 4.7 | 2.0 | 3.1 | | 24.2 |
| Seepage Q2 (m³/s) | 7.5 | 9.7 | 10.8 | 1.3 | 3.8 | 6.3 | 11.6 | 39.4 |
| Seepage Q3 (m³/s) | 11.1 | 12.1 | 14.8 | 9.1 | 1.1 | 15.2 | | 63.4 |

measurements (Q3) were organized in similar hydrological conditions to those in case Q2 but the discharge was increased to 150 m³/s. The observed maximum losses were 63.4 m³/s because part of this flow inundated areas of the polje outside of the river channel and lost water directly into a series of ponors there.

The best results were obtained where the hydrological station for the simultaneous measurements were accompanied by a hydrogeological monitoring station (Fig. 4.25).

**Figure 4.25.** Temporary hydrological gauging station with adjoining piezometric borehole in the Trebisnjica River bed (dry conditions).

The hydrogeological monitoring station consists of a piezometric borehole in the river bed and piezometers located in both adjoining river banks. The number of these piezometers will depend on the hydrogeological conditions, i.e., on intensity of karstification in the local area. Usually one to three piezometers were installed on each side of the river. One of the hydrogeological/hydrological stations is shown in Fig. 4.11. This station consists of seven piezometric boreholes: one in the middle of the river bed and three on each side of river, at distances of approximately 30 m between boreholes (piezometers).

# CHAPTER 5

# METHODS TO DETERMINE THE PERMRABILITY OF DAM FOUNDATION AND RESERVOIRS

## 5.1 GEOLOGICAL SETTINGS OF DAM SITES AND RESERVOIRS

Investigations and analysis of seepage from reservoirs and by-pass around dam sites are crucial matters in dam design. They are particularly important for dams situated on karstified rocks because the necessary preventive measures can be unacceptably risky and expensive.

The nature of karst presents a great variety of risks for dam and reservoir construction. Almost every dam site in karst has its distinct geological and hydrogeological features. A special approach has to be applied in order to counteract seepage from the future reservoir. To deal with karst problems successfully, innovation, engineering practice, its feasibility, and commercial understanding all have to be considered. Adequate investment in investigations is required. *Any restriction on investment in comprehensive site investigation usually results in expensive failure.*

The commonly applied, standard hydrogeological models for seepage analysis and computation of possible leakage from reservoirs are of no use at many dam sites in karst. Hydrogeological singularities, in the form of karst conduits of large aperture, are the main potential leakage pathways. These features are rarely predictable from surface investigations and limited numbers of boreholes. The most modern geophysical methods still have a hard time detecting the spatial location of a cavern or conduit at a depth of more than 20 m. Because these karst singularities are randomly distributed, standard mathematical models cannot be used as reliable tools to solve the seepage problems.

Detailed geological mapping of the dam foundations is crucial for completion of the final geological model for the dam site. In many cases there are caverns in the bedrocks below the body of the dam: Grabovica Dam (Herzegovina), Ourkiss (Algeria),

Salman Farsi (Iran), Logan Martin (U.S.A.), and Boqaata (Lebanon). The chance of these caverns being discovered by investigations from the surface, including boreholes, is less than 10%. In all the cases mentioned here, except the Ourkiss dam site, only dental treatment (exploration/filling/grouting from surface locations) was undertaken.

A detailed geological map and high quality photo data base taken during investigations and construction will be very important if problems appear later during operation of the dam.

The hydrogeological map should contain the following basic information: hydrogeological properties of each lithostratigraphic unit; tectonic (structural) features; estimated position of the groundwater divide with adjacent catchment areas; borehole locations; hydrological monitoring stations; listing of all of the hydrogeological features (springs, estavelles, ponors, collapses, caves, shafts, dry valleys); and results of tracer tests.

One of the crucial geological features in karst hydrogeology is the position of the groundwater divide (watershed). This will generally coincide with the highest elevations of the base of karstification. Unfortunately, because this feature is usually deep underground and not sharply delineated, it is extremely difficult to determine its spatial location. Geo-electrical methods (particularly geo-electrical sounding), using deep boreholes for a variety of different logging methods, and water level monitoring, are the most effective methods to determine the approximate base of karstification.

From the perspective of dam stability, karst features in carbonate rocks are not as hazardous phenomena as is the impermeability (i.e., the hydrogeological properties).

## 5.2  SEEPAGE CONDITIONS IN KARSTIFIED ROCKS

To analyze the natural seepage potential at the dam site knowledge of the following properties is important:

- Lithological composition.
- Structural properties (characteristics and density of discontinuities).
- Rock mass permeability (water pressure tests, pumping tests, slug tests).
- Groundwater regime (groundwater levels, gradients and flow velocities).
- Presence and properties of caverns and conduits.

Lithological composition and structural properties should be determined by detailed surface mapping and analysis of borehole cores. The data collected by the cores are crucial. Obtaining the proper rock mass permeability based on test checking and correct interpretation of collected data is difficult due to the large number of natural and technical factors, most of which are unpredictable. Analysis of the groundwater regime (piezometric data) is crucial because it incorporates all important rock properties, particularly parameters related to the depth of karstification. For a large number of authorities the groundwater level data will be the most important because they represent the hydrogeological properties for a much larger rock volume than is the case with laboratory or downhole permeability tests.

## 5.3 INVESTIGATION BOREHOLES

Karst features, particularly their position and contours, are scarcely predictable from the surface even when the best methods of investigation are applied. As a result, boreholes with core extraction are still the most important and frequently used investigation tool in karstified rocks. Their role is multipurpose: to collect borehole core to determine the lithology and structural properties; to define the quality of the rock mass; to determine the permeability of the surrounding rocks by applying a variety of methods; to monitor water level fluctuations; to be used for geophysical investigations (for application of different logging and cross-hole exploration techniques) and TV camera (video-endoscopy); to take water samples for different analyses; to be used as an injection or observation points for groundwater tracer tests; and to be used for grout mix injection and filling any cavernous spaces. Large diameter boreholes (wells) are used to define hydrogeological parameters by pumping tests, to fill (plug) large caverns beneath the water table, and for the construction of cut-off walls.

The location and depth of the investigation boreholes depend on the surface geology and its interpretation in the form of cross-sections, geophysical data and the characteristics of the proposed dam structure. Proposed borehole depths are not necessarily the final depths because the latter must depend on findings during the drilling. If a supposedly final run encounters a karstified zone or water pressure tests show huge leakage (the required pressure is not reached or the permeability is too high to be measured) drilling should continue until at least 10 meters of solid rock mass is encountered.

If there are upward flows at the dam site, particularly hydrothermal flows, deep boreholes are always required. Pressure, temperature and chemistry of the upward flow should be subject to permanent monitoring.

However the capacities of boreholes to detect karst caverns is still limited. In many projects in karst, large caverns have gone undiscovered in spite of substantial numbers of investigation boreholes and application of the most advanced geophysical methods in them. Due to the fact that average karst porosity is less than 2% in most cases, the likelihood that boreholes will penetrate into the caverns in the rock is negligible. *Even if boreholes are spaced no more than 3 m apart, the probability of detecting a channel with diameter of 0.8–1 m is less than 50%.* Normally, investigation boreholes are spaced 50, 100, 200 or more meters apart. Numbers of karst conduits, even large caverns, can go undetected in between them.

Absence of karst features at the surface does not mean that deep and large karstic features are not developed underground. In the case of the left bank of Sklope Dam, Croatia, drilling failed to detect any large cavities. After excavating just one short gallery, a cavern with a volume of 25,000 m³ was discovered. During investigations at the Salman Farsi dam site, 3,800 m of boreholes were drilled but no conduits or caverns were detected. The results of the excavation of three short galleries in the right bank showed even more advanced karstification than had been assumed earlier. On the basis of these findings the entire design of the grout curtain was changed. In spite of 36,000 m' of boreholes at the Keban dam site, a huge cavern with a volume of 600,000 m³ was not discovered.

## 5.4  PERMEABILITY CHARACTERISTICS DERIVED FROM PRESSURE TESTS AND GROUNDWATER REGIME ANALYSIS

To evaluate the permeability of the rock mass at dam sites and reservoir banks, the Lugeon test and groundwater regime analysis are most widely applied. Among the authors advocating the test are Lugeon, 1933; Wittke, 1968; Houlsby, 1977, 1992; Ewert, 1992; Lombardi, 1996, 2003, and many others. Greater attention was paid to a permeability of karstified rocks perspective by Ivanov, 1976; Selimović, 1977; Božović, 1985; Nonveiller, 1989; Zogović, 1990; Bruce, 2003; Riemer, 2015.

Other tests such as pumping tests, slug tests and air pressure tests are not so frequently used as the Lugeon test. The pumping test is limited to the saturated sections, i.e., it is applicable chiefly below the dam foundations or local river bottom. Results of water pressure tests are expressed as Lugeon units or specific permeability.

The ***Lugeon unit*** represents the quantity of water adsorbed through a one meter section of the borehole at a pressure of 10 bars over a time interval of one minute (Lugeon, 1933).

***Specific permeability*** is defined as the quantity of water adsorbed through one meter of the borehole at pressure of 0.1 bar over a time interval of one minute. Techniques of water pressure testing have been explained in detail by many of the above-mentioned authors.

On the basis of thousands of tests using these different methods, it appears clear that all of them are imperfect, including the widely used water pressure test.

More than 75% of Lugeon tests in karstified rock mass find permeabilities of less than 3–5 Lu. The number of sections with moderate or higher values is much less frequent. Two common examples of the permeability distribution down investigation boreholes are shown in Fig. 5.1.

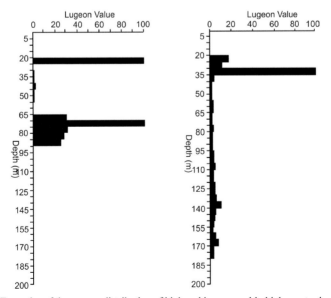

**Figure 5.1.** Examples of the uneven distribution of high and immeasurably high grout mix consumption along boreholes.

In karstified sections of boreholes this method needs some adaptation. Frequently, during Lugeon testing in cavernous zones it is not feasible to reach the largest planned pressure (usually 10 bars). In such cases lower pressure steps (1, 2 and 3 bars) should be applied. If it is still not feasible to reach the desired pressure, the length of test sections should be reduced by using the double packer method (Fig. 5.2). Very often the test of a 5 m section produces unrealistic results because the obtained value for permeability is cumulative (Fig. 5.2(1)). By testing in one m increments (Fig. 5.2(2)) the cavernous zone can be determined much more precisely. It is a very important task, particularly if the cavern is in the grout curtain route and so needs specific treatment.

The relationship between pressure (P in bars) and flow (Q – L/min) provides useful guidance on the permeability of a rock mass. These relationships are presented

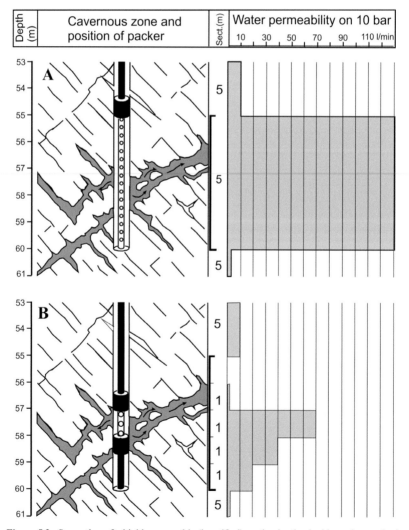

**Figure 5.2.** Separation of a highly permeable (karstified) section by the double packer method.

and analyzed in many publications related to grouting technology, and so need not be elaborated in detail here. Just the generalized pressure-flow relationship is presented in Fig. 5.3.

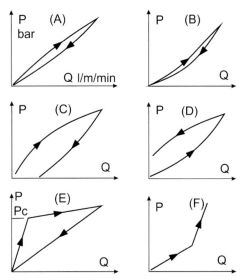

**Figure 5.3.** Generalized relations between pressure and absorption rate (flow): (A) laminar flow; (B) turbulent flow; (C) washing process of joint fill; (D) plugging of joints as a consequence of the test; (E) hydraulic fracturing; and (F) abrupt flow increasing due to technical problems (flow by-pass packer).

In karstified rocks, however, there are some specific features that warrant more consideration. One feature seen in many P/Q graphs in a wide variety of rocks is a sudden increase in water absorption that is due to hydrofracturing. In karstified rock this effect in the P/Q graph can be a consequence of hydrofracturing but, also, may be a result of the presence of a cavern close to the grouting hole. If a bedrock joint involved in an incremental pressure test is connected to a nearby cavity, a pressure step of 25 or more bars (P critical) can open it physically or wash out sediments filling the joint, and so make a strong hydraulic connection with the cavity. As a consequence, an instantaneous increase in grout mix consumption (increasing the flow rate) will be registered (Fig. 5.4).

A well-known condition observed in engineering practice in karst is that the water absorption rates during the pressure test will, in many cases, represent the permeability along the borehole only. In many cases the water pressure tests data indicate conditions directly contrary to the true permeability of the surrounding rock mass. For instance, down a 110 m deep borehole located inside a large ponor zone with a swallowing capacity of 15 m$^3$/s (Cernica Polje, Herzegovina), the Lugeon values were constantly less than 1 Lu. In this case there was no indication that large karst conduits existed close to the borehole. Ground water level fluctuations in this same borehole (a piezometer) clearly showed evidence of a vigorously fluctuating ground water regime in surrounding karst channels.

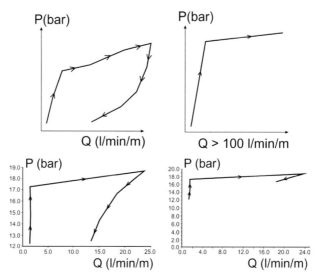

**Figure 5.4.** Relationships between pressure and absorption rates that indicate hydrofracturing or the presence of cavernous spaces close to the grouting borehole.

The commonly used packers are not always watertight. To avoid that problem double packers or special, much longer, packers ('balloon' packers) should be used. The stress field induced by the axially symmetrical outflow from the borehole into the limited rock mass enclosing it is quite different from the real flow that occurs within the surrounding rock mass.

The Lugeon test in karstified rock is not the perfect method according to many authors but it still "remains the main engineering tool in assessing the permeability of dam foundations and in evaluating the achieved efficiency of the grout treatment" (Božović, 1985). Increasing pressure during the test has a tendency to open existing joints, to create new joints or wash out sediments from cracks and cavities. This means that by the standard applied pressure (10 bars) the natural rock permeability can be disturbed. To avoid the destructive influence of a 10 bars application some authors suggest five bars for the nominal pressure.

Interpreting the correlation between Lugeon test data and grout mix consumption is particularly controversial. Discrepancies between these data has been discussed by many authors. For example, Ivanov (1976) studied relationships between the Lu test data and grout mix consumption at Špilje dam (FYUR Macedonia). In 70 sections with permeabilities more than 10 Lu the grout mix consumption was less than 33 kg/m. In contrast, in other 23 sections with permeabilities less than 2 Lu, consumption was more than 500 kg/m. This author also compared Spilje with the grout curtains at the Rama Dam (Bosnia and Herzegovina), which is in a rock mass with similar geological properties. The average permeability of the Rama dam site was 12.6 Lu and average grout mix consumption was 109 kg/m'. The average permeability of the Špilje dam site was 10 Lu and average grout mix consumption was 571 kg/m'.

The grouting records in the case of Karun 3 dam (Iran) show similar relationships between permeability (Lugeon value) and the grout mix comsumption. Sections with high Lugeon values accepted very small amounts of grout mix. Figure 5.5 shows: Left – results of water pressure tests along one borehole, and, Right–a graph comparing the Lugeon test data with grout mix consumption down three boreholes (100 m deep), all situated in karstified limestone. The permeability in the 10 meters section (50–60) is only 3 Lugeon but the grout cement take is 2300 kg. Section 35–40 m has a permeability that is almost four times higher (11 Lugeon units) but the cement take was only 75 kg. These data show different permeability and groutability results in the same sections of borehole when exposed to the pressure of two different fluids: Newtonian (water) and Binghamian (grout mix). This situation is confirmed at many other sites in other karstified rocks.

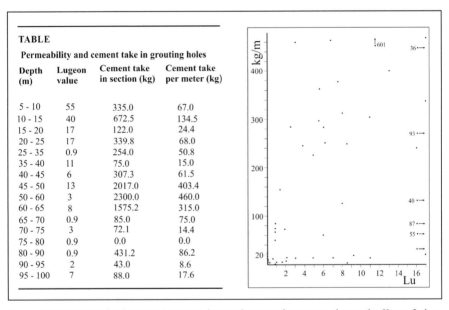

**TABLE**

Permeability and cement take in grouting holes

| Depth (m) | Lugeon value | Cement take in section (kg) | Cement take per meter (kg) |
|---|---|---|---|
| 5 - 10 | 55 | 335.0 | 67.0 |
| 10 - 15 | 40 | 672.5 | 134.5 |
| 15 - 20 | 17 | 122.0 | 24.4 |
| 20 - 25 | 17 | 339.8 | 68.0 |
| 25 - 35 | 0.9 | 254.0 | 50.8 |
| 35 - 40 | 11 | 75.0 | 15.0 |
| 40 - 45 | 6 | 307.3 | 61.5 |
| 45 - 50 | 13 | 2017.0 | 403.4 |
| 50 - 60 | 3 | 2300.0 | 460.0 |
| 60 - 65 | 8 | 1575.2 | 315.0 |
| 65 - 70 | 0.9 | 85.0 | 75.0 |
| 70 - 75 | 3 | 72.1 | 14.4 |
| 75 - 80 | 0.9 | 0.0 | 0.0 |
| 80 - 90 | 0.9 | 431.2 | 86.2 |
| 90 - 95 | 2 | 43.0 | 8.6 |
| 95 - 100 | 7 | 88.0 | 17.6 |

**Figure 5.5.** Relationships between Lugeon values and grout mix consumption at the Karun 3 dam. Left: Table; Right: Graph (results from three investigation boreholes).

Similar discrepancies between Lugeon values and grout mix consumption have been reported at many other dam sites in karstified rocks. One of the examples is the results obtained during construction of a grout curtain in the reservoir bank of the Elati basin, Ilarion Reservoir, Greece (Fig. 5.6). It is clear that in the sections with 10 Lu or less, consumption can be 1000 kg/m' and between 20 and 100 Lu can be less than 5 Lu.

According to the experience gained at many dams in karst this unexpected relationship "*low Lugeon value/huge grout mix consumption, and vice versa*" should not be considered unusual: rather, it is a frequent and common finding.

While dam sites usually have geologically and geometrically identical permeability on both banks, in many cases this will not be true. Results of water pressure tests at the Bakhtiary dam site (Iran) from an elevation 940 m down to 385 m showed that permeability in the right bank is quite different from that in the left bank (Fig. 5.7).

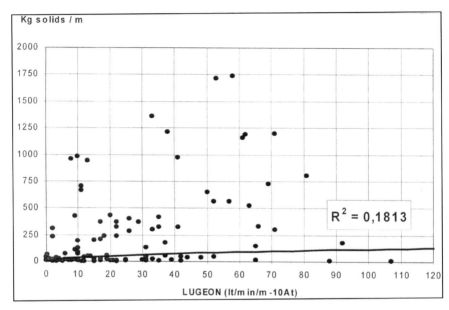

**Figure 5.6.** Correlation between Lugeon values and grout mix consumption, Ilarion Reservoir, Greece (Public Power Corporation, Athens).

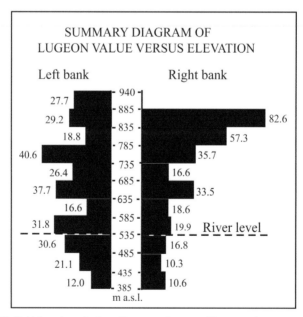

**Figure 5.7.** Bakhtiary dam site, Iran. Summary diagram of Lugeon values versus elevation.

The relatively high permeability on both sides is a consequence of an extremely tectonized rock mass with a number of wide open discontinuities; the intensity of karstification is low.

## 5.5  KARSTIFICATION AND HYDRAULIC CONDUCTIVITY

To analyze rock mass permeability at dam sites, a key parameter is the coefficient of hydraulic conductivity. In karstified rock masses this parameter cannot be defined by adopting the common hydrogeological approaches associated with the Darcy Law.

On the basis of pressure tests (Lu tests) an approximate hydraulic conductivity is:

- 1 Lu (lit/min/m'/10 bar)  =  1.3 x $10^{-5}$ (cm/s)
- 10 Lu                            =  1.3 x $10^{-4}$
- 50 Lu                            =  6.5 x $10^{-3}$
- 100 Lu                          =  1.3 x $10^{-3}$

From this scale, a permeability range between 1 and 10 Lu would appear to imply the absence of karstification in the surrounding rock mass. Only the permeabilities between 50 Lu and 100 Lu may be the consequence of an intensively fractured or a slightly karstified rock mass.

However, intensive or high degrees of karstification cannot be gauged by the Lugeon values, as has been shown clearly in the examples cited above. The Lugeon test does not offer the technical possibility of determining the exact value of the hydraulic conductivity in cases where there is intensive karstification. In these cases the permeability should be expressed as > *100 Lu* or as ***undefined high permeability***. The presence of karst conduits in the vicinity of the boreholes can only be suspected but not unambiguously determined.

The key hydrogeological feature of karst is the existence of '***hydrogeological singularities***' (solution conduits and cavities). In the case of karst channels with apertures of more than 20 cm the velocity of the water flow is usually more than 0.5 cm/s.

## 5.6  ADITS (INVESTIGATION GALLERIES, DRIFTS)

Adits are the unavoidable and most effective method of investigation when the dam site is on or in karstified rocks. A large number of important geotechnical parameters of the rock mass at such dam sites cannot be defined by surface observations and borehole investigations alone. Discontinuities such as bedding planes, joints and faults are essentially planes of weakness in the rock. Solution features are developed along these discontinuities. However, on the surface significant properties of discontinuities are mostly masked by different weathering processes. Deduction of their likely properties at depths of 100 m or more is not easy and is usually questionable. Boreholes represent the rock mass properties only in their close vicinity and in many cases do not provide enough good quality data. Adits alone allow direct observations of the geological properties of the rock mass as a whole and permit use of different methods of investigation deep below the surface. In karstified rocks the key problem (hazard) is the cavern or conduit, its spatial location, size and properties, empty or filled.

At many dam sites, in spite of large numbers of boreholes and the use of a variety of different geophysical methods of investigation, large caverns were not discovered before the adits were excavated (Salman Farsi Dam, Iran; Sklope Dam, Croatia;

Wudeongde dam site, China; Wulichong Reservoir, China; Khao Laem Dam, Thailand; Ombla underground dam project; Karun 4, Iran).

In many cases after the adit excavation, speleological investigations are also required. Along and between adits different geophysical and geotechnical methods are applied. The locations of the adits (elevation, orientation and length) should be based on the detailed geological mapping, geophysical investigation data and findings in boreholes.

The number and position of adits may differ, from routinely two or three in each bank of the dam, up to a large number of them. Usually the target is detection of cavernous spaces. After such a space is detected, the adit should follow the discontinuity along which the cavern is developed. In many cases in karstified rocks the findings in one adit will require construction of a further one or more of them. Another approach is the excavation a few mutually connected adits in a regular geometrical arrangement at different levels. Besides the detection of caverns, such arrangements are suitable for detailed geological mapping, the application of many different inter-adit geophysical methods, local tracer tests and borehole construction.

At the Sklope Dam site and Ombla underground dam site, large caverns were discovered after excavation of adits no longer than 10 m each. At the Keban dam (Turkey) the experience was different. In spite of cutting 11 km of galleries, a large cavern (volume 600,000 m$^3$) was not discovered: it was detected only during the first attempts to fill the reservoir! During investigation of the Oymapinar dam (Turkey) 33 adits with total length of 10 km were excavated.

Investigation adits are the basic tools in the Wudongde project (China), which is located in heavily karstified Middle Proterozoic rocks—limestone, metamorphic limestone and marble. Of 154 investigation boreholes, only nine of them detected any small caverns. 186 investigation adits (total length 16.516 m) were then excavated in addition (Fig. 5.8). In 77 adits 90 small size caverns and one huge cavern (on the right bank) were discovered.

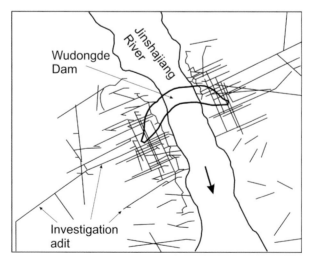

**Figure 5.8.** Wudongde dam site, China. Position of the investigation adits. Changjiang Survey, Planning, Design and Research Co., Ltd. Wuhan, 2011.

This cavern was investigated by eight levels of adits between elevations of 860 m and 1060 m above sea level. Based on this dense grid of adits, a cavern volume of 369,000 m³ was determined. This cavern is now filled with limestone blocks, crushed limestone, silty clay and sand (data by Three Gorges Geotechnical Consultant Co. Ltd., 2011).

In the case of dam sites where there is a cover of deep overburden, adits are also used as locations for the boreholes and for geophysical investigations (Fig. 5.9).

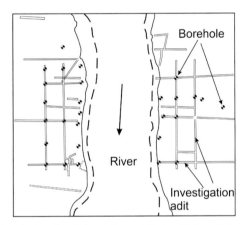

**Figure 5.9.** Wudongde dam site, China. Adits used for installation of piezometric boreholes.

One uncertainty in using adits is their potential destructive role within the dam site rock mass. Each adit may become an artificial karst channel. The surrounding rock can become highly disturbed due to the blasting. Almost always during dam construction all adits have to be completely plugged. This procedure requires, sometimes, additional complicated grouting works. The worst situation occurs if an adit is left unplugged. It then can play the role of an untreated karst channel and become a source of major leakage. Its treatment may need reducing the water table in a reservoir in order to plug it. In cases where the dam and all other structures have already been constructed, this operation can be very complicated and expensive.

# CHAPTER 6

# WATERPROOFING IN KARST

## 6.1 GENERAL ASPECTS

As a consequence of the nature of karst, it is practically impossible to select a dam and reservoir site that will be acceptable for construction without any application of waterproofing technologies. Due to the distinctive nature of karst porosity in the form of hydrogeological singularities, seepage occurs through ponors, shafts, caverns and solution-widened (karstified) joints at the surface. In many cases, in this environment grouting alone is not an adequate geotechnical measure. To reduce or eliminate seepage or greater leakage, specific waterproofing technologies and a certain flexibility during the building of the remedial structures are needed.

The earliest grouting, using clay as grout mix, was undertaken in France more than 200 years ago (1802). Since that time modern grouting techniques were developed, particularly during the second half of the 20th century. Many articles and books have been published on the subject of rock mass grouting. A wealth of information about rock properties, technologies, materials and numbers of case studies have been presented and analyzed. However, the majority of these publications are focused on the grouting of granular and fractured rocks.

One of first cases of grouting and plugging of karst caverns occurred during construction of the Hales Bar Dam (U.S.A., 1905–1913). Asphalt, cement and concrete were placed in the cavities. Basic experience in grouting was obtained during the construction of many dams and reservoirs as part of the Tennessee Valley project (U.S.), where many of the large structures were located in karstified rocks. Massive grouting works and plugging of karst caverns were carried out in the period, 1922–1945.

The first karst treatment in Russia (Soviet Union) was done during the construction of the Volhovskaya Dam in 1926. A few more dams in karst were constructed in 1955–1957. A number of water proofing methods in karst were evaluated during construction of the Kahovskoy Dam (1956) on the Dnieper River, Ukraine.

Due to high water potential in the Dinaric karst (a major area of former Yugoslavia) grouting technology became an important issue in the successful construction of dams and reservoirs. In 1948 a national board for the investigation and selection of proper grouting technologies in karst was founded. The first conference on 'Consolidation in

Karst' was organized in 1961 in Split, Croatia. More than 30 reservoirs in karstified rocks were constructed, largely between 1952 and 1975. The largest are Peruča, Krupac, Slano, Grančarevo (Bileća Reservoir), Špilje, Buško Blato and Mratinje (Piva). One of the largest and most successful reservoirs that is located entirely in highly karstified rocks is Bileća Reservoir (1.27 billion m$^3$).

In the early 1960s the construction of a number of dams in the Zagros Mountains of western and southern Iran was begun. This mountain chain, rich in groundwater, is composed mostly of karstified carbonate rocks. A large part of it belongs to the Karun River catchment. The most prominent dams are Karun 1, 3 and 4, Dez, Gotvand, and Rudbar Lorestan. Well known dams in Iran with karstification problems are: Lar, Salman Farsi, Marun, Seymareh and few more.

During the second half of the 20th century massive grouting was undertaken at many dam sites located in karst in other nations: U.S.A., Turkey (particularly Keban Dam), China, Honduras, Thailand, Vietnam, Switzerland, France, Spain, Iraq, Greece and others.

Based on this experience, a substantial number of monographs and articles have been published. A selection of the books focused on grouting in general are by Lugeon (1933); Ewert (1985); Houlsby (1989); Nonveiller (1989); Lykoshin et al. (1992); and Lombardi (2003). Grouting and plugging problems and methods in karst have been presented primarily in individual papers and reports relating to particular dams and reservoirs, rather than in dedicated volumes.

The seepage problem is of lasting interest for dam geologists and engineers and has become one of the primary topics at many ICOLD Congresses. Karst as unfavorable for water storage has been the subject in many reports. Particularly important is Question 58, 'Foundation Treatment for Control of Seepage' presented at the Lausanne ICOLD Congress (Božović, 1985).

The cost of attempting construction in karst is sometimes an expensive failure due to massive seepage. Some modern dams and reservoirs have been abandoned entirely: Hales Bar and Center Hill Dam, U.S.A.; Montejaque, Spain; Vrtac Reservoir and Liverovići Dam, Montenegro. Other dams operate with great leakage: Lar, Iran; Salakovac, Buško Blato and Višegrad in Bosnia and Herzegovina; Samanalawewa, Sri Lanka; May, Onac, Keban and Ataturk in Turkey; and Mosul in Iraq. Reservoirs in karst that were impossible to impound have become frightening memories for dam engineers. As a consequence, the term 'karst' has a bad connotation—it implies a likelihood of failure. Long-time dam engineers have tended to avoid karst areas in spite of the fact that many offer sites that are excellent from the topographic and geotechnical (mechanical strength) perspectives. However, the period of avoiding karst has been of a relatively short duration; with the increasing demands for water resources in karst areas, particularly for irrigation and power production, the issue has become one of knowing how to select safe and watertight dam and reservoir sites in them. The need for efficient waterproofing techniques has increased tremendously. Due to increasing understanding of the distinctive geological properties of karstified rocks and improving remediation concepts and technologies, a number of recently constructed dams and reservoirs are operating successfully.

## 6.2 KARST FEATURES AND PROCESSES – SOURCES OF LEAKAGE

Because of the distinct hydrogeological and geotechnical properties of karst, it is practically impossible to foresee all of the possible reactions and disturbances that may arise as consequences of building artificial (engineering) structures: dams, grout curtains, underground excavations and reservoirs. Problems will appear at the surface or underground during the construction period but most tend develop during the operating time afterwards.

Caverns, empty or filled, are a major obstacle in the dam foundation area, along the grout curtain or tunnel route and on the bottom or banks of reservoirs. Filled and empty caverns and channels exist in the temporarily (or seasonally) saturated zone and also in the zone of base karst flow (phreatic zone).

Cave fillings differ but mostly contain fallen carbonate rock blocks, with sandy clay fillings, or very plastic clay only. In undisturbed natural conditions a small proportion of clay may be eroded in instances of extreme rains that generate some turbulent flow but, due to the natural cohesive qualities of clay, underground erosion by flowing water tends to be negligible. However, after construction of dams, grout curtains or grouting galleries, this natural equilibrium condition can be drastically changed.

Every manmade structure provokes drastic disturbance of a number of parameters in the karst aquifer, particularly its mechanical erosion capabilities. As a consequence of dam construction, differences in height between water upstream and downstream of the structure become very great in a (geologically very short) period of time. The pressure in the karst aquifer behind the dam is dramatically altered. The velocity and energy of turbulent flows are greatly increased. Reactions can range from catastrophically rapid erosion and sudden increases in leakage to long-term progressive erosion that may not be immediately noticeable.

In the case of the Samanalawewa Reservoir in Sri Lanka, during reservoir filling "a sudden burst originating at the leakage outlet resulted in washing away of about estimated 25,000 m$^3$ of the right bank earth material within 10 hours" (Laksiri et al., 2005); see more in Chapter 13.

During excavation of the Kuhrang III Tunnel in Iran, water inflow from a small cavern at the tunnel face began at a rate of less than 10 L/s. After four hours the rate had increased to 1160 L/s; within 24 hours, more than 1000 m$^3$ of sediments (boulders, gravel, sand and silt) had been washed out from an undetected cavernous system ahead of the tunnel face.

The progressive type of erosion has been observed in the rock foundations of many dams situated in karstic areas. During the first filling of the Višegrad Reservoir (Bosnia and Herzegovina) in 1986, seepage of 1.4 m$^3$/s was found deep beneath the dam foundation. In spite of intensive re-grouting, the leakage rate gradually increased to 6.5 m$^3$/s (1996), 9.4 m$^3$/s (2003) and 13.92 m$^3$/s (2008). During this period more than 50,000 m$^3$ of clay has been washed out from joints and caverns; see more details in Chapter 13.

In some other cases the effects of progressive erosion have not been so massive and destructive, despite continuing for a long period. Seepage below the Špilje Dam of about 340 L/s began in 1987; up to 1995 it increased to only 380 L/s.

Due to progressive erosion at the Gorica Dam site (Herzegovina), seepage of 1.5 m³/s in 1966 increased to 4.5 m³/s in 2003. A sudden squeezing out of clay from karstified joints below the dam foundation occurred after there was a large earthquake (magnitude 4.8) on the Adriatic coast with the epicenter at a distance of 80 km. Muddy water discharges 10 to 50 m downstream from the dam looked like boiling water, with chunks of compact clay (1–1.5 kg in weight) swirling in them.

After construction and remedial works at the El Cajon Dam (Honduras) initial seepage of 1.65 m³/s was reduced to 0.1 m³/s (1995/96). However, due to progressive erosion, seepage had gradually increased to 2–3 m³/s by 2015.

To prevent seepage from reservoirs in karst there are three waterproofing alternatives:

- *Underground alternative* by constructing grout curtains, cavern plugs and cut-off walls,
- *Surface alternative* by applying various structures and techniques on the bottom and banks of the reservoir.
- *Coupled Surface and Underground.* In many cases one of the above approaches cannot be effective on its own. To achieve acceptable seepage in such cases, coupled surface and underground treatments are necessary, e.g., at Buško Blato Reservoir, Bosnia; Akköpru Reservoir, Turkey.

## 6.3 UNDERGROUND WATERPROOFING

### 6.3.1 Waterproofing Structures

The basic objective when designing a proper waterproofing structure is to block seepage from the reservoir by intersecting the rock mass and occasional karst conduits and cavities in the most efficient and economic way. To achieve this, the rock mass should be grouted with a suitable grout mix, and all caverns plugged down to a certain depth to achieve the desired permeability.

The common materials for sealing the underground by grouting and plugging are cement, clay, and bentonite, in the form of a thick or thin grout as the basic suspension, and in particular cases polyurethane foam, asphalt, sand, gravel synthetic sponge, rock blocks, self-compacting concrete, reinforced concrete plugs, cut-off (diaphragm) walls and a range of chemicals.

There are no general rules established for foundation treatment, grout curtain treatment design, or cavern plugging technique in karstified rocks. Each dam has therefore to be evaluated by the particular conditions prevailing at the site and to be treated according to general experience and engineering judgment. It is in the nature of the karstic environment that the design work is only finished when the construction is finished. *Modifications to the design during waterproofing construction in karst are therefore the rule, not the exception.*

Basic underground sealing structures are: the grout curtain (cut-off); positive cut-off; cut-off wall; bath-tub structure; underground retaining walls and concrete plugs.

*A grout curtain (grouted cut-off)* is constructed by injecting the grout suspension through injection boreholes, into the pores, joints and cavities in any kind of rock. This structure should be thin but as watertight as possible. An absolutely watertight grout curtain is not realizable, however. The location and extent of the grout curtain (laterally and in depth) depends on local conditions in the rock mass.

*A Positive cut-off* is a grout curtain that is extended tightly into adjoining impervious geological formations where the hydrogeological properties and geological structure allow this.

*A Cut-off wall (diaphragm wall)* is a linear and compact body (clay, cement or concrete) constructed from drilled overlapping piles or along deep excavated trenches.

*A 'Bath tub structure'* is an underground structure that usually consists of mutually connected vertical, inclined and sub-horizontal grout curtains (without an open bottom) that is emplaced to completely isolate a hazardous part of the rock mass, i.e., to prevent or minimize filtration into or out of the enclosed rock mass.

*An Underground water retaining structure* is a dam constructed by reinforced concrete in a karst channel or large cavern in order to block groundwater flow.

*A Concrete plug* is a large cavernous space entirely filled by concrete (self-compacting concrete) or with pre-compacted concrete. Pre-compacted concrete (grouted aggregate) is very useful for plugging caverns and channels in saturated (sub-watertable) karst rocks and in open conduits with active stream flow.

All of these waterproofing structures are widely applied in karstified rock masses. However, the most important is the optimum grouting of solution-enlarged joints and cavities, together with plugging any karst channels and caverns along the curtain route.

## 6.3.2  Determining the Grout Curtain Route

The key parameters for selection of the routes for grout curtains at dam sites in karstified rocks, are careful interpretation of the following categories of information:

- the general structural and hydrogeological properties of the rock mass.
- the position of the minimum groundwater level and the regime of GWL fluctuations.
- water pressure test data in boreholes located beneath the dam and in both dam abutments.
- water pumping or infiltration tests at selected locations.
- local tracer tests.
- detection (as much as possible) of the position of caverns in the zone around the potential course of the grout curtain.

Determination of structural and hydrogeological data and their interpretation are the key starting points in selecting the proper alignment of a grout curtain. Geological parameters include: lithology, inclination of strata, dip direction and properties of discontinuities, rock mass strength and deformation behavior. Geological structural relationships between the karstified and any impermeable rocks (if the latter exist) may play a decisive role in selecting an efficient grout curtain route and its lower limit underground. In such cases, use of a positive cut-off structure is the best solution.

A water pressure test (Lugeon test) is the method routinely applied to determine the permeability of rock masses at dam sites. Due to the double porosity nature of karst permeability (karst channels situated inside a compact rock mass of low permeability) about 80% of Lugeon values are usually less than 3 Lu. In a karstified but lithologically uniform and very deep rock mass, a suspended curtain is the only solution. In this case, the water pressure test data are the main engineering tool in assessing permeability along the curtain route, particularly for determining the depth of the base of karstification.

Groundwater level data also play an important role in determining the depth of the grout curtain. Usually the minimum groundwater level will be in the zone of the base of karstification, i.e., close to strata with low (unkarstified or much less karstified) permeability. For suspended grout curtains that zone is usually accepted as sufficiently watertight at a permeability of less than 3 Lu. For many dam geologists working in karst, the groundwater regime will take priority over the Lugeon tests data when selecting the basal contour line.

Generally speaking, grout curtains in dam abutments should be oriented in the upstream direction. However, because there are numerous uncertainties in karst terrains, the alignments in left and right dam banks can be quite different. Figure 6.1 shows a

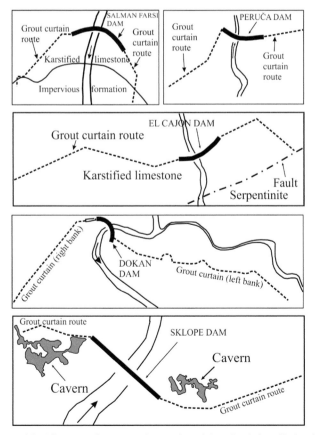

**Figure 6.1.** Alignment of grout curtains at selected dam sites in karstified rocks.

few simplified sketches where grout curtain alignment has accommodated particular structural and hydrogeological conditions in some dams in karst.

In Fig. 6.1 the curtain alignments were selected on the basis of geological structural and permeability data and the groundwater fluctuations. In some cases intensive karstification played the decisive role. For example, at Sklope Dam the grout curtain was redesigned to pass downstream of one large cavern on the left bank and upstream of a similar cavern on the right bank. In the case of Salman Farsi Dam grout curtain in both abutments is orientated in a downstream direction to be tight in impervious formation.

### 6.3.3  Grouting Galleries

At a number of sites in karst with hanging curtains down to depths of 200 m or so, successful grouting has been achieved using open courses along the perimeter and on top of the dam. However, due to the rugged nature of karst topography and the great height of many dams, construction of grout curtains from the surface alone is not possible at many sites. Another uncertainty is the possible (and often expected) occurrence of caverns along the curtain route. To detect and plug caverns along grout curtain routes in the abutments and dam foundations from the surface only becomes impossible. Grouting galleries must be installed. For detection and easy access into each cavern, the vertical distance that is selected between galleries is also of great importance.

For instance, in the case of Khao Laem Dam (Thailand) total gallery length is 22 km and vertical distance between galleries is 14 m. At Keban Dam (Turkey) 11 km of galleries have been constructed; two of them were driven beneath the bed of the Euphrates River. 11.5 km of galleries was excavated for the grout curtain at the El Cajon dam (Spain) and 10 km of galleries for the Oymapinar Dam (Turkey). To achieve the required water tightness beneath and in the flanks of the Salman Farsi dam site 5.82 km of galleries were excavated: the length of grouted sections in the galleries is 3.75 km and their vertical separation is between 18 and 33 m.

The routing and vertical separation between galleries depends on numerous geological and hydrogeological findings during investigations so it is not possible to set down strict rules. However, on the basis of current experience, the vertical distance between grouting galleries in highly developed karst should be not more than 30 m.

In some cases a designed grout curtain route can be extended as a consequence of new geological findings during the underground curtain construction: often the route will need modification (realignment).

### 6.3.4  Grout Curtains – General Design Criteria

The basic objective when designing a proper grout curtain is to block seepage from the reservoir into the river valley(s) downstream by intersecting the rock mass and occasional karst cavities in the most efficient and economic way.

1. To achieve the desired impermeability, the rock mass should be grouted with suitable grout mix to a determined depth below the dam and into both abutments, based on geological and hydrogeological assessments of the rock mass.

2. In general, criteria are related to the objectives of a project. In the case of a small reservoir for potable water, particularly if is situated in an arid or semi-arid area, requirements for seepage limitation will be extremely strict. If some 'Guaranteed Ecological Flow' is required downstream of the dam, this reduces the need for the curtain to be highly impermeable. If a reservoir has a large volume or there is continuous flow into it, comparatively high leakage rates (5–10% of inflow) can be accepted, provided (of course) that the stability of the dam is not endangered, i.e., if there is no risk of progressive erosion along the leakage routes.

3. As noted, the vertical distance between grout galleries should not exceed 30 m, in order to intersect as many karst features as possible with local galleries or shafts.

4. The one-row curtain, i.e., the single plane curtain, is suggested unless exceptional conditions call for a multiple row curtain in sections with high grout take and high radius of penetration (i.e., in heavily tectonized or karstified rock).

5. To justify the advantage of a one-row curtain, high pressure and stable grout mixes should be applied.

6. Orientation of grouting holes should be angled to intersect the orientation of the dominant bedding planes and joint systems in the most efficient way; i.e., to cross them at right angles where possible. Therefore, grouting holes must be designed on the basis of detailed structural analysis of the dam site. Most important are the systems of discontinuities (bedding planes, joints, faults) along which the principal karst channels have developed. Usually, the curtain is a combination of vertical and upstream-inclined planes. Sometimes it is difficult to apply this criterion due to a desire to orient all curtain planes with an upstream inclination.

7. Grouting procedure will be done by the 'split spacing' method with a staged execution of, first, P (primary), then S (secondary), T (tertiary), Q (quaternary) and (B) quinternary holes; (C) – check holes. The standard spacing of P-holes is 4 m, with a proportional decrease of the spacing for lower category holes: if the geological properties permit, spacing of P holes can be increased to 8 m.

8. A stable grout mix (decantation of less than 5% after standing for two hours) with a ratio of water/dry component or water/cement of 0.6 to 1.0 is recommended. In many cases in karst a ratio of 1.0 has been successfully applied as the starting ratio. Application of thicker suspensions (1.5:1) has also given good results.

9. High grouting pressures are suggested (max 40 to 50 bars).

10. In the case of large joints and cavities with apertures less than 20 cm, sand (up to 4 mm diameter) has to be added to create a grouting mortar.

11. A permeability of 3 Lu below the dam and 5 Lu in the lateral parts are the general criteria to be applied, according to suggestions presented in Q 58, Foundation Treatment for Control of Seepage (ICOLD 1985).

12. The objectives of the usual closing criteria (criterion of water permeability and curtain sustainability) are 3 Lugeons for 90% of check holes and grout take less than 50 kg/m.

13. Comprehensive cylinder strength after 28 days shall be 10 MPa.

## 6.3.5  Exceptions in the Case of Hypogene Karstification

The possibility that there is simultaneous karstification by both meteoric water and water of hypogene origin has to be anticipated at many dam sites. To apply successful measures against leakage, correct diagnosis of the origin of any caverns is essential for proper design of the grout curtain (alignment, depth, inclination, properties of the grout mix). In particular, the existence of karst features of hypogene origin at dam sites makes construction of watertight barriers more complicated.

From the practical dam engineering view point two general questions arise in the case of hypogenic karstification at dam sites:

1. How to determine proper alignment of grout curtains (length, inclination and, particularly, the depth); and,
2. How to select the proper technology for grouting and cavern plugging beneath and above the water table.

Essential for selecting the proper depth and alignment of the grout curtain is the determination of the three-dimensional spatial locations of the zones with karst channels that have a role of principal ways of ascending hydrothermal and/or $CO_2$-rich or $H_2S$-rich, waters. To create watertight reservoirs, all of these flow routes should be blocked by the grout curtain. With respect to chemistry of these flows, the grout mix has to be resistant to pressure and the destructive action of $H_2S$ and its products such as native sulfur or $H_2SO_4$. Plugging weakly consolidated cave sediments below the water table is an important but technically complicated task. Particularly sensitive is the case where there is high pressure with rapidly ascending hot and mineralized water. At the Salman Farsi dam site, there are two deep rising waters, one thermal and one cold. Sometimes, even where the grouting boreholes are very closely spaced, one hole will intercept hot water, the next will be cold. The full shape and depth of Golshan's Cavern, a probable hydrothermal cavity, was never completely established at this site.

The verticality of rising flows and the (mostly) unknown depths and locations of their sources are unfavorable conditions because the ordinary purpose of grout curtains is mainly to prevent sub-horizontal groundwater filtration below the dam. Grout curtains should be deep enough to block rising flows.

## 6.3.6  The Final Layout of the Grout Curtain

In a karstic formation with a highly random distribution of solution features some uncertainties will always remain. Conventional investigation methods are not powerful enough to detect the spatial locations and sizes of all caverns and karst channels during the initial investigations. Due to this characteristic of karst, the design work is only finished when the curtain construction is completed. Modifications based on findings during construction are therefore the rule and not the exception, as emphasized above. Extending the grout curtain beyond the designed limit, increasing its depth, changes of numbers of rows and grouting holes patterns, are routine practices in many curtains in karst.

One of the important questions in grout curtain layout is whether to construct the curtain from grout one gallery tunneled into the banks at the dam site, or to do it using open courses on top of the surface, or from a few closely spaced galleries. From experience at a number of dam sites, successful grouting was achieved from top galleries down to depths of 150–200 m. However, in heavily karstified rock with numerous large caverns, using galleries with a vertical separation of not more than 30 m is the best practice, as suggested in 'Grout Curtain – General Design Criteria' above. Closely spaced galleries make for easier detection of caverns along the curtain route, access to the caverns (by shafts or adits), and cleaning and plugging each cavern separately. The Khao Laem Dam in Thailand and Salman Farsi Dam in Iran are the most significant examples. The vertical separations between the six galleries in the Khao Laem Dam was 14 m. The Salman Fasi grout curtain was constructed from five galleries with vertical separations of 18–36 m.

In some cases the size of the karst features discovered by the galleries is too large and the problem is too complex or expensive for remedial works. In such cases, modification of grout curtain alignment is the only solution. An example is the modification of alignment in the right bank of the Salman Farsi Dam, Iran. Due to a large cavern detected in the proposed curtain route (chamber length 130 m, height more than 70 m and width 15–25 m, plus a number of secondary caverns) the original curtain route had to be redesigned. A new upstream bypass route (between the reservoir and cavern) was selected as the technically and economically most feasible solution (Fig. 6.2). In the original tendering design the length of the grouting section was to be 2,697 m. After the large cavern and others were found along that proposed route, the length of new grouting sections, redesigned and executed, was 3,752 m.

Curtains required also considerable modification and adaptation in the cases of Sklope Dam (Croatia), Seymareh Dam (Iran), Clarence Cannon Dam (U.S.A.), Chichik Dam (Uzbekistan) and others.

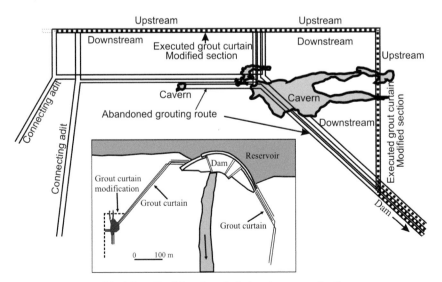

**Figure 6.2.** Salman Farsi dam, Iran. Redesigned grout curtain alignment.

## 6.3.7  Grout Mix (Suspension)

A grout mix consists of several constituents combined in different proportions. The key properties of the mix are: mixing ratios, stability, resistance against mechanical erosion and chemical corrosion, water permeability and thixotropy. The properties of flow of the grout mix (a Bingham fluid) are defined by its stability. The terms *stable* and *unstable* suspension describe the structure of the resulting sedimentation. A grout mix in which the distribution of the particles is very uniform once they have settled is considered a stable suspension. Different kinds of admixtures (bentonite or plasticizers) are used to increase stability of the mix, which can be determined by simple tests in the laboratory. In the example presented in Fig. 6.3 it is clear that 2% of super-plasticizer increased the stability of the grout much better than in the experiments where no super-plasticizer was added.

From its major constituents a grout mix may be: a *cement suspension, clay-cement suspension, clay-cement-bentonite suspension, clay-bentonite suspension and chemical grout.*

The common ingredients for grout are the dry components (cement, clay, bentonite, sand, fillers, additives/stabilizers) and water. The grout mix is defined as the proportion of the dry component by weight divided by the quantity of water needed to make a suspension (DC/W).

Grouting technology is based, chiefly, on use of cement as the key component. Cement-based grout mixes consist of conventional or microfine cement, aggregates (fly ash, bentonite, sand, slag, sawdust), additives and water. Because cement is the base of the mix, the cement suspension is defined as the water/cement ratio by weight (W/C). The roles of the W/C ratio and superplasticizers to achieve the required strength, viscosity and stability of grout suspension is presented in the examples in Fig. 6.3. It is evident the settling of a simple mixture of water with cement is much faster than the suspension with an additive (super-plasticizer).

Clay-cement, clay-cement-bentonite and clay bentonite grout mix are clay-based suspensions. Bentonite is impure clay consisting mostly of montmorillonite—usually more than 70%. Montmorillonite is a clay mineral that weathers from igneous rocks (chiefly, volcanic ash) in warm wet climates. Particles of montmorillonite are 1–2 μm in plain view, thickness ranges from $10^{-3}$ μm to 2 x 10 μm, the specific surface area is 50–120 $m^2$/g when dry and 700–840 $m^2$/g after swelling and dispersing in

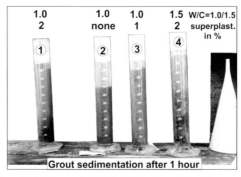

**Figure 6.3.** Samples of grout mix (suspension) with different quantities of superplasticizer and W/C ratio.

water. Montmorillonite is used for the preparation of stable cement suspensions, for plasticizing thick mortar for grouting large fissures and cavities (Nonveiller, 1989). The properties of swelling on contact with water make bentonite a good sealing material.

In a clay-based grout mix the content of clay or clay + bentonite is more than 50%. This grout mix has high thixotropy, i.e., it develops strength by forming soft gels when at rest. In a clay-cement suspension, the percentage of cement is 35–40%. Clay-based suspensions have been very successfully applied in some grout curtains in karst (La Amistad, Rio Grande, Mexico; Gallo Matese, Italy; Limmernboden, Switzerland). In the case of the Canelles Dam, Spain, the suspensions were: clay 70 to 60 kg and cement 30 to 40 kg, water 100 L. In the Dinaric karst clay-based suspensions have been used for at least 18 grout curtains. A clay-cement stable grout mass was used in the Peruča Dam, 75% clay; Buško Blato, 55 to 75%; Rama, 45 to 57%; and Slano Reservoir, 50 to 65%. For waterproofing cavernous spaces at the 123 m high Grančarevo Dam site (Bosnia and Herzegovina) the following grout mix suspension was successfully applied: clay – 66%, bentonite – 33% and Sodium Hydroxide – 1%; DC/W = 1:2.5. For special cases here, a mix: cement – 35%, clay – 43%, bentonite – 21% and $Na_2CO_3$ – 1% was used (Karamehmedović, 1965).

**Polyurethane foam** is a gaseous emulsion of uniform, very small diameter bubbles. It has a very high coefficient of expansion—the final volume can be more than 20 times larger than the initial one. After being mixed with a cement suspension the foam expansion is drastically lower.

**Diesel oil** and **cotton flocks** were used to plug the main leakage zone in the Dokan Dam (1955–1960) in Iraq.

Dry **sawdust** is sometimes used. If added immediately before pumping, the sawdust absorbs water and in this way reduces the fluidity of the grout mix.

**Asphalt** grouting was used for sealing leakages at the Hales Bar dam site (1919), Clayton Dam (1939/41) and Great Falls Reservoir, all in the U.S.A. It is not a common injection material in modern grouting practice.

Within the last few years **hot bitumen** has been used in U.S.A. for plugging large underground water flows. The bitumen is pumped continuously, through specially installed pipes, into the karst stream channels to be sealed off. According to Bruce (2003)—"The hot bitumen encounters the water which quickly removes the heat from the material (injected at temperatures of 200°C and over). The material begins to gel and congeal and thus, when pumped at sufficiently high rates, will begin to overwhelm the water flow".

The hot bitumen technology was successfully applied in Canada to plug the karst channels with active flow in the case of Stewartville Dam, Ontario (see Chapter 13).

**Molten asphalt** was successfully used to seal off fractured zone and karstic voids encountered at depths of 80–86 m (Abkemeier and Stephenson, 2005).

### 6.3.8 Grout Mix Penetration

The characteristic double or triple porosity that plays such an important role in the groundwater regime of karst aquifers is necessarily very significant when grouting.

Due to the multi-porosity properties of karstified rocks the behavior of the grout mix flow within the rock mass is not predictable and cannot be strictly controlled. In spite of detailed surface investigations, the nature and location of numerous solution-widened joints, caverns and varying quantities and types of infilling sediments cannot be determined with adequate precision.

The extent of penetration of a grout mix along the joints (for a given pressure and grout mix cohesion) depends on many different factors: first of all of its opening, change of aperture, roughness of joint walls, and natural secondary filling. Lombardi (2003) presented a theoretical model which gives the relationship between grouting pressure, aperture of joints, cohesion of slurry and maximal reach of the grout (Fig. 6.4). "Accordingly, the higher the pressure, the wider the joint and the lower the cohesion, the greater will obviously be the reach." In general this conclusion presents relations between key parameters of grout mix penetration. This model may be appropriate for open planar, smooth-walled fractures of constant thickness ('pull apart' fractures). However, in reality, particularly in cases where joints are significantly karstified, the parameters are different and more complex.

If the grouting borehole intercepts a joint that has a cavern developed along it, the grout mix can penetrate to the karst void only after the joint is opened sufficiently. To achieve this, increasing pressure is needed to widen the joint and force grout along it into the cavern. This situation requires hydraulic fracturing ('fracking'—Fig. 6.5). The graph displayed is of the relationship between pressure and mix consumption in the case where a grouted joint is connected to other cavities. Due to the joint being comparatively tight, grout penetration starts only after pressure has become high enough to force it open. Immediately after the grout mix begins to fill the cavern beyond it, consumption (Q) increases and pressure (P) stabilizes or is slightly decreased until the cavern is fully filled. A renewed abrupt pressure increase indicates that the cavern is now filled. If a cavity is far from the grout curtain route (5 or more meters) economic effects can be negative due to unnecessary waste of grout.

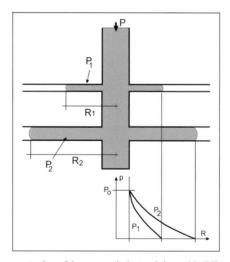

**Figure 6.4.** Simultaneous penetration of the grout mix in two joints with different apertures, with a graph of the injection pressure distribution (Lombardi, 2003).

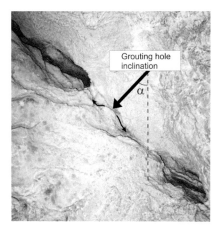
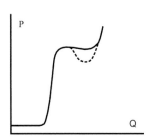

**Figure 6.5.** The pattern of pressure increases that is observed when connecting a point of injection to a cavern.

Inter-relationships between joints and viscosities of the suspensions play important roles also. The viscosity depends on the grout ingredients and mixing ratio. In reality all parameters of joints change erratically, sometimes within short distances, particularly where there has been karstification. The common situation is one of frequent changes from wide openings to very narrow cracks. The common cement-based suspension consists of conventional (Portland) cement and admixtures. Approximate relationships between sizes of conventional cement (blain No. 3500), microfine cement (> 7000) and clay particles are shown in Fig. 6.6.

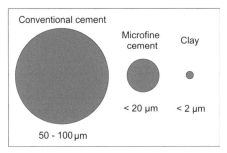

**Figure 6.6.** The relationships between approximate sizes of conventional and microfine cement and clay.

It is clear that slurries with microfine cement and clay-based slurries (thixotropic suspensions) are able to penetrate into much narrower joints than grout mixes with conventional cement.

Numerous discontinuities in karstified rocks or cavities are filled with clayey sandy sediments or broken fragments of the surrounding carbonates. Where the filling is of rock fragments, the grout mix can penetrate between them to make a compact and impervious mass (Fig. 6.7A).

In many cases all of the fractures enlarged by karstification will not become fully filled with the grout mix, as seen in Fig. 6.7B: after analysis of the check hole core, additional grouting was required in this section.

**Figure 6.7.** Examples of penetration of grout mix in karstified rocks. (A) Thick grout mix in a cavernous space with rock fragments; (B) Partial penetration of grout mix along joints only weakly enlarged by karstification.

According Evert, 2005, "the tests revealed that an planar shaped conduit of 0.3 mm in width and 20.8 mm in breadth yielded water takes between 0.9 and 1.3 LU, depending on the length of the path: 300 mm and 900 mm, respectively. A grouting pressure of 12 bar was needed to penetrate that path using a suspension of W/C-ratio = 1. The flow through that path stopped once that the pressure dropped below 9 bars".

## 6.3.9 Groutability

The groutability of a karstified rock mass depends on many of its characteristics, the first being its hydraulic properties. These depend on the presence and spatial patterns of the fractures (orientation and frequency of any joints, faults and bedding planes that are solutionally opened); their aperture and roughness; the frequency of interconnections between them; properties of any infillings; presence of caverns (size, empty or filled, below or above the water table); type of karst porosity; properties of cavern sediments (granulometry, stiffness and degree of lithification); connections between discontinuities and caverns; water pressure; presence of very soluble rocks (evaporites); velocity of any water flow; water temperature and chemistry.

If a cavern along a grout curtain route is filled with unconsolidated materials, a high pressure grout mix may penetrate preferentially along the contact between solid rock and non-groutable deposits (Fig. 6.8A). When this part of the grouted rock becomes exposed to groundwater pressure, mechanical erosion of the unconsolidated fill is to be expected. Where a fractured block of limestone in a cave is surrounded by sandy-clayey sediments, the grout mix tends to penetrate through the joints in the limestone only (Fig. 6.8B). The clayey-sandy sediment mass will be eroded rapidly when the water pressure increases.

Another case occurs where the mix forms 'nests' of grout within the unconsolidated deposits (Fig. 6.9). Once again, if such caverns become exposed to high water pressure, all deposits will be eroded and removed easily. To achieve the required water tightness

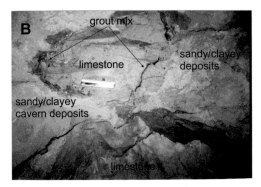

**Figure 6.8.** Penetration of grout mix in cavernous spaces filled with unconsolidated deposits.

**Figure 6.9.** Grout mix 'nests' in sandy clayey deposits in a cavern.

in the grout curtain, all sandy-clayey deposits must be removed from the cavern and replaced with Self-Compacting Concrete.

The groutability of volcanic ash is also questionable. If a cavern filled with volcanic ash is grouted, large parts will retain their natural consistency and be prone to erosion.

## 6.3.10  Grouting Pressures

From common practice and experience, it can be concluded that the minimum pressure should be greater than the weight of the overlying rock. The standard procedure is to perform stepwise pressure increases, generally down to 80 m. In deeper sections constant pressure is maintained. The maximum pressure rarely exceeds 50 bars. Applied grouting pressures usually range between two and six times the overlying rock weight $(2 \gamma h + 6 \gamma h)$, as shown in Fig. 6.10.

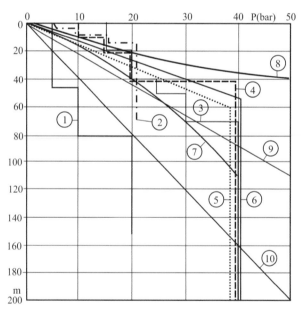

Figure 6.10. Relation between grout pressure and grout curtain depth (examples). (1) Dokan Dam; (2) Gorica Dam; (3) Grančarevo Dam; (4) Rama Dam; (5) Piva Dam; (6) Peruča Dam; (7) $P = 1.5\ h^{0.7}$; (8) Task Committee on Cement Grouting; (9) $P = 0.45\ h$; (10) $P = 0.25\ h$.

## 6.3.11  Grout Mix Consumption

According to the Deer classification, specific consumption based on 38 cases is: (chiefly volcanic and sedimentary rocks)—Very low < 12.5 kg/m'; Low 12.5–25; Medium low 25–50; Medium 50–100; Medium high 100–200; High 200–40; and Very high > 400 kg/m. In karstified rock the classification is different.

As a consequence of the nature of karst, the grout mix consumption in karstified rocks can vary from borehole to borehole and from grouting section to grouting section. In the majority of dam sites in about 80% of the grouting sections (often more) the average grout mix consumption is very low, indicating an almost impervious rock mass. However, along the karstified sections, grout consumption can increase enormously, making the average consumption at the site much higher than in insoluble rocks.

For instance, in Iran, at the Salman Farsi, average grout mix consumption was 79.193 kg/m'. However, the reported grout mix consumption in the right abutment varied between 6 kg/m', and 234,048 kg/m' per section, and in the left abutment between 4 kg/m' and 81,000 kg/m'. The average consumption at the Lar Dam site was 44 kg/m' but in some parts of deeply karstified zones it was between 1000 and 40,000 kg/m'. At Seymareh Dam average grout consumption in fractured vuggy zones was about 500 kg/m'. This was considerably higher than average consumption in the non-fractured, vuggy limestone (100–200 kg/m') or in the crystalline limestone (50–80 kg/m').

Average consumption in the Sklope Dam (Croatia) grout curtain was 53 kg/m'; however, in one 5 m section consumption was 5,454 kg/m' (in fact 27,270 kg per section) and in the next borehole 3440 kg/m' without any pressure (Božičević, 1971).

The average grout mix consumption of the Berke Dam grout curtain (Turkey) was 167 kg/m'. At a section which consisted of slightly karstified limestone, the average grout mix consumption was 44 kg/m' and 80 kg/m', depending on the position (riverside or mountain). In a heavily karstified blocky limestone, average consumption was 518.1 kg/m', with that of P holes being 3270.0 kg/m' (Altug and Saticioglu, 2001).

The bar chart presented in Fig. 6.11 summarizes results of the average grout mix consumption for 93 grout curtains from different karst regions of the world.

Based on the above chart, the classification of grout mix consumption in kg/m' is shown in Table 6.1.

Due to extreme inhomogeneity of the karstified rock porosity the grout mix consumption vary in short distances from a few kg/m' up to a few 10 of tons per

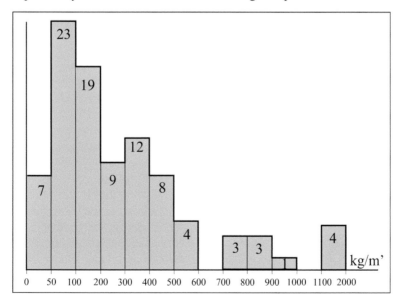

**Figure 6.11.** Bar chart of the distribution of average grout mix consumption per meter in karst.

Table 6.1. Grout Mix Consumption in Karstified Rock.

| Classification of Grout Mix Consumption | Quantity of Dry Mix (kg/m') | Distribution (%) |
|---|---|---|
| Very low | up to 50 | 7 |
| Low | 50–100 | 24 |
| Medium (most frequent) | 100–400 | 44 |
| High | 400–1000 | 20 |
| Very high Undefined high | > 1000 | 5 |

grouted meter. Because of that an average grout mix consumption is much higher than in the not karstified rocks.

In the Deer classification for non-karstified rocks, a grout mix consumption up to 12.5 kg/m', classified as Very low, is four times less than Very low consumption in karstified rock (up to 50 kg/m').

Average grout mix consumption per length meter in karstified rocks (classified as Very low) is rarely less than 50 kg/m' (Hiawase, Medvode, Gallo Matese, Panix dams). Even if almost the entire rock mass is compact and with an average consumption of 10–15 kg/m', two or three small cavernous zones alone, with consumption of a few hundred to a few thousand kg/m', can change (increase) the average consumption tremendously. Because of that, average consumption less than 25 kg/m' in karst is exceptional (Guntersville Dam).

Consumption rates between 50 and 100 kg/m' were found in 25% of the examples analyzed. Weakly fractured and mostly thick-bedded and slightly karstified carbonate rocks belong to this category. This grout mix consumption is categorized as Low (Pueblo Vijeo, Salman Farsi, Geheyan, Elati, Piva, Karun 3).

The most frequent values for grout mix consumption in karstified rocks ranges between 100 and 400 kg/m'. This was found in 43% of the cases. In them, the rock was frequently intensively tectonized and with karstified zones. In non-karstified rocks, such consumption is rated Medium High to High. In karst, it is only Medium. This consumption was found at 51 of 84 examples analyzed (El Cajon, Keban, Grančarevo, Ataturk, Wujiangdo, Punta del Gall, Oymapinar).

Grout mix consumption between 400 and 1000 kg/m' is rated as High; found in 20% of 93 examples analyzed (Berke, Dokan, Camarasa, Sklope, Slano, Great Falls). Average consumption of more than 1000 kg/m is rare (Fort Loudon, Lourus, Krupac).

From the data in Fig. 6.11 and Table 6.1, it is clear that grout mix consumption in karstified rocks is quite different from that in non-karstified rocks. Figure 6.11 does not include consumption of the grout mix used for filling karst cavities, i.e., where grout consumption is more than 2000 kg/m'.

Depending on the nature of the karstification, grout mix consumption is sometimes considerably different beneath the dam foundation (central part) and in the banks and lateral curtain sections. This difference is particularly high in cases of hypogene karstification (Višegrad Dam, El Cajon Dam, Salman Farsi Dam) where, due to the upward water flow, deep karst channels had developed. In the case of El Cajon, grout mix consumption below the central dam section was more than twice as much as in dam site banks (Fig. 6.12).

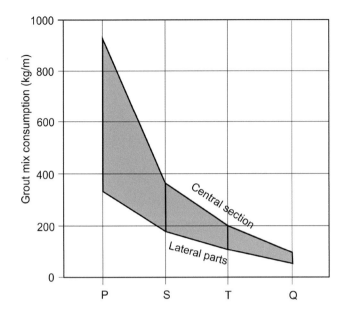

## 6.3.12  One, Two Row or Multi-row Curtains?

There are no widely accepted rules for the number of curtain rows in karstified rocks. Depending on the geological conditions (particularly, the intensity of karstification), type of dam and its geometry, the crucial parameters for curtains (length, depth, number of rows) differ from site to site.

Two principal parameters that are basic for grout curtain design are the efficiency (water tightness) and the cost. Grouting is expensive. With a two row design the curtain will be wider and grout consumption generally higher than in a one row curtain. Using closer holes, the latter will be thinner but its water tightness may be same or even better.

In many cases, due to the karst, modification of the curtain design and adaptation to new findings during its execution are necessary. Usually, in heavily karstified zones, an additional row or a few rows are needed. At some sections of the Berke Dam (Turkey), there was a five row curtain, and in some sections of the Wulichong Dam (China) a seven row grout curtain 10 m wide was used. In both cases there was a combination of vertical and inclined grouting holes.

The example presented in Fig. 6.13 is representative of the findings at many sites in karst. Inside the left bank of the Karun 4 Dam (Iran), a two-row grout curtain (separation 0.9 m) along a 320 m long curtain section was constructed. The split spacing method was applied. From the graphs it is clear that the desired permeability was achieved by the first curtain row alone. Grout consumption along the second row was negligible. The split spacing used along the first row was very effective.

The vertical distribution of the grout mix consumption at Karun 4 is presented in Fig. 6.14. Grout mix consumption along the first row ranged between 50 and 300 kg/m', but locally up to 700 kg/m'. Particularly large consumption occurred in the P and S boreholes. To a depth of 130 m grout consumption along the second row

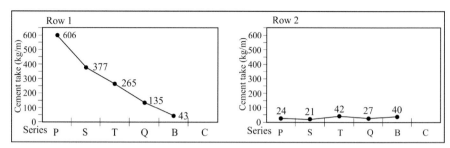

**Figure 6.13.** Difference of average grout mix consumption along the first and second rows of a grout curtain.

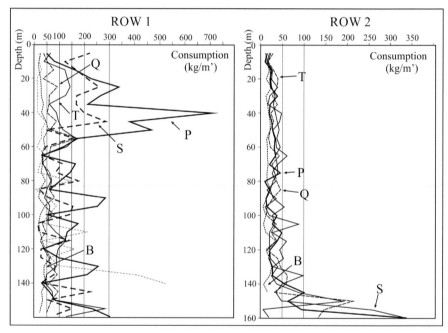

**Figure 6.14.** Distribution of grout mix consumption by depth at Karun Dam. Difference between consumption along the first and second rows.

was always less than 50 kg/m' but there was a sharp increase below 140 m due to local cavernosity at the bottom of row 2.

Usually, a multiple row concept is understood to use bigger distances between grouting holes and lines, in order to construct a wider curtain. However, the actual distances between boreholes tends to increase with a depth to form a frayed fringe at the curtain bottom, i.e., some karstified joints may be left untreated. Those are weak and easy erodible places. In the case of suspended grout curtains, their degradation usually begins in these zones.

Based on similar experiences at many dam sites, a one row grout curtain with split spacing closure is recommended by many engineers. The separation between grouting boreholes should be reduced progressively until the target permeability is achieved or the consumption of grout mix falls to the acceptable level—less than 50 kg/m'. If at some sections consumption increased considerably additional investigations should be done. If the reason for difference in consumption is the cavern, the plugging technology should be applied. If huge consumption is a consequence of tectonized and karstified rock mass the second row should be locally constructed.

## 6.3.13 Grout Curtain Depth

Where the geological structure allows a positive cut off (i.e., extending the tight curtain into an impervious formation at reachable depth, especially beneath the dam itself), the desirable curtain depth is easily defined. If the depth of the karstified formation is a few hundred meters or more, however, hanging curtains should terminate in the less pervious rocks below the minimum water table level. In this case it is a difficult matter to decide the lower contour of the curtain, particularly where the karstification is a consequence of hypogene action. Earlier in the history of the design and execution of grout curtains, general practice presumed a linear relationship between the height of the dam and the depth of the curtain needed. Based on empirical data for dams that were mostly built on non-karst rocks, this relation gave the range:

$$h = 0.4 \, H – 0.6 \, H$$

where h is the grout curtain depth (m), and H is the height of dam.

The U.S. Bureau of Reclamation recommends the following common relation:

$$h = H / 3 + c$$

where c is a constant between 8 and 25 m.

Later experience with a larger number of completed curtains suggests that there is not a linear relationship, especially where the grout curtain is in karstified rocks. The empirical relationship between depth of the curtain and the depth of reservoirs situated in karst terrain extends over the range:

$$h = 0.3 \, H – 8.0 \, H$$

In general, the frequency of karst features decreases with depth. However, in soluble rocks (carbonates and evaporites) with thicknesses of 1000 m and more, solution features have been detected much deeper than the expected base of karstification. Cavities are often detected by deep boreholes, during excavation of tunnels where the bedrock overburden is more than 1000 m, and in deep mines. Minimum water levels of more than 300 m in deep boreholes is a clear indication of how deep karst processes can penetrate. Due to these facts, the depths needed for grout curtains in karstified rocks is quite different from that in non-karst rocks. At the Ataturk Dam (175 m high) the grout curtain below the foundation is 300 m deep; at the Canelles Dam, Spain (151 m high) the curtain below the foundation is 180 m deep, which is 340 m below the reservoir level; at Višegrad (the dam rises 50 m above the river bottom), grout curtain depth is more than 160 m; Sklope Dam, Croatia (74 m high) grout curtain depth = 120 m; El Kansera Dam, Morocco (57 m high), grout curtain depth = 120 m. An extreme case is the Župica Dam in Bosnia. The height of the dam is only 24.75 m, however the average depth of the curtain is 185 m, its length 692 m, and its surface area is 127,777 m².

Data collected for 30–40 other dams on karst foundations show a typical depth of about 1.5 times the height of the dam (Fig. 6.15; Riemer, 2015).

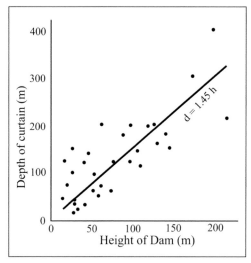

**Figure 6.15.** Dam height versus depth of curtain of selected dams in karst areas (Riemer, 2015).

## 6.3.14  The Flushing Method

The flushing method is commonly used to improve the geotechnical conditions and stabilize the permeability in dam foundations. The clayey and sandy materials commonly found in karstified joints and other small cavities are flushed and replaced by cement slurry. Water and pressurized air are applied through specially designed arrays of boreholes used for the washing (Fig. 6.16).

Key steps for the flushing method presented in Fig. 6.16 are:

- The cluster of three boreholes in a cell has to be used simultaneously—one for air injection, the second for water, and the third for discharging the air, water and flushed material.
- By changing the extent and location of water and air packers in the injection holes, flushing should be continued until the outflow water becomes clear. This indicates that all cracks and cavities have been washed out.
- After washing a 4 or 5 m cell, pressure grouting has to be done from top to bottom.
- To prevent uplift of the foundation surface during washing and grouting, continuous and very strict monitoring is necessary.

In cases where vertical or near-vertical fractures have been enlarged by solution, grouting boreholes should be angled to intersect these fractures as much as possible.

For foundation treatment this flushing method was used successfully at Khao Laem Dam (Thailand), Grančarevo Dam (Herzegovina) and Piva Dam (Montenegro).

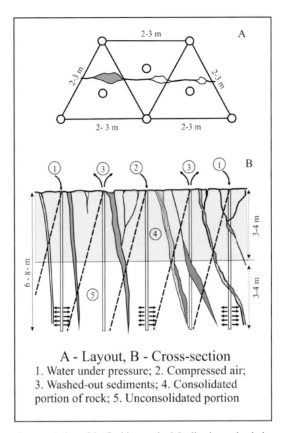

Figure 6.16. Schematic presentation of the flushing method. Inclined grouting holes are shown by dashed lines.

## 6.3.15 Grout Curtain Surface Areas

Grout curtains in karstified rocks are much larger than curtains in other geological formations.

| | |
|---|---|
| Ataturk (Turkey) | 1,200,000 m², length 5,5 km, depth to 300 m. |
| Dokan (Iraq) | 755,000 m², length 2541 m |
| El Cajon (Honduras) | 610,000 m², depth to 250 m |
| Berke (Turkey) | 533,000 m², depth to 235 m |
| Limmernboden (Swiss) | 544,740 m² |
| Buško Blato (BiH) | 475,000 m² |
| Khao Laem (Thailand) | 437,000 m² |
| Slano (Montenegro) | 404,224 m², length 7.011 km |
| Wulichong (China) | 262,000 m² |
| Salman Farsi (Iran) | 261,000 m², depth below dam foundation 140 m. |
| Peruća (Croatia) | 242,000 m² |
| Karun 3 (Iran) | 200,000 m², depth below dam foundation 200 m |
| Geheyan (China) | 180,000 m² |

## 6.3.16 Treatment of Caverns and Conduits

The greatest impacts on dams and reservoirs occur when there are natural caves along the grout curtain routes. Caves and caverns function as giant sediment traps, accumulating samples of all clastic, chemical and organic debris mobile in the local environment during the life of the cave (Ford and Williams, 2007). The volumes of caverns can vary between a few decimeters to more than one million cubic meters. The likelihood of detecting and characterizing most caverns with remote sensing methods deployed on the surface is negligible despite intensive investigations, particularly where the cavities are at depths of more than 20 m.

Honeycomb Cave, in the bank of the Guntersville Reservoir (Tennessee River) was one of first caverns successfully plugged to prevent potential seepage in karstified rocks (Tennessee Valley Authority projects, 1949). Among Dinaric karst projects, underground channels were plugged successfully at Krupac Reservoir (Montenegro, 1957) and Buško Blato Reservoir (Bosnia). A few experiments with karst channel plugging have also been carried out in eastern Herzegovina.

Table 6.2 gives the recorded volumes of caverns in some other dam and reservoir banks in different karst regions.

In many cases filling a cavern has thus been a successful solution, particularly where the cavity is above the pre-dam groundwater level. Plugging karst stream channels carrying rapid water flow, however, is technically extremely complicated. Different kinds of grouting mortar, self-compacting concrete and pre-packed concrete have to be used. At the Keban dam site about 30 caverns were treated. According Ruichun and Fuzhang (2004) in the Geheyan Dam and reservoir area more than 600 caverns with a total volume of 60,000 m$^3$ were detected. At the dam itself, as many as 79 caverns were discovered, with an aggregate volume of 22,000 m$^3$.

**Table 6.2.** Cavern volumes at selected Dams and Reservoirs.

| Dam and Country | Cavern volume (m$^3$) |
| --- | --- |
| Keban (Turkey) | 6600,000 (Petek Cavern) |
| | 150,000 (Crab Cavern) |
| Wudongde (China) | 369,00 filled with blocks, sandy material, |
| Salman Farsi (Iran) | clay |
| | 150,000 (Golshan Cavern) |
| Power plant Čapljina (Herzegovina) | 150,000 |
| Maotiohe, 4th Cascade (China) | > 100,000 |
| Lar (Iran) | 90,000 |
| Pueblo Vijeo (Guatemala) | 60,000 |
| Sklope (Croatia) | 25,000 |
| Khoubin (Vietnam) | 17,600 |
| Canelles (France) | 10,000 |
| Geheyan (China) | 60,000 |
| Diyawini (Sri Lanka) | 12,000 |
| Slano (Montenegro) | 6,000 |
| El Cajon (Honduras) | 5,500 |
| Dokan (Iraq) | 5,000 |
| Ourkiss (Algeria) | Unknown volume |
| Gergebilsky (Dagestan) | Unknown volume |

As noted, in other cases, due to technical and economic issues, grout curtains have been rerouted to by-pass caverns either on the upstream side (Salman Farsi Dam, Iran) or downstream side (Sklope Dam, Croatia) instead of attempting to infill them.

Many of these caverns were not discovered during very detailed investigations from the surface, including closely spaced boreholes and use of different sophisticated geophysical methods. Because of this detection problem, the existence of undiscovered caverns (their location, size, form, empty or filled) must be allowed for even where the surface-based methods have found none, or detected only erratic and minor showings. In practice, many cavities have been discovered and investigated during the excavation of investigation adits or grouting galleries.

Fallen or transported rocks, sandy clay, silt and pure clay are the most common clastic deposits in caverns. As in the case of joints, caverns can be partially or entirely filled with clastic sediments transported in suspension. Facies that are frequently deposited in caverns vary from almost fluidized clay (Wujiangdu Dam, China, with a water content of 56%), through weakly cemented and erodible sandy clay (Salman Farsi Dam, Iran), to well lithified siltstone of low permeability and resistant to erosion (Piva Dam, Montenegro). From the point of view of water tightness, the weakest places are usually the contacts between the natural deposits and the cavern walls.

Particles of different granulometry are often deposited in caverns in rhythmic sequences, parallel or sub-parallel to the depositional surface. For example, Fig. 6.17 shows a cavern fully filled with a range of clayey sediments. Over a basal thick and dark plastic clay containing falle limestone clasts from the cave roof. As a sequence of brown to yellow clay to sandy sediments were laid down in at least four different phases. All phases are separated with thin layers of calcite precipitated from films of static or slowly flowing water rich in dissolved calcium carbonate. Frequent changes of the contents in suspension (different minerals) have resulted in many thin accretionary laminates. In many other cases the deposits will consist of pure plastic clay layers without any rock fragments, pebbles or sand/silt grains (Fig. 6.18).

Due to sedimentation and precipitation from the solutant is mostly discontinued process the laminated sequences are different in granulometry, color and degree of lithification. Mostly the thin laminated layers of hypogene origin are separated (Fig. 6.19A). Some of them are of precipitated origin (broken flowstone, a vadose zone deposits) and looks as accrete mass however are not resistant against hydro-jacking (Fig. 6.19B).

Gravel and boulders are common deposits in caverns and conduits. They can be transported from the surface through large swallow holes (allochthonous sediments) or created in the cave from breakdown blocks transported by rapid stream flow (autochthonous). In the course of evolution in a karst aquifer, hydraulic activity gradually decreases in the upper channels as the lower ones are enlarged because the springs shift downwards in tandem with valley entrenchment. Net transport of clastic sediments is gradually replaced by net sedimentation. The basal unit often consists of large boulders, and is usually succeeded by gravel, sand and finally, when velocity is too low, silty clay and pure plastic clay are deposited (the typical 'abandonment facies'; Ford and Williams, 2007). If a grout curtain route crosses such filled caverns, application of common grouting technology can be problematic.

**Figure 6.17.** A natural cavern filled with rhythmic sequences of waterborne sediments – intercepted in a grout curtain route.

**Figure 6.18.** Layer of pure, dark and plastic clay, overlying with some coarser laminations.

In some cases sedimentary breccia that is cemented by calcite may plug an entire cavern (Fig. 6.20).

The particles that are encountered in sedimentary breccias tend to be poorly sorted to chaotic mixture of all particle sizes from a few millimeters up to a few decimeters. From the grouting viewpoint, sedimentary breccia have the same properties as other sound carbonates.

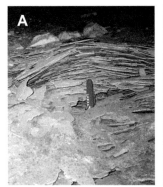

**Figure 6.19.** Cavern deposits as consequence of hypogene origin.

**Figure 6.20.** A cavern naturally plugged with sedimentary breccia.

If there is tectonic breccia in a cavern that consists of pebble-sized and larger clasts only without any fine-grained matrix, the porosity is high and the permeability immeasurable, > 100 Lu (Fig. 6.21). In this case, any clayey matrix will have been washed out by (mostly) turbulent water flow.

In many large passages and caverns the breakdown blocks present serious difficulties during construction of grout curtains. The instability of enormous blocks in the cavern ceiling is an extremely dangerous hazard during remedial works and for any structure built inside the cavern. Piles of breakdown can be more than 20 m high and the sizes of blocks can range from decimetric to tens of cubic meters

**Figure 6.21.** An example of a very porous tectonic breccia in a core.

(Fig. 6.22). Most often the full depth of such piles, until contact with solid rock, is not known and thus cannot be made the foundation for any waterproofing structure.

The most complicated situation arises where a large cavern is filled with big breakdown blocks that are infilled with and completely buried by clay matrix. It is a considerable grouting problem if the cavern is situated in the saturated zone (sub-water table).

**Figure 6.22.** Breakdown carbonate blocks in a cavern in the grout curtain route.

### 6.3.17 Design Criteria for the Treatment of Caverns and Conduits

The following treatment covers the methods for plugging larger cavities or chimneys encountered in dam excavation or along the grout curtain plane. They may be empty or filled with clastic deposits. Larger karst voids should not be treated by grouting alone, or not grouted at all. Plugging techniques may be applied, as described below:

1. The location (spatial position and extent) of each cavernous space has to be carefully determined in order to connect it properly with the grout curtain.
2. The larger cavernous spaces (metric size) detected by drilling, have to be connected to the grouting gallery(s) by an adit or a shaft that is accessible for people. Speleological investigations should be arranged if necessary.
3. Every open karst channel at a horizontal distance up to 10 m upstream and 10 m downstream of the grout curtain plane should be plugged.
4. For filling solutional openings and channels with apertures less than 20 cm, grout mix and grout mortar (up to 4 mm) should be used. For filling channels with apertures of more than 20 cm, a SCC has to be poured, using pumps.
5. Caverns or channels filled with sandy-clayey material in the grout curtain route have to be cleaned at least 5 m in either direction from the grouting plane. If possible, all caverns above water level should be cleaned manually. The cleaned space will be subsequently filled (plugged) with Self Compacting Concrete (SCC) or pre-compacted concrete. The contact between the plug and limestone (particularly, in the roof area) should be grouted.
6. Particular care has to be given to proper grouting of the discontinuities along which the karst features were developed, because there may be further undiscovered caverns along them.
7. Treatment of caverns and channels has to be completed before any drilling and grouting starts in a given section of the gallery. The same applies for the concrete lining of grout galleries: lining has to follow karst treatment.
8. Treatment of filled caverns encountered below the GWL (curtain sections below the dam) needs a special approach. Those caverns are not accessible. The common plugging technologies cannot be applied. Because of these circumstances, the air/water flushing method has to be used to remove any silty-clayey material. This method is also used successfully for consolidation grouting.
9. Particular problems arise where the conduits and channels with underground flows are under pressure. The crucial requirement is to stop, or to considerably decrease, the velocity of flow by all available technical measures.
10. One means of successfully plugging deep-seated caverns below the water level is the 'stop-start procedure'. If, in the process of drilling (above or below the GWL), a bit enters a small and empty cavern, regardless of the depth drilling should be stopped immediately and the cave filled by gravity (without pressure) with five tons of grout mix. After an 8-hour delay, an additional five tons should be pumped (without pressure also) and the entire procedure repeated until the borehole is completely filled. The next stage is redrilling the borehole through this grouted cavern section down to the designed depth. To complete the task, bottom-up grouting with the selected cement mix should be done, including the already-filled section.

11. The above criteria have to be interpreted in a flexible way and adapted to any particular treatment situation.

## 6.3.18  Treatment Procedures for Caverns

The greatest hazards for engineering structures, particularly dams, occur when caverns are present. If a cavern is located beneath the dam (in the bedrock foundations) or along the grout curtain route, two main problems must be solved: the dam stability and the potential seepage losses. Detection and treatment of each cavern is the chief prerequisite for the construction and sustainability of the project. The hazards of cavernosity on dam stability are much less than the consequences of major seepage. The most common treatment approaches are to fill the entire cavernous space, construct individual plugs in the karst channels, and/or the construction of cut-off walls.

Detection and characterization of caverns remains very difficult despite intensive attempts at improvement. Current methods are still not able to detect their spatial positions from the surface and so geotechnical prevention measures at the design stage are usually based on a limited amount of good quality information. Even if borehole permeability is found to be not more than one or two Lugeons, this does not mean that the surrounding rock is compact and impervious. Clay deposited in discontinuities can block test water penetration even under high pressure.

*The general objective when constructing a grout curtain with the required tightness is to detect caverns along the grouting route as far as possible and to fill all of them prior to starting curtain construction.* Of course, some of them will remain undiscovered and should be detected during grouting procedure. To detect and investigate caverns the most effective approach is: closely distanced investigations and grouting galleries; orientation of grouting galleries into the assumed domains of the largest caverns; extensive investigation of all detected karst features (geophysics from the galleries, speleological investigation, boreholes, auxiliary adits and shafts) is needed, plus analysis of the karst evolution at the site and analysis of sediments deposited in the caverns.

The range of conditions encountered in the caverns and conduits to be plugged is wide: location - in the aeration or in saturated zones; properties - empty or filled; geometry - isolated single cavern or highly tectonized and karstified rock mass; hydraulic properties - from stagnant pools to flowing water under high pressure.

In most cases the caverns have developed along major fractures, vertical, sub-vertical or horizontal. Many of these cavities are only metric in size but their detection and sealing is not easy, particularly if they are beneath the dam foundation and under the water table, as were the cases of the Kentucky Dam (USA) and the Višegrad Dam (Bosnia and Herzegovina). The cavity beneath the Kentucky Dam foundation and beneath the groundwater level was 21 to 30 m wide with a height of 1.8 m. A vertical shaft was dug to seal it with concrete.

For grouting cavities in both aeration and saturated zones, where their volumes are believed to be small (a few cubic meters up to a few tens of cubic meters) successful plugging can usually be achieved by injection of grout through grouting holes. For these cases a cement-based grouting mortar is suggested, with the ratio: sand – 65%; cement – 34% and bentonite – 1%.

If a cavern is **above the groundwater level** (in the aeration zone) and large enough to be accessible for people, the general guidelines are:

a. Excavation of an access adit or shaft from the grouting gallery to the cavern.
b. Detailed geological and speleological mapping of all accessible parts of the cavern.
c. Excavation of all sediment deposits and washing of cavern walls.
d. Construction of wooden formwork to isolate a section for plug construction, i.e., to prevent unnecessary consumption of the concrete in grout mortar.
e. If necessary, instead of wooden formwork, reinforced auxiliary plugs can be built.
f. Filling of the cavern or selected parts of it with SCC, mortar or grout mix.
g. Grouting the contact between plug and natural rock walls of the cavern.

For proper treatment, speleological investigations are an important source of information about the cavern morphology (geometry) and properties of the deposits. Excavating the sediments and washing the cavern walls for better contact between concrete and rock walls is very unpleasant and time-consuming work; however, it gives the best results (Fig. 6.23).

After speleological investigation and careful cleaning of a cavern in the curtain route it has to be filled with Self Compacting Concrete—SCC (Fig. 6.24).

Auxiliary offsets or shafts must often be constructed for efficient cavern treatment (Fig. 6.25). To prevent unnecessary consumption of SCC in a karst channel a concrete plug should be installed approximately 10 m from the grout curtain. Caverns more than 10 m upstream and downstream of the curtain do not require treatment. To achieve better contact between concrete and cavern walls, an additional second grouting row is suggested.

Figure 6.26 shows the filling procedure used for a very complicated system of karst caverns and channels discovered between two galleries along a grout curtain route.

**Figure 6.23.** Manual cavern cleaning.

**Figure 6.24.** Karst cavern filling with CCT, after cleaning and careful washing.

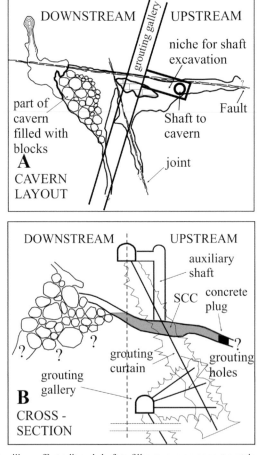

**Figure 6.25.** Auxiliary offset adit and shaft to fill a cavern on a grout curtain route with SCC.

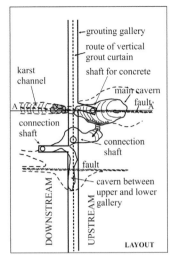

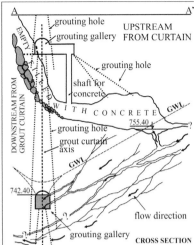

**Figure 6.26.** Construction of the plug in a cavern detected between two galleries on the grout curtain route.

For easier cleaning and filling, the cave system was connected to the grouting galleries via shafts. The bottom of the cavern was covered with 0.5 m to 2 m of sandy-clayey deposits. Before filling with SCC, these were removed and the solid rock was washed. After filling all empty pockets in the ceiling, the caves were grouted from the gallery and an offset adit. A large void downstream of the grout curtain was left untreated.

In the Geheyan Project (China) "sixty caverns were indentified along the proposed grouting curtain route, with total volume of over 6000 m³. Before curtain grouting, those revealed karst caverns were back-filled, and the karstified rocks thus become a kind of fissured medium that made grouting much easier and faster" (Ruichun and Fuzhang, 2004). For filling nine caverns in the Salman Farsi dam site (Iran) more than 3.100 m³ of Self Compacting Concrete was used prior to execution of the grout curtain.

For plugging karst channels **below the water table**, particularly if they carry flowing water, the key prerequisite is to stop or considerably reduce the velocity and quantity. To plug a cavern or large channel with active flow in the saturated zone the following procedure is suggested:

Contours and properties of cavern have to be determined using:

a. small diameter boreholes to define the approximate contours of the cavern;
b. cross-hole seismic methods to provide additional data;
c. analysis of sediments if the cavern is filled;
d. echo-sounding to define contours and the volume of the submerged cavern from inside it;
e. measurement of flow velocity using tracer tests and/or hydrological equipment;
f. depending on cavern size and flow velocity, large diameter boreholes (from 128 mm up to 1000 mm) should be used to access the cavern.
g. depending on findings from the above methods, selection of the appropriate grain size for the aggregate for cavern filling (usually it is from a few millimeters up to a few decimeters in diameter).

h. when a cavern is filled to its top with crushed stone and aggregate, consolidation grouting should be performed from the bottom of the cavern upwards until all pore space (standard porosity of 35 to 45%) is filled with grout mix. It is particularly important to grout the contacts between the plug and the cavern walls.

Where very large caverns are discovered in a permanently saturated rock mass, either a complete filling or a rerouting of the grout curtain are the only solutions. One of the largest submerged cavities was the Crab Cavity, 320 m below the crest level of the Keban Dam (Turkey). Its volume was approximately 105,000 m³. The cavern was detected and treated during the investigation and construction period. The cavern was filled with 64,000 m³ of concrete and injected solids introduced through a large borehole (Božović et al., 1981). During the filling of the reservoir, when the water level was approximately 10 m below the design maximum, a huge leakage suddenly occurred (26 m³/s). The large 'Petek Cavern' was discovered, and explored by speleologists (Fig. 6.27).

For the filling, 605,000 m³ of limestone blocks, gravel, sand and clay were dumped into the cavern through an excavated shaft 2.5 m in diameter. After cavern filling, the leakage decreased to 9–10 m³/s.

In many cases the remedial works are expensive and time-consuming and produce only questionable results even where the best engineering practices are followed. To stop leakage from Slano Reservoir in Montenegro, a grout curtain seven kilometres

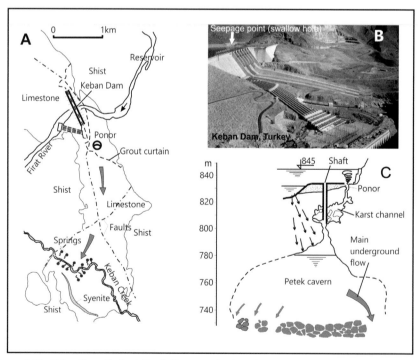

**Figure 6.27.** Keban Dam (Turkey). Petek Cavern at the dam site. (A) Layout, with main leakage route; (B) Keban Dam (photo) with location of seepage collapse; (C) Petek Cave, cross-section (Milanović, 1979; Božović et al., 1981).

long was constructed. A large cavernous zone, 'Koštrikova luka', was discovered in one section at depth of about 90–100 m (Fig. 6.28). Grouting was only possible during the periods of lowest water table because, when the water table was high, the groundwater flow was too fast. Plugging this cavernous zone took four years and five months to complete. During this period 4,124 m³ of gravel and 4,500 tons of grout mix (dry component) were injected into the cavernous zone but were nearly instantaneously washed out again—a failure until the full grout curtain was built (Vlahović, 1981).

To fill a cavern at a depth of 210 m below the Lar Dam foundation, boreholes of 214-mm-diameter were drilled. The cavern was filled with about 38,000 m³ of gravel (grading 5 to 50 mm), 34,000 m³ of crushed rock and 13,000 m³ of cement mortar (Djalaly, 1988).

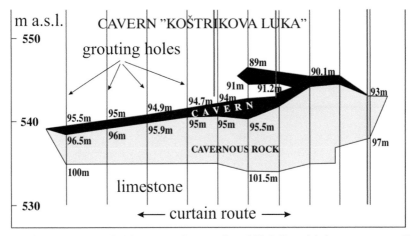

**Figure 6.28.** Slano Reservoir, Montenegro. Cross-section of 'Koštrikova luka' cavernous zone (From Vlahović, 1981).

## 6.3.19 Plugging Underground Karst Water Flow

The most challenging remedial work in karst is to plug active channels carrying significant quantities of flowing water in the submerged (sub-water table) parts of an aquifer. This problem was crucial in many dam projects: Krupac Reservoir, Montenegro; Charmine Reservoir, France; El Cajon, Honduras; Canelles, Spain; Višegrad and Čapljina PPP, Bosnia and Herzegovina.

Usually these flows are under high pressure, fast and turbulent. After reservoir filling, the problem is increased tremendously. In some cases pressure is increased by 10 bars or more and the velocity of turbulent flow in narrow sections increased up to 8–16 m/s. In these circumstances standard grouting technology is useless. The crucial problem is how to reduce the velocity and destructive water energy. It is of highest importance to plug the karst flow before filling the reservoir because pressure and water velocity will be much lower than afterwards.

Many different materials and techniques have been tried to reduce or stop karst flow: different types of cement mortar with sand, gravel, aggregate, chemicals, polyurethane foam, synthetic sponge, asphalt, hot bitumen and, for large caverns, rock blocks to form precompacted concrete.

- Foam products based on polyurethane have a very high expansion coefficient when in contact with water. During the foaming process the final volume can be more than 20 times larger than the initial volume. The strength of foam is not high enough to be applied as the final solution, however. It is used to slow or stop the flow in order to make a final plugging by cement-based grout mix or mortar possible.
- Synthetic sponge (cut into pieces of 1–3 cm) combined with sand as an additive to the grout mix can be a very efficient means to plug karstified joints with apertures of a few cm.

Beneath the foundation of the Višegrad Dam, in 2013/2014 more than 37.000 m$^3$ of crushed limestone was inserted into the cavernous spaces, mostly at depths between 100 and 200 m. Leakage of 15 m$^3$/s was decreased to 4.5 m$^3$/s. To plug an underground river in the subsurface curtain route of Wulichong Dam (China) a concrete plug 33.46 m high, 13.9 m long and 2–10 m wide was constructed, using 4811 m$^3$ of concrete.

## 6.3.20 Groutability of Clay-Filled Caverns

Caverns filled with clay, particularly with soft clay and fine sand, are frequently present in dam sites in karstified rocks. Due to their very low density these sediments are readily subject to intensive erosion and piping. In geotechnical practice such sediments are usually declared to be non-groutable. The common practice is to remove the sediments and replace them with concrete. However, this method cannot be used where the caverns are deep beneath the water table.

Most residual soils in karst are classified as plastic clay or sandy clay. Important properties of these deposits are:

- the plastic limit – the content of water below which the clay becomes brittle and crumbly,
- the liquid limit – water content above which the clay acts as a liquid, and
- the plasticity index – where water content is greater than the plastic limit but less than the liquid limit, the soil will act as a plastic material.

In general, these properties will differ depending whether the hosting discontinuities and cavities are above the water table or if they are submerged. The best case is where soil (clay) deposits have become much denser as consequences of prior stress history and a long duration of filling, i.e., the clay deposits have become almost lithified. The worst case is that of the very large, tectonized and karstified section of a rock mass where large blocks and mid-sized fragments float in a clay matrix (Fig. 6.29). In natural conditions this formation is mostly impermeable. However, if this mass is located along the grout curtain route the effectiveness of the grouting will be dubious, with a lot of technical problems. If it is situated above the saturated zone, direct access is necessary and mine stabilizing technology should be used. When such masses become exposed to turbulent flow under high pressure, movement of the blocks

**Figure 6.29.** Huge blocks in a cavern 'float' in a matrix of clay.

is to be expected. Large empty spaces will be created between blocks and considerable turbulent flow through them can have disastrous consequences.

If there are large cracks and caverns filled with clayey deposits, their location, extent and properties must be investigated and analyzed in detail. Where clay fillings have a water content near to the liquid limit ($\sim 58\%$) or more under natural conditions, problems become serious. If clays with this consistency are found deposited in karstified joints, artificial washing is suggested. A problem with such soft clay grouting was crucial in the case of the Wujiangdu Dam in China. A lot of caverns filled with soft clay and fine sand were discovered in foundation area at a depth of about 250 m beneath the river bed (Zhang and Huo, 1982). Due to high water content, near the liquid limit, clay density was very low (Table 6.3).

To improve resistance to high water pressure and prevent erosion of the clay two, locally three, rows of boreholes were used. Spacing between boreholes was 1.5 to 2.5 m. A descending method with maximum pressure of 60 kg/cm$^2$ was applied. Grouting started with a thin grout mix (water cement ratio 8:1 by weight), gradually increasing to a maximum of 1:0.8, depending on the grouting conditions experienced.

From an analysis of samples taken from the grouted clay, Zhang and Huo (1982) concluded that during grouting under high pressure the clay filling was subject to four

**Table 6.3.** Physical Properties of soft clay (Zhang and Huo, 1982).

| | |
|---|---|
| Natural water content (%) | 56.0 |
| Natural dry density (g/cm$^3$) | 1.10 |
| Clay fraction (%) ($< 0.005$ mm) | 54.0 |
| Liquid limit (%) | 57.5 |
| Plastic limit (%) | 32.4 |
| Plasticity index (%) | 25.1 |

conditions: fracturing, extrusion, consolidation and chemical hardening. Before the grouting, the empty space between the clay deposits and the cavern roof was filled with grout mix (Fig. 6.30A and B). Afterwards, due to the high grouting pressure, there was hydraulic fracturing of the clay fillings in both radial and circumferential directions (Fig. 6.30C).

By comparing results before and after the grouting, the improvement of the mechanical properties and permeability of the soft clay was clear. For instance, the dry density of 1.1 g/cm$^3$ in the natural clay was increased to 1.26 g/cm$^3$ after grouting and the average unconfined compressive strength was increased to 21.2 kg/m$^2$.

Cases where large caverns or long and wide karstified cracks exist in dam abutments are also particularly complex. In the case of the gravity-arch Gergebilska Dam (Dagestan, Russia) a problem with dam stability and with filtration (leakage) were tackled together. To improve the geotechnical and hydrogeological properties of the rock mass in the right bank, a wide and 50 m deep tectonic discontinuity was cleaned and filled with concrete (Semenov et al., 1962; Fig. 6.31).

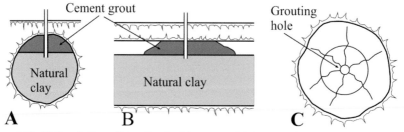

**Figure 6.30.** Wujiangdu Dam, China. Sketch of the cavern with its portion filled by grout. (A) Transverse section, (B) Longitudinal section, (C) Patterns of hydraulic fracture lines.

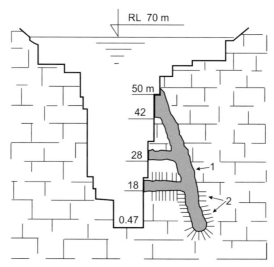

**Figure 6.31.** Schematic presentation of the Gergebilska Dam site, Russia. Semenov et al., 1962, from Lykoshin et al., 1992.

### 6.3.21 Plugging of Large Karst Channels and Caverns – Overview

The plugging of individual cave passages with underground flow to stop seepage losses has been applied effectively in a number of cases. Examples include: Honeycomb Cave in the Guntersville Reservoir, USA; Krupac ponor (4 m³/s) in Krupac Reservoir, Montenegro; karst channel (0.8 m³/s) in the Charmine Reservoir, France, reducing loss from 800 l/s to 20 l/s; a number of karst channels in the foundation of the Khao Laem Dam, Thailand; in the case of Wujiangdu Dam, China (early 1980s, concrete gravity, 165 m high), a specific grouting method was used to plug cavities filled with soft clay (pressure up to 60 kg/cm²); Salman Farsi dam site, Iran, to plug six large caverns 3,125 m³ (7,500 tons) of concrete was used; Khoabin Dam, Vietnam (17.600 m³ of caverns have been cleaned for the construction of a concrete curtain wall 10.5 km long); Buško Blato Reservoir, Bosnia, three large plugs along one channel (a former underground river). In the case of PP Čapljina (BiH) plugging of a karst channel with a flow 300 l/s at a depth of 32 m below zero level, was undertaken using a 146-mm diameter borehole. To block an underground river in the curtain route of Wulichong Dam (China), a concrete plug 33.46 m high, 13.9 m long and 2–10 m wide was constructed.

### 6.3.22 Cut-off Diaphragm Structures (Piles, Overlapping Piles, Trenches, Mine Structures)

Cut-off diaphragm walls are the most effective but expensive watertight structures. Among the few methods used the most frequent are overlapping piles, trenches and mines. Cut-off walls are needed in cases where groundwater seepage occurs through wide tectonized and/or karstified zones in which the common grouting techniques are ineffective.

The overlapping piles method has been chosen at a number of dam projects as the less risky solution: Khao Laem Dam (Thailand); Wolf Creek, Beaver, Walter F. George and Patoka Dams (U.S.A.); Wujiangdu and Wulichong dams (China); Pavlovska Dam (Russia); Akköprü Dam (Turkey); and La Loteta Dam (Spain). Cut-off wall technology is particularly successful in the case of well developed, but shallow karstified formations, which is the common situation in the U.S.A. This technology can be efficiently applied in saturated rocks (sub-water table).

The overlapping piles technique consists of a series of vertical holes (shafts) drilled in two phases. Primary holes are drilled on centers a standard distance apart – L, and separated from each other by untouched rock - D, Fig. 6.32(A). The secondary holes are then drilled between them to overlap the primary holes (B). The thickness of the wall – T, is the distance between two secant points. In some cases, where the geological conditions permit, the, secondary piles are not necessary because some separation between boreholes can be tolerated. But in such cases the distance between borehole centers should not be more than 1.5 times borehole radius. The selected diameter of holes and distance between centers depends on geological conditions, water pressure in the reservoir and the technology available. Usually the borehole diameter – R is between 150 and 600 mm, however, in some cases diameters of 1000 mm or more (1200 mm) have been adopted. The open holes are backfilled using mostly bentonite,

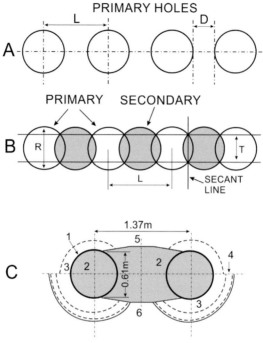

**Figure 6.32.** Schematic presentation of the borehole arrangement for cut-off construction using piles: (A) Primary hole arrangement; (B) Final cut-off (secant) wall. (C) Wolf Creek cut-off wall: Plan of primary and secondary elements. 1. Primary element, permanent casing; 2. tremie concrete; 3. weak grout; 4. centerline wall; 5. section in the rock; 6. section in the embankment and overburden (Fetzer, 1979).

cement bentonite and tremie concrete. If frequent caverns are expected or detected along the cut-off wall, backfill with concrete is the mandatory solution. If a drill hole enters a cavern this space should be filled with concrete up to the collar and, after reaching the required density, should be redrilled down to the designed depth. This section needs local grouting along the wall.

Figure 6.32(C) is the original design for a cut-off wall at the Wolf Creek Dam (Fetzer, 1979). In the embankment above the water table, the primary holes (1.29 m diameter) were excavated to a depth of about 21.3 m and cased with 1.19 m diameter temporary steel casing. Excavation was continued inside the casing to a depth of 42.6 m. A bentonite mud and airlift technology was used to remove the cuttings.

To prevent leakage from the Akköpru Reservoir (Turkey) a double row cut-off wall (concrete piles) was constructed along the edge of a thick horizontal blanket of reinforced concrete for a distance of 800 m and to a depth of about 40 (Fig. 6.33). Throughout its length the cut-off wall is locked tightly into a weakly pervious flysch formation to prevent any leakage from the reservoir via overlying highly permeable conglomerates into extremely karstified limestone. This very conservative and expensive solution was adopted with the ultimate aim of eliminating the seepage risk almost entirely. The first filling of Akköpru Reservoir was successful, justifying the investment in the cut-off structure.

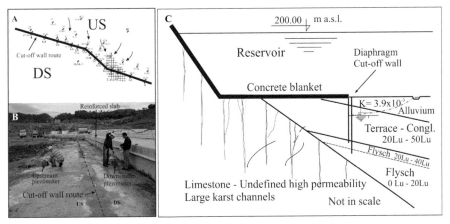

**Figure 6.33.** Akköpru Reservoir, Turkey. Protective measures (reinforced concrete slab and vertical cut-off) to prevent connection of reservoir water with karstified marblized limestone. (A) Layout of cut-off wall route. (B) Photo—cut-off wall route and reinforced concrete slab. (C) Schematic cross-section with position of the cut-off wall. Design: DSI Ankara.

In the case of La Loteta Dam the cut-off wall is 1567 m long, 0.8 m wide, with an average depth of 23 m, Gutierrez et al., 2015, see Chapter 7.

In the case of the Khao Laem Dam (Thailand) a concrete diaphragm wall was constructed in three sections containing karst caverns (Fig. 6.34). The wall consists of overlapping 762 mm piles drilled on 615 mm centers. Maximum depth was 50 m. The total surface area of the wall is 15,900 m².

A few different types of cut-off technology using overlapping piles are discussed here:

1. Where there were karst channels between 0.2 and 10 m in diameter, the diaphragm wall of intersecting concrete piles (762-mm diameter) was adopted and drilled on 615 mm centers using the following procedure:
   a. Detailed investigation
   i. Three-dimensional geological mapping
   ii. Closely spaced exploratory drilling (6 m between holes)
   iii. Water pressure tests in the exploratory holes
   b. Low-pressure grouting of exploratory holes to reduce the absorption of concrete during the wall construction
   c. Drilling primary holes (762 mm) and concreting using the tremie pipes method
   d. Secondary holes drilled and concreted three days later
2. For treatment of minor karst porosity (open widths of 5 to 200 mm) below the galleries, 300 mm-diameter diaphragm walls were constructed under water, using the same overlapping pile method.
3. A positive cut-off wall (165 mm in thickness) was used where the small karst features and fissures were filled with clay and silty material in amounts that were acceptable for conventional grouting.
   a. 165 mm diameter holes on 150 to 250 mm centers were drilled from the inverts (troughs) of the upper galleries to the crowns of the lower galleries. The vertical distance between the galleries was 14 m.

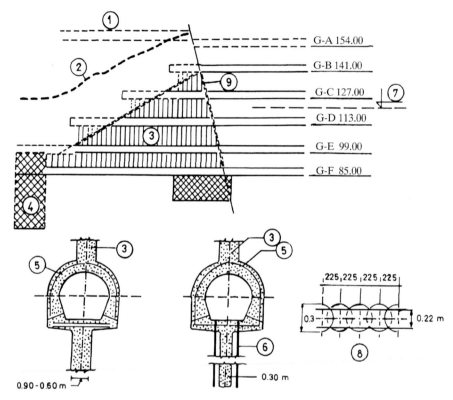

**Figure 6.34.** Khao Laem Dam, Thailand. Cut-off wall: (1) Elevation of the dam crest; (2) ground surface; (3) cut-off wall; (4) diaphragm wall; (5) steel rib support; (6) grouting hole; (7) dry season water table; (8) plan view of the positive cut-off; (9) fault zone (Bergado et al., 1984).

    b. The holes were flushed with high-pressure water.

    c. Effectiveness of the flushing was checked with a TV camera.

    d. The holes were filled with grout mortar using a 60-mm flexible pipe and needle vibrator.

The design of the cut-off wall at the right bank of the Pavlovskaya Hydroelectric Power Station consists of 22 concrete boreholes 915 mm diameter to drilled depths of 12 to 15 m (Kagan and Krivonogova, 1999).

Application of mining technology is possible where the bedrock is above the water table. Usually the structures are large and require specific methods for the underground excavation. In the example of Wulichong Reservoir, Yunnan, China, the reinforced cut-off wall was constructed inside a huge cavern (Fig. 6.35). The wall is 100.4 m high, 50–30 m long and 2–2.5 m thick. The amount of used concrete was 15.152 m³.

In the right abutment of the Karun I (Shahid Abbaspoor Dam) in Iran a wide clay-filled zone was weak and permeable. To prevent leakage through this zone, a massive concrete cut-off structure, 100 m high and about 30 m wide, was constructed (Fig. 6.36). The cut-off was protected by a grout 'curtain box' (A). Mining technology

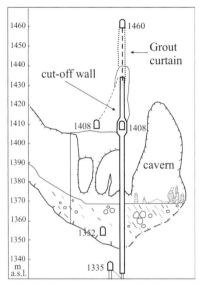

**Figure 6.35.** Wulichong Reservoir Project (China). Large cavern treatment, cut-off wall, cross-section (Kang and Zhang, 2002).

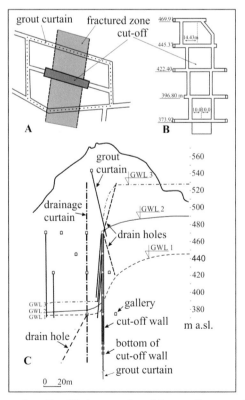

**Figure 6.36.** Karun I Dam, Iran. Simplified presentation of the cut-off structure in the right bank, A-B–layout. C–cross-section (design by Harza).

(galleries and vertical shafts) was used to install intervening concrete panels (B). A cross-section perpendicular to the cut-off structure, with the effects on ground water in the right bank (C).

In the case of the Walter F. George Dam, in addition to an overlapping piles cut-off structure 410 m long and 30 m deep set into the karstified limestone beneath the reservoir level, two more lateral trench cut-off walls (lengths of 100 m and 55 m) were constructed by applying hydromill technology; for more information, see Chapter 13.

Trench cutting technology (hydromill and clamshell methods) was successfully applied to construct watertight underground walls at a number projects in karst: Mississinean, Walter F. George and Centre Hill Dams (U.S.A.), Gotvand Dam (Iran), and in non-karst rocks at Peribonka Dam (Canada) and many others. Some of these structures (Gotvand and Peribonka) are as much as 150 m deep. This technology can be applied from the surface only. The cut-off diaphragm wall structure consists of primary and secondary panels (panel wall technique).

The clamshell technology is simple and less expensive and has been applied in a number of projects (Fig. 6.37). However its efficiency depends on the physical properties of the rock mass.

To improve the water tightness of the trench filling (clay, clay/cement, SCC concrete) the hydro mill technology was improved with an additional geomembrane installed along the upstream trench wall (Fig. 6.38).

**Figure 6.37.** Rudbar Lorestan, Iran. Cut-off wall construction using 'clamshell' technology.

**Figure 6.38.** Gotvand, Iran. Cut-off wall construction, hydro-mill technique.

## 6.3.23 Drainage of Dam Foundations and Grout Curtains in Karst

It is well established that the rock beneath dam foundations and behind grout curtains should be drained, particularly in the case of high concrete dams. Usually, systematically arranged drainage holes are constructed from grouting galleries.

Due to the special nature of karst porosity, a karstified rock mass can be considered to be self-drained. Because the density of this porosity is not uniformly distributed, however, the draining efficiency of karstified rock will also be irregularly distributed. Zones with many joints and bedding planes tend to have the highest concentrations of karst conduits, i.e., best-drained. The effectiveness of drainage outlets in such zones can be problematic. During the construction of grout curtains and sealing of caverns along a curtain route particular care should be taken to leave the downstream karst channels free for the discharge of water, i.e., the location of drainage points should be determined by the karst features in the bedrock rock, not by any prior engineering design. Locations of drainage holes have to be selected on the basis of the karst geological mapping and other data collected during the excavation of the grouting galleries.

In some cases, a drainage curtain is constructed separately, 30 to 40 m downstream of the main grout curtain. To build it, additional drainage galleries should also be constructed (Fig. 6.39). Based on current experience, separate drainage curtains can be sources of additional groundwater inflow from areas outside of the reservoir as well. The need for this kind of extra drainage structure should be analyzed carefully because of the differing nature of the karstification at each and every dam site.

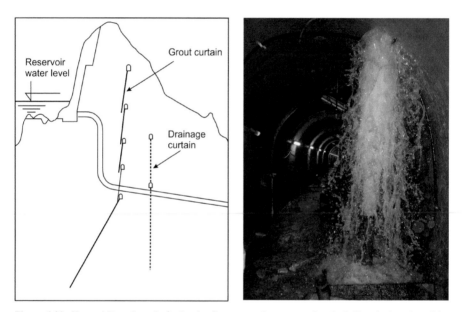

**Figure 6.39.** Karun 4 Dam, Iran. Left–sketch of grout curtain cross-section, including the location of the drainage curtain and galleries. Right–water erupting from a borehole in the drainage gallery.

## 6.4  SURFACE WATERPROOFING

Manmade reservoirs change the regime of water on the surface and underground, provoking a number of different destructive reactions: erosion, suffosion, strong uplift, air hammer and water hammer effects. The consequences can be leakage from the reservoirs, sometimes catastrophic. The successful operation of many existing reservoirs in the world is due to efficient measures of prevention and remediation against the destructive effects of creating artificial bodies of water in karst areas.

Surface waterproofing technology is required in cases where the water table is permanently deep below the reservoir bottom, as well as when it temporarily rises to the bottom or above it (the 'estavelle regime'). There are two general technological approaches to surface prevention and remediation—creating impervious blankets or building preventive structures. Such solutions usually require installing a drainage layer or drainage pipe network below the watertight layer.

Commonly used blankets include compacted natural soil, clay layer(s), geomembranes, geosynthetic clay liners, asphaltic screens, shotcrete, reinforced shotcrete, reinforced concrete slabs, different kinds of geotextiles, and grouting carpets. To isolate concentrated seepage zones the common technologies and structures are dental plugging of individual openings in the carbonate rocks, construction of dikes and dams in front of large caves and ponors, construction of cylindrical dams around ponors and estavelles, and construction of one-way valves.

## 6.4.1 Damming Ponors and Estavelles

Since ancient times inhabitants of karst terrains have tried to control water losses into ponors and estavelles (Milanović, 2003). A variety of different structures has been devised to prevent degradation or erosion of arable land or protect against natural obstruction of karst channels, i.e., to minimize the duration of natural flooding. Similar structures to keep ponors operational or prevent seepage from reservoirs are still used frequently today. Figure 6.40A and B shows two different designs to keep ponors operational, and C and D to prevent seepage losses from reservoirs.

A rare circular dam to prevent water losses from the naturally flooded area of Nikšićko Polje (Montenegro) was built around Slivlje Ponor (Fig. 6.40C). It has a diameter of 50 m and a swallowing capacity of around 120 m³/s. There is a similar but smaller structure at the perimeter of Vrtac Reservoir to prevent leakage through the Opačica estavelle (Fig. 6.40D).

**Figure 6.40.** A and B. Two structures built to control stream sink losses at Peloponnesus, Greece. (C) Cylindrical dam around a large ponor. (D) Cylindrical dam around a large estavelle at the perimeter of a reservoir in Nikšićko Polje, Montenegro.

If there is an estavelle in the bottom of a reservoir, a non-return (one-way) valve allows inflow of water even when the reservoir is full, and prevents loss underground when the estavelle regime changes to become a swallow hole (Fig. 6.41). Due to a variety of reasons this particular structure was not successful; after the first impounding, a number of new sinkholes were created in its vicinity.

Similar structures have been used to prevent seepage from reservoirs in the Chinese karst (Fig. 6.42A and B). In the example of reservoirs in Cili County, Hunan Province, one-way valves (A) were constructed to prevent water loss through estavelles in Cuanhu Reservoir. The circular dam (B) was constructed in Balyangwang Reservoir to prevent water loss and to permit air to circulate in an aeration zone beneath the reservoir floor.

Dikes and dams in front of large sink holes or areas of seepage are frequently used to obtain the necessary water tightness of reservoirs in karst. A few examples are presented below.

To prevent leakage from the man-made Buško Blato Reservoir (Bosnia), two different approaches were applied: construction of grout curtains behind ponor inlets (including the plugging of karst channels) and closing off the largest sinkhole area by building the three km long and 19.3 m high rock-fill 'Kazaginac' Dam (Fig. 6.43).

A similar solution (dike) was applied to eliminate loss to a big zone of ponors in marblized limestone in Mavrovo Reservoir (FYUR Macedonia).

The Anchor Dam (Wyoming, U.S.A.) was constructed "during the dam-building boom of the late 1950s and early 1960s" (Jarvis, 2003). To prevent major loss from the reservoir, a sinkhole 100 m in diameter and 15–20 m deep was walled off by an earth dike in 1976 but the height of the dike was below the crest of the dam.

Concrete dams in front of entrances of large karst channels at the edges of storage reservoirs are used also to prevent water losses (Fig. 6.44).

**Figure 6.41.** Non-return valve (A) installed in the reservoir in Nikšićko Polje, Montenegro. (B) New sinkholes that developed after the one-way valve was installed.

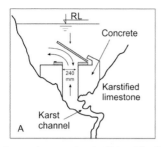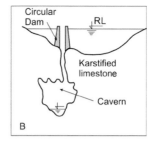

**Figure 6.42.** Reservoir protection in China. Simplified sketch of one-way valve (A), and circular dam (B). Hydroelectric Survey and Project Institute of Hunan Province, 1979.

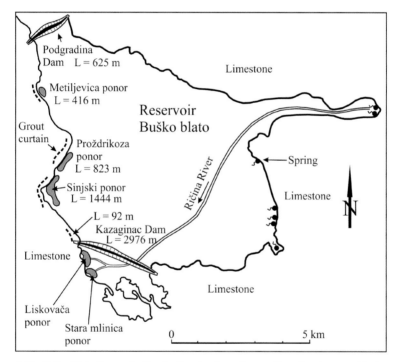

**Figure 6.43.** Buško Blato Reservoir, Bosnia. Construction of an auxiliary dam to prevent leakage from the reservoir by blocking flow to the largest ponor zone (Nonveiller, 1989).

A bolder design was used to create an artificial reservoir upstream of the sink of the Tianqiao River, Guangxi Province (Fig. 6.45). Because the design level for the proposed reservoir was much higher than the cave entrance the entire opening was sealed off by a concrete slab, including the limestone cliff above it.

## 6.4.2 Dynamic Compaction and Clay Blanket

Development of subsidence (collapses) is a common occurrence where the reservoir bottom consists of alluvial or other unconsolidated (granular) sediments deposited on top of karstified rocks. In natural conditions the process of piping, a consequence

**Figure 6.44.** China, Guangxi Province. Dam in the front of the entrance into a large karst channel. It protects storage in a large reservoir.

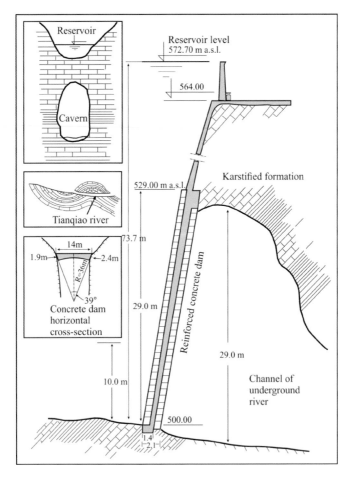

**Figure 6.45.** Tianqiao sinking river, China. Concrete dam at channel entrance of Tianqiao underground river (Dai Jingchun, 1984).

of vigorous ground water level fluctuations coupled with surface water percolation, provokes collapse sinkhole development. The collapse typically begins at the contact between the karst rock and the unconsolidated sediments. Depending on the thickness of the sediment and intensity of the piping or arch propagation this process sometimes takes many years before a collapse appears at the surface—or it can develop with catastrophic rapidity. A few stages of collapse propagation are sketched in Fig. 6.46. During reservoir operation the magnitude of this process can increase tremendously. Subsidence (collapse) induced by reservoir operation has resulted in considerable leakage in many cases: Lar Reservoir, Iran; Mavrovo Reservoir, FYUR Macedonia; Perdicas Reservoir, Greece; Kamskaya Reservoir, Russia; Hutovo Reservoir, Herzegovina; North Dike, Florida; Hamam Grouz, Algeria; Samanalawewa, Sri Lanka, and many others.

Frequently, the 'cavity' takes the form of a stopping-upwards arch in the soil at a shallow depth below the surface that is undetectable during preliminary investigations at a proposed reservoir (Situation No. 3, Fig. 6.46).

The role of dynamic compacting is to destroy the natural porous structure of granular soils and to increase the impermeability of the compacted layers. Compaction is very efficient if the compacted layer consists of more than 15% of clay. The standard technique is compaction by vibrating rollers. The impermeability of large parts of the reservoir floor is improved during compaction by vibrating rollers and, in addition, at many points the roofs of shallow soil arch 'cavities' will be collapsed. This is an unintended but useful means of searching for and detecting shallow but concealed soil arch cavities.

For large natural collapses in a reservoir floor, including collapses induced by vibrating rollers, point-positioned dynamic compaction can be applied. By using the repeated high-energy drop of steel tampers weighing 5–20 tons (i.e., pile driving equipment), onto the area of collapse, the strength and consistency of sediments in zones weakened by piping is considerably improved; the depression created should be backfilled by material with similar properties to the natural soil, and a geomembrane added for a final cover and additional support.

The clayey blankets are not frequently applied to prevent seepage through the karstified reservoir bottom. It usually consists of one or two clay layers with a drainage system beneath the blanket. Due to different influences of underground water

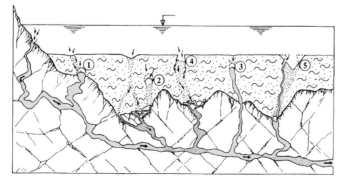

**Figure 6.46.** Schematic presentation of a few stage of collapse genesis.

(uplift and suffusion) and reservoir water (pressure and degradation), as well as clay chemistry, the blanket is prone to be demolished. Degradation is a long process, starting with local sinking, but it is not neccessary to ending in collapsing. In the case of the Hammam Grouz Reservoir, Algeria, a number of small depressions were observed after 17 years of operation (Figs. 6.47 and 13.36). However, the collapse occurred at the perimeter of reservoir.

Particularly after the experience with Tarbela Dam in Pakistan this kind of reservoir waterproofing became questionable. Geosynthetic clay liners became accepted as much reliable technology for waterproofing reservoir bottoms is situated in karstified rocks.

**Figure 6.47.** Hammam Grouz Reservoir, Algeria. Above - empty reservoir; Below - collaps at perimeter of reservoir

### 6.4.3 Grouting Carpet (consolidation)

When large and deep sinkholes in a reservoir floor are connected directly to open karst conduits with high flow capacity, treating only the alluvial sediments above the karst can be insufficient and too risky. Figure 6.48 shows the case beneath the bottom of the Hutovo Reservoir, where the former sink point of a river was discovered, but now buried beneath about 50 m of alluvial sediments. Water sank through a few collapses and channels in the alluvium to collect in the karst channel in the paleorelief. To minimize the risk of seepage from the reservoir the entire rock mass surrounding the sinkhole (ponor) and all karst channels were grouted. The grid spacing of the grouting holes and the paleotopography beneath the alluvial sediments are shown in Fig. 6.48(a). Via the grouting holes (70 m deep) all karst channels and joints were plugged. Before grouting, sinkhole funnels and channels in the alluvium were backfilled with clay/cement mixtures. After grouting, the entire area was blanketed with a geomembrane for additional support.

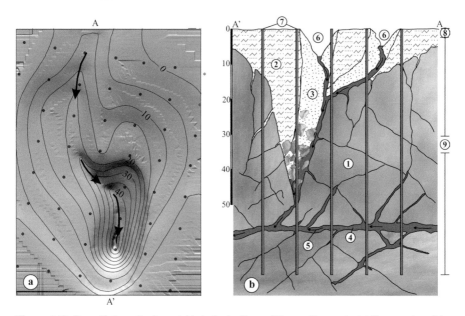

**Figure 6.48.** Remediation of a deep sinkhole in the floor of Hutovo Reservoir. (a) Topography of the limestone bedrock with the grouting holes grid. (b) 1. Karstified limestone; 2. Alluvial deposits; 3. Sandy deposits with boulders; 4. Karst channel; 5. Direction of underground flow; 6. Collapse preceding the remedial works; 7. Grouting boreholes; 8. Surface zone compacted and covered with geomembrane; 9. Depth of grouted rock mass.

### 6.4.4 Shotcrete

Shotcrete (sparry concrete) is an efficient material to achieve the necessary level of water tightness for reservoirs and karstified river beds. For full efficiency, the shotcrete should be applied after cleaning the rock of vegetation (tree stumps and roots), with clay from the cracks removed, caverns dentally treated and any sections prone to uplift protected by one-way valves.

The average thickness of the shotcrete has been 5 cm. To prevent micro-cracks, steel reinforcement meshes (dia. 3 mm) or different types of fiber (steel, plastic, glass) are being used. The length of the fibers is from 3 to 30 mm, i.e., at least three times the diameter of the largest grain in the aggregate. In order to assure satisfactory anchoring, ends of steel fibers should be thickened or bent.

To prevent losses (75 m³/s) along 62.5 km of the Trebišnjica River bed more than $2 \times 10^6$ m² were blanketed with shotcrete (Fig. 6.49).

This solution was very effective overall. However some local problems appeared immediately after the shotcreted river bed became operational. Due to strong uplift, there was local destruction, as seen in Figs. 6.49B and 6.50A. These destructions happened when sinkholes are not discovered and properly treated during the shotcrete blanket construction.

After 35 years of operation, the structural integrity of the shotcrete in some sections has been destroyed because it was of low quality, not meeting the specifications sufficiently strictly (Fig. 6.50B).

Shotcrete was applied to prevent seepage through the karstified bank of the Hammam Grouz Reservoir (Algeria). In the case of Salakovac Dam, Herzegovina, efficiency of shotcrete was negligible due to vegetation, particularly roots, that were not completely removed.

**Figure 6.49.** Trebišnjica River bed protected by shotcrete.

**Figure 6.50.** Damages at shotcrete lining due to (A) underground water uplift, and (B) during operation due to (locally) low quality of shotcrete.

## 6.4.5 Reinforced Concrete Slabs

Akköpru Dam (Turkey) is an embankment structure rising 110.5 m above the river channel. A small area of the reservoir (2 km upstream from the dam site) is in direct contact with heavily karstified limestone. Several 10s of wide and deep karst shafts were discovered in an area of 250.000 m$^2$ (Fig. 6.51). The water table is very deep, 100–116 m below the reservoir bottom (close to the sea level).

To prevent leakage a thick reinforced concrete slab was laid down on the reservoir floor and up to an elevation of 100 m on the adjoining left bank (Fig. 6.52). This engineering solution was designed by DSI Ankara.

**Figure 6.51.** Akköpru Reservoir, Turkey. Karst shafts in the floor.

**Figure 6.52.** Akköpru Reservoir, Turkey. Protective reinforced slab in the reservoir bottom and along the bank.

## 6.4.6 Geomembranes

Geomembranes are very useful covers, frequently applied for waterproofing reservoirs and the bottoms of tailings ponds. In karstified rock this kind of prevention generally requires additional special structures (different kinds of sub-membrane drainage and aeration installations). In order to design an efficient reservoir waterproofing solution, the key requirement is to prevent any contact between reservoir water and the karstified limestone at the open ground surface or at the bedrock contact underneath any alluvial deposits.

A smooth limestone or soil surface is needed to support any of the protective measures. If there are wide openings, karstified cracks or voids, special approaches and technologies should be planned as part of the design to reinforce the resistance of the waterproof layer. Most types of geomembranes are efficient up to maximum pressures of 10 bars in reservoirs. If they have to span gaps (wide open cracks or circular openings such as ponors), then their tensile strength and puncture resistance must be increased.

After reservoir impounding, the alluvial sediments will be saturated and vertical (sometimes also horizontal) circulation of water can occur in them. The effectiveness of the surface treatment will depend on the treatment of the real sources of any problems. Almost always these real sources are in the karstified substratum beneath the unconsolidated deposits. If they are not properly treated any waterproofing structure in the reservoir bottom can be ineffective or fail. If it is under strong pressure from above, or is exposed to any strong upward pressure from below, a geomembrane alone will not be sufficient to resist these forces. In some cases geomembranes have not prevented the creation of long and wide cracks in reservoir floors.

Geomembranes are products manufactured with rigid quality control, and are easy and quick to use during construction. If they are used to protect the upstream face of a dam, consolidation grouting of dam foundations is not necessary, as it is in the case for a dam with a central clay core.

An example of the design of an impermeable structure with geomembranes is presented in Fig. 6.53 (Šumarac, 2008). Geomembranes were installed in Ourkiss

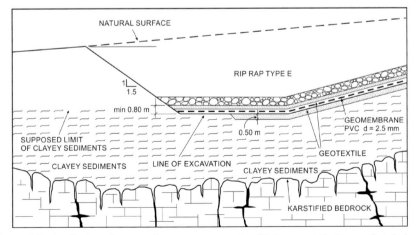

**Figure 6.53.** Application of geomembrane to protect a karstified reservoir bottom against leakage. Šumarac, 2008.

Reservoir (Algeria) with a deep groundwater table (40–60 m below the dam foundation). A few meters of clay sediments covered karstified limestone containing many empty caverns, and strong air currents were detected in the investigation of boreholes. A PVC geomembrane, 2.5 mm thick, between two layers of geotextile, was laid over the natural clayey sediments. A filter-drainage bed beneath the geomembrane drains any percolating water downstream of the dam. In this case any increase of hydrostatic pressure beneath the impervious carpet and the dam is prevented.

To protect seepage from the Ourkiss reservoir the impermeable carpet of 520,000 m² was constructed. Geomembrane of 550,000 m² was protected from both sides with geotextile. A total of 1,1000,000 m² of geotextile was applied.

Geomembranes have been used successfully in tailings ponds on karstified rocks: Cerro Corona, copper/gold mine, Peru (Fig. 6.54); Gacko coal mine, Herzegovina; and many other tailings.

Geomembranes can be breached for several different reasons: however, most are due to defects in the basal layer(s)—usually due to collapse in the underlying alluvial deposits.

The efficiency of geomembranes depends on proper treatment of the soil and rock underneath. If the surface of the reservoir floor is not properly designed and constructed, the protective measures are exposed to destruction. The most dangerous are undiscovered sinkholes in the alluvium close to the reservoir floor (Fig. 6.55).

Due to compaction of the surface soil layers wide cracks can be created (Lar Dam, Hutovo Reservoir). In some cases compaction is a consequence of the use of mobile heavy machinery (tracked or similar).

Long and wide cracks can be created even if the reservoir bottom is protected with geomembranes. Large cracks usually occur during the impoundment of the reservoir, due to differences in soil mechanical properties between the natural sediments and the artificially compacted upper layers.

**Figure 6.54.** Cerro Corona, Peru. Andean karst. Tailings waterproofing protection by geomembrane.

**Figure 6.55.** Geomembrane damage due to collapses and wide cracks in unconsolidated sediments. (A) Collapse sinkhole below the geomembrane, (B) wide crack in alluvial deposits below geomembrane, and (C) large crack developed in spite of the reservoir bottom being protected by a geomembrane.

Frequently geomembrane destruction occurs due to the roof of an undiscovered suffosion 'cavern' (soil arch) collapsing and the geomembrane above it losing support. Two examples are seen in Fig. 6.56. This type of accident occurred because the tensile strength and puncture resistance of the geomembrane was not strong enough to withstand the pressure of the reservoir water column coupled with the diameter of the collapse. However, the basic cause of the problem was an undiscovered shallow empty void beneath the reservoir bottom.

Aeration equipment to evacuate the air from cavities during a rise in groundwater level is not commonly installed beneath geomembranes because the natural aeration will be adequate. Laying a PVC geomembrane cover directly over compacted alluvial soil is usually very effective, particularly where the reservoirs are in rocks not prone to karstification.

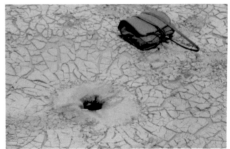

**Figure 6.56.** Destruction of the geomembrane in a reservoir bottom due to collapse in the supporting layer.

## 6.4.7 Dental Treatments

'Dental treatment' means plugging just the karst openings or cracks seen at the surface by different means, mostly concrete today. It is synonymous with filling a cavity in the enamel of a tooth, without considering the health of the root. It is a very local measure only. In ancient times people tried to prevent seepage through karst openings by using soil fillings or wooden structures. Their lifetimes were short—after the first flood, most were destroyed.

From the very beginning of reservoir construction, this limited approach has been adopted:

- To improve the geotechnical strength of dam foundations by filling caverns, and
- to prevent seepage losses from reservoirs.

For the first case dental treatment can be an efficient measure; however, in the second case dental treatment alone will (mostly) end in failure. Every open joint is only one part of a larger system of similar joints which can cover thousands of square meters. Many of them are masked by soil and visually not detectable. Following reservoir impounding and water level fluctuations, large numbers of new seepage points are likely to be created. In some cases dental treatment can be locally effective (Fig. 6.57).

Only a dental treatment approach to the estavelle type of karst opening is particularly questionable. Plugging alone will not be effective. Due to strong uplift forces during rising water stages, new seepage loss points will be created. To improve the efficiency of dentally treated karst openings, local grouting or shotcrete blankets can be applied. However, this is not usually enough. Construction of uplift-releasing structures is necessary.

**Figure 6.57.** Dental treatment in the bank of Keban Reservoir. A wide crack is filled with concrete.

## 6.4.8  The Destructive Role of Air-hammer and Water-hammer Effects

One of the important properties of the karst aquifer is the presence of a distinct unsaturated (vadose) zone. In karstified rocks, caverns and conduits above the water table contain air at atmospheric pressure. In many different karst regions of the world strong air currents have been encountered in innumerable piezometric boreholes. The direction of the air current is often reversible, depending on external changes of air temperature and pressure. Clearly, the presence of air currents of some magnitude is one of the important features of the karst unsaturated zone. Many different kinds of protective blankets in reservoir floors can be destroyed due to it.

As has been stressed above, many dams and reservoirs in karst areas are located in valleys or poljes where the water table lies deep beneath the floor. This means that below the dam sites and reservoirs there is an unsaturated zone where conditions can change rapidly. In some cases after heavy rains, this zone becomes completely saturated in a very short time. Emptying (dewatering) of the zone is also fast. As a consequence of this behavior, the integrity of reservoirs depends on air flow dynamics in this zone. From the point of view of water tightness, during construction and operation stages the properties and behavior of the aeration zone between the reservoir bottom and water table should be carefully studied.

Fluctuations of water tables can be fast and with high amplitudes. At times of sudden heavy rains, vertical and turbulent water flows saturate karst channels and caverns. Rates of the water table rise can be 80–100 m/24 hour, implying that large volumes of karst porosity is saturated extremely quickly and water occupies space previously filled with air. Due to the geometry of many karst channels, particularly inverted siphon forms, air can become trapped in air pockets. Increasing pressure can then expel the trapped air very suddenly into the upper channels. Usually this process is followed by marked by a strong explosion and destruction of remedial works (protective blankets and plugs). Air-hammer and water-hammer effects have been noted and reported in reservoirs in Dinaric and Chinese karst regions (Milanović, 1981; Chen, 1986). Accordingly only a verbal record (D. Ford) in exploration drilling into a metamorphic marble body with iron ore in the Canadian Shield in Ontario, the borehole hit a closed cavity where air pressure was so high that it blew 50+m of drilling rods back out of the hole, and continued to blow for hours afterwards.

Experience with many reservoirs and other projects in karst confirm that the (usually unexpected) problems with pressurized air and strong groundwater uplift are serious. Because explosively pressurized air is a consequence of the groundwater regime, the two determinants (water and air) are closely coupled and in many cases it is not easy to distinguish which played the decisive role, water or air.

Rapid groundwater rise with increasing air pressure in the aeration zone can produce air bubbles directly below the protective structures, with destruction of geomembranes as a consequence. If the air pressure reaches the limit of the plastic deformation, the geomembrane explodes (Fig. 6.58). After destruction (explosion) the torn opening becomes a channel connected to karst conduits in the bedrock and so becomes a sink point. To protect geomembranes against such damage an effective solution is the construction of an aeration pipe (Fig. 6.59). This has to be installed

**Figure 6.58.** Destruction of a geomembrane due to air pressure: (A) Geomembrane balloon due to strong air pressure, (B) a geomembrane after exploding.

directly into a major karst conduit opening in the bedrock in order to discharge the air under pressure into the air above water level in the reservoir. Aeration pipes can be efficient only when properly inserted into such karst channels. Because of this factor, karst channel openings in paleorelief should be uncovered by excavating the unconsolidated sediments burying them in likely areas. Detection of channels by drilling boreholes alone is not suggested because chances of finding them are less than 10%. The best method is to follow any channel in the buried relief to the karst opening at its contact with the alluvial deposits, and excavating there.

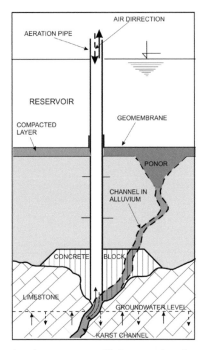

**Figure 6.59.** Aeration pipe. Right: cross-section of an aeration pipe structure, and left: an aeration pipe in operation in a reservoir.

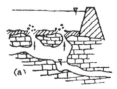
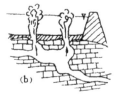
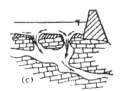

**Figure 6.60.** Process of exploded collapse by water and air. (a) Send up bubble; (b) Break through the cover; (c) Collapse and leakage. Jian, 1986.

Similar problems with increasing air pressure after dam construction, and seepage from reservoir, have been registered in Chinese karst (Fig. 6.60).

In order to counteract the destructive role of pressurized water (water-hammer) and air (air-hammer) similar structures have been built in some Chinese reservoirs. A few examples are presented in Fig. 6.61. In the case presented in 6.61(A) one vertical pipe and one inclined up the reservoir bank provide air escape outlets for two different caverns that are separated by a probably airtight shale unit. Pipe diameter is between 100 and 200 mm. Figure 6.61(B) shows the structure at Mata Reservoir, Cili County, Hunan Province. The aeration pipe is connected to an underground stream channel and anchored to the inclined reservoir bank. The seepage measures at Yintaoao Reservoir, Fenghuang County (Fig. 6.61C), show the use of both an aeration pipe and a natural

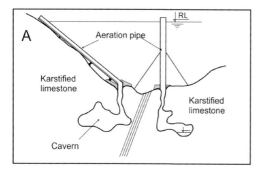

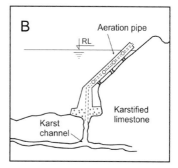

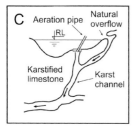

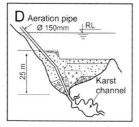

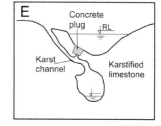

**Figure 6.61.** Different anti-seepage and pressure releasing structures in reservoirs in Hunan Province, China. Survey and Exploration Team, Hydropower Institute of Hunan Province, 1979 (in Chinese).

opening (a sinkhole) connected to the karst channel for the release of air and water overflow. A second aeration structure in the same reservoir is shown in 6.61D. A natural cave channel coupled with dental plugging of a branch passage beneath the Loubokeng Reservoir, Hunan Province, was used as for the releasing in a similar manner (6.61.E).

# CHAPTER 7

# DAMS AND RESERVOIRS CONSTRUCTION ON EVAPORITES

## 7.1 EVAPORITES – RISKY ENVIRONMENT FOR DAM CONSTRUCTION

Numerous dams and reservoirs all over the world that have been built on rocks containing evaporites have encountered dissolution problems during their construction and/or first filling, or later during the operation. The highly soluble nature of the common evaporite rocks, chiefly gypsum and salt, sometimes carnallite, sylvite and glauberite, is known to create a highly hazardous environment despite massive protective measures being applied to prevent dissolution and failures.

In 'Gypsum-karst problems in constructing dams in the USA' Johnson (2008) concluded that if gypsum karst is present at a dam site or reservoir it can compromise the ability of the dam to hold water in a reservoir, and can even cause its collapse. "Gypsum karst in the abutments or foundation of a dam can allow water to pass through, around, or under a dam, and solution channels can enlarge quickly, once water starts flowing through such a karst system". These conclusions are confirmed at number of dams and reservoirs built in evaporites around the world.

For instance, the worst American civil engineering failure of the 20th century was that of the St. Francis Dam (California, USA) that killed 450 people along the St. Francisco Canyon and Santa Clara valley. The foundation of the 60 m high concrete gravity dam was a conglomerate with gypsum intercalations. On March 12/13 1928 the dam collapsed abruptly, due to weakening of a paleo-landslide feature in the left abutment. According to Cooper and Calow (1998) the strong uplift was partly attributed to gypsum dissolution. Once rapid solution has started at a site due to the hydraulic head created by filling the reservoir, additional protective measures become technically extremely complicated.

Evaporite rocks are widespread. Practically every dam and reservoir constructed on strata containing them will face at least one of the three principal problems: seepage loss, stability, and water pollution. Recent authors reporting the problems of dams and reservoirs in evaporites include Brune, 1965; Pechorkin and Pechorkin, 1979; James and Kirkpatrick, 1980; Anagnosti, 1987; Ford and Williams, 1989, 2007; Guzina et al., 1991; Lu and Cooper, 1997; Cooper and Calow, 1998; Pearson, 1999; Gutierez et al., 2003, 2015; Johnson, 2003, 2004, 2008; Lu and Zhang, 2006; Maximovich, 2006; Barjasteh, 2012; Sissakian et al., 2015; Mahjoob et al., 2014.

There have been many presentations focusing on dam problems in evaporites at different international conferences. Particularly important was the International Engineering Geology Symposium on the topic 'Engineering geological problems of construction on soluble rocks' held in Istanbul, Turkey, 1981. A total of 55 papers, divided into five different thematic sections, were presented.

## 7.2 EVAPORITES – GENERAL CHARACTERISTICS

Evaporites belong to the class of minerals that are formed chemically. Precipitation, i.e., deposition of a mineral from an ionic solution, takes place due to oversaturation of the solution with a given component. This is usually caused by evaporation of the solvent, commonly water, but can also be due to the addition of other components that increase or decrease the solubility of the given component (e.g., reducing $CO_2$ in solution renders the calcium-carbonate less soluble), fluctuations in temperature and pressure. In such ways, but chiefly aqueous evaporation, salts and gypsum are precipitated from an over-concentrated solution.

> Chemically, *gypsum* is hydrated calcium-sulfate, bonded with two molecules of crystalline water ($CaSO_4 \cdot 2H_2O$). Its solubility in standard conditions is ~ 2.4 g/L. Anhydrite is calcium-sulfate ($CaSO_4$). Absorbing water, anhydrite converts into gypsum.
> Glauberite is the monoclinic mineral $Na_2 Ca (SO_4)_2$ with a solubility of 118 g/L. Halite is the mineralogical name for rock salt, NaCl. It is the principal deposit from the evaporation of sea water. Its solubility is 365 g/L. As the last of a body of sea water evaporates, thinner deposits of sylvite (KCl) and carnallite (KCl. $MgCl_2 . 6H_2O$) may be precipitated on top of the salt; their solubility is slightly higher.

Among the number of gypsum compositions the most frequent are: solid and compact, as seen in Fig. 7.1(A); tectonized and karstified (B); and as a gypsum breccia (C). In some cases where it is fine-grained and homogenous in texture (*alabaster*), it can be easily mistaken for carbonate rock in a borehole core.

The intensity of karstification in evaporites depends on their solubility and the rate of removal of the dissolved solids by flowing water. Salt is about 160 times more soluble than gypsum. As the rate of flow increases the dissolution rate increases rapidly, sometimes catastrophically. Gypsum dissolution in flowing water can be ten to one hundred times faster than limestone but is still only about one thousandth the rate of halite.

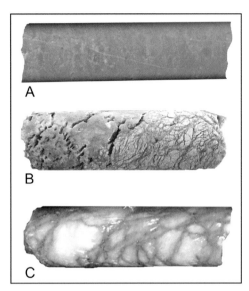

**Figure 7.1.** Gypsum borehole cores. (A) solid gypsum core; (B) tectonized and karstified core; (C) compact gypsum breccia.

From investigations made in Spain, water flowing at 0.35 m/s dissolves 0.393 m of gypsum per square meter of surface per year. Flow of 0.99 m/s dissolves 0.853 m/m² per year (Cooper and Calow, 1998). Calculations show that approximately one m of gypsum a year can be removed from a gypsum surface by a water flow of about one m per second. In the Kamskoye reservoir the rate of dissolution of karstified gypsum banks is 1.5–2.0 m/year (Pechorkin and Pechorkin, 1979). According to investigations by Filippov and Shkolnik (1988) the time necessary to create karst channel one meter in diameter and one km in length in gypsum, can be from 12 to 40 years. For the same initial conditions in limestone ~ 10,000 years is required and in dolomite, 35700 to 52600 years (from Kozyreva and Trzhtsinsky, 2004). There has been a great deal of computer modeling as well as field experiments, etc. Rates vary according to initial aperture, hydraulic gradient, etc.

From laboratory and field studies by Mancebo Piquera et al. (2012), in the case of evaporites the water-mineral interaction is complex and may have significant effects on the porosity. Changes of groundwater flow induced by construction of dams in rock that contains gypsiferous deposits along with halite, thenardite, mirabilite, glauberite, epsomite and anhydrite provokes a series of processes that will increase porosity: (a) mineral dissolution; (b) increase of void volumes by the transformation of glauberite to gypsum; (c) increase of voids by physical erosion. The processes that reduce porosity are: (a) processes of expansion (as it hydrates in the presence of water, anhydrite can increase its volume between 30–67%); and (b) crystallization of gypsum and other sulfates (increase in volume and pressure of the crystallization of these sulfates is comparable to the expansion of anhydrite).

Osipov (1981) analyzed soluble salts (sulfates, carbonates and chlorides) in clays and their properties. These studies have shown that salts of syngenetic origin have less influence upon the behavior of clays under loading than do secondary salt inclusions that form during diagenesis. The uni-axial compaction strength of these clays can reach some hundreds of kg/cm$^2$. However, the presence of secondary salt inclusions can be most unstable; with vigorous exchange of water, they can be dissolved and removed. The process of salt dissolution in clays is in general of an osmotic character and can lead to the formation of a fairly porous clay structure. As a consequence, the permeability increases and physical collapses will occur.

In flowing water solution of salt in a reservoir bank can be as fast as one cm/hour or more. Figure 7.2(A) shows part of a river bank (the proposed reservoir bank, before impoundment) after it was exposed to flood water. In less than 10 hours more than 10 cm of salt was dissolved (Milanović, 2004).

**Figure 7.2.**  Gotvand Reservoir (Iran). (A) High water salt solution notching created by a river flood of less than 10 hours; (B) Sinkhole and morphology of a fresh underground channel in halite.

Evaporite solution is particularly fast where vertical or sub-vertical discontinuities cross each other. Figure 7.2(B) shows the underground morphology developed by dissolution in halite beneath a shallow sinkhole depression on the surface.

In some cases, large collapses have propagated from salt beds located 1000 m or more beneath the surface (Ford and Williams, 2007). Even where the salt deposits are very deep beneath a reservoir bottom, the possible risk of collapses and seepage loss cannot be ignored.

A specific problem arises when the alluvium and material used for the foundations of an earth dam contain substantial quantities of disseminated gypsum grains. Preferential solution of the gypsum may induce piping which creates huge porosity and, finally, subsidence within the earth dam structure, Gutierrez et al., 2003.

Thick deposits of crystalline salt are prone to sliding, even where they are massively bedded. One of a few slides in salt noted in the banks of Gotvand Reservoir is shown in Fig. 7.3. After the filling and submergence of the bank here, rapid movement of this mass into the reservoir is to be expected.

**Figure 7.3.** Gotvand Reservoir, Iran, showing a displaced salt mass (rotational slide) before reservoir impoundment.

## 7.3  LIST OF DAMS AND RESERVOIRS IN EVAPORITES WITH CONSTRUCTION PROBLEMS

Evaporites are particularly vulnerable if they are present in the foundation of dams or in abutments of reservoirs. Worldwide more than 60 dams have been affected by gypsum and salt dissolution problems and needed rehabilitation - in the U.S.A., China, Iran, Iraq, Argentina, Algeria, Guatemala, Russia, Switzerland, Peru, Spain and Venezuela.

According Cooper and Calow (1998) and Johnson (2004) several dams in the U.S. were seriously endangered or were abandoned because of solution problem in gypsum: **Upper Mangum Dam (**Oklahoma, abandoned before construction); **McMillan, Avalon**, and **Rio Hondo** dams in New Mexico; **San Fernando, Dry Canyon, Buena Vista, Olive Hills, Rattlesnake** and **Castaic** dams in California; **Stanford Dam** (Texas); **Quail Creek Dike** (Utah, catastrophic failure, 1989); **Red Rock Dam** (Iowa); **Fontanelle Dam** (Oklahoma); and **Moses-Saunders Power Dam** (New York/Ontario).

Dams reported to have problems or concerns because of gypsum dissolution elsewhere include: **El Isiro Dam** (Venezuela); **Poechos Dam** (Peru); **Beninar Reservoir, San Juan Reservoir, La Loteta, Caspe, San Loran, Estremera** and **Alloz Dam** (Spain); **Joumine Dam** (Tunisia); **Moncenisio Reservoir** (border Italy/France); **Hessigheim Dam** (Germany); **Birsfelden** (Switzerland); **Farkhad Dam** (Uzbekistan); **Casa de Piedra Dam** (Argentina); **Puebla de Pava** (head race tunnel, Guatemala); **Kamskaya Dam** (Russia); **Tannur Dam** (Jordan); **Tbiliskaya Dam** (Georgia); **Erevanskaya Dam** (Armenia); **Mignachevir Dam** (Azerbaijan); **Irganayskaya Dam** (Dagestan); **Lower Kafirniganskiy Dam** and **Baipazinskaya** (Tajikistan); **Chardarinskaya Dam** on Syr-Daria River (Central Asia); **Huoshipo, Yangmazhain,**

**Langdai, Baiyanjiao and Mahuangtian Reservoirs** (China); **Haditha, Mosul** and **Hatra Dam** (Wadi Tharthar – Iraq); **Tang-E-Shemiran Reservoir, Khordad 15th Reservoir, Marun Reservoir, Gotvand Dam** and **Reservoir, Khersan III, Jarreh, Gheisaragh** and **Nargesi Dam** (Iran); **Djedra Dam, Foum El Gerza Dam** (Algeria); and a few others elsewhere in the world.

Salt is present in the foundation rocks of the **Nurekskaya** and **Rogunskaya** dams in Tajikistan. In the foundations of the world highest **Rogunskaya Dam** (height of 335 m, earth-filled) in Tajikistan, a thick salt layer was treated by a combination of grout and salt curtains. The **Gotvand Reservoir** (Iran) suffers many problems due to salt and gypsum, as seen in the illustrations above.

## 7.4 SELECTED CASE STUDIES

The worst American civil engineering failure of the 20th century was the failure of the **St. Francis Dam** (California, USA) killing 450 people along the St. Francisco Canyon and Santa Clara valley. The bedrock of the concrete gravity dam (60 m high) was a conglomerate with gypsum intercalations. On March 12/13 1928, the dam collapsed due to rejuvenation of a paleo-landslide in the left abutment. According to Cooper and Calow (1998) strong uplift was partly attributed to gypsum dissolution.

Another example of catastrophic failure is that of the **San Juan** earth dam (Spain). During the first filling of the reservoir in 2001 there was intensive dissolution of gypsum and a portion of the dam collapsed, sending a very big flood downstream.

A 6.6 force earthquake in February 1971 caused failure of the **Lower San Fernando Dam** in California. After the impounding, there was seepage through gypsum intercalated in shales, siltstones and sandstones. Remedial grouting has been applied. To avoid the possible danger, about 80,000 people were temporarily evacuated.

According Johnson (2004) recent projects related to gypsum karst and dam-construction problems in the United States include: (1) abandonment of the **Upper Mangum Dam** site in southwest Oklahoma; (2) failure and collapse of the **Quail Creek Dike** in southwest Utah; (3) leakage from **Horsetooth Reservoir** and **Carter Lake Reservoir** in north-central Colorado; and (4) leakage from Anchor Dam in northwest Wyoming. One 0.6 m bed of gypsum in a 50 m thick dolomite formation has been a recurring source of problems in the **Moses-Saunders hydropower dam on the St Lawrence River** (U.S.A./Canada border).

Proposals for a 33 m high earth-filled dam at the **Upper Mangum** site in Oklahoma began in 1937. The bedrock foundation consists of 60 m of gypsum with thin interbeds of dolomite and shale. The decision to consider the site was based only upon its favorable topography, without proper respect for the geology despite the predominance of gypsum in the abutments (Fig. 7.4). Only in 1999 was it recommended that further examination of the Upper Mangum site be abandoned. "Engineering measures needed to remediate karstic foundation construction here would add greatly to the cost of construction, and still would not assure tightness of the reservoir or integrity of the dam" (Johnson, 2004).

**Quail Creek Dike** is a 25 m high earth-fill embankment constructed in 1985 over well-jointed and karstified strata consisting of dolomite, gypsiferous siltstone, gypsum and sandstone. Many karst features were detected during the site investigations. When

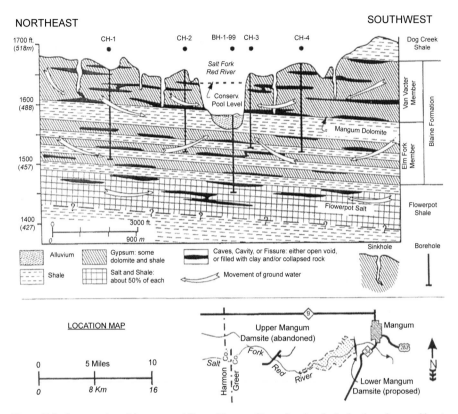

**Figure 7.4.** Cross section of the proposed Upper Mangum Dam, showing the bedrock geology and karst features in gypsum, dolomite, and salt (Johnson, 2004).

filling operations began, seepage beneath the Dike caused local collapse and failure, provoking considerable damage and evacuation of people downstream. As a remedial measure, a 25 m deep and 600 m long cut-off wall has been constructed through the gypsum-karst zone (Payton and Hansen, 2003). The earth-fill embankment is now replaced with a roller-concrete structure. After a second-time filling, seepage losses had reached about 340 L/s in 2002.

Due to outcrop of gypsum beds and the likelihood that they would be exposed to solution in the bottom of the reservoir, the original site proposed for the **Cedar Ridge** dam in Texas was declared to be too risky. "The presence of gypsum beds in this portion of reservoir means that there may be karst pathways whereby impounded water could escape a reservoir" (Johnson and Wilkerson, 2013). Based on lengthy studies, a new location was suggested approximately 8 km upstream where an impervious formation separated the reservoir floor from underlying gypsum.

Failure of the **Anchor Dam** "is a textbook example of where not to build the dam" (Jarvis, 2003). The dam is a 66 m high concrete arch dam situated near Thermopolis in Wyoming (U.S.A.). The reservoir area (Owl Creek) is underlain by red beds, dolomite and gypsum formation. Solution cavities in the dolomite beds were encountered during excavation at the dam site and more than 50 sinkholes were documented in the reservoir area. A year after it started filling, a sinkhole nearly 100 m in diameter and 20 m deep developed approximately 350 m upstream of the dam. To prevent water loss into it, an earthen dike was constructed around the sinkhole "but the height of the dike was below the crest of the dam". Since the 1960s (first filling) the reservoir has never filled to more than just a small pond.

**The Mosul Dam** (Iraq) is located on the River Tigris approximately 60 km upstream of Mosul, the second city of Iraq. It is a multipurpose project for irrigation, flood control and hydropower generation. The dam is 113 m in height, 3.65 km in length, 10 m wide at its crest. The design surface area of the reservoir at the beginning of the undertaking was 380 km$^2$ with a storage capacity of 11.11 billion m$^3$.

It is an earth fill dam constructed on bedrocks of the Fatha Formation (Middle Miocene), which consists of cyclically deposited gypsum, gypsum breccias and anhydrite beds, alternating with marl and limestone. The aggregate thickness of the gypsum beds is 18 m in average. Initial permeability in the massive gypsum and anhydrite, measured by pressure test, was close to zero. High initial permeabilities, ranging from 10 to 50 Lu units or more, were measured in the fissured limestones and brecciated strata. The highest measured permeability was found at the boundaries of or within the tectonized and brecciated gypsum and anhydrite beds. Depths for the grout curtain were selected mainly on the basis of initial permeabilities. The greatest depth of the curtain was 120 m on average (Anagnosti, 1987).

The dam design was based on the assumption that impermeability of the foundations and prevention of any dissolution of gypsum and anhydrite would be achieved by implementing a deep grout curtain strengthened by a grout blanket under the clay core. For grout curtain construction (4,454 + 1,504 m long) 732,000 m$^3$ of grout mix was used. However, during excavation of the dam foundations, a number of sub-vertical ruptures were discovered (Guzina et al., 1991). They were intensely karstified, even in the foundation rocks, and caused intensive gypsum and anhydrite dissolution and seepage (Fig. 7.5). This has created numerous problems during the construction, the impounding and subsequent operation of the dam.

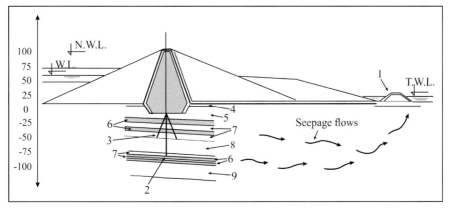

**Figure 7.5.** Mosul Dam, Iraq. Cross-section perpendicular to the axis of the dam. (1) Cofferdam; (2) deep grout curtain; (3) piezometer; (4) river alluvium; (5) marly breccia; (6) gypsum/gypsum breccia/anhydrite; (7) marl/marly limestone; (8) chalky series (vuggy limestone); (9) limestone (hard and dense) Guzina et al., 1991.

The first partial impounding of the reservoir began in February 1986. Shortly after it started seepages were observed through a 1000 m long section of the dam foundations. Chemical analyses of seepage water showed that increases in dissolved solids were consequences of fluctuations in the reservoir level, i.e., consequences of varying pressure and velocity of groundwater flow (Fig. 7.6).

By the month of June in 1986 leakage through the dam site area had increased to 1,400 l/s. The leakage paths are located at average depths between 60 and 70 m. The seepage increased as a consequence of enlargement of flow routes due to the intensive dissolution of the gypsum and anhydrite. By measuring the total dissolved solids, average amounts of gypsum solution of 42 to 80 t/day were found (Guzina et al. 1991). During the six month observation period 13,000 t of gypsum and anhydrite were dissolved beneath the Mosul foundations. As consequences of such massive dissolution, large voids were created, mostly between the evaporites and less soluble (limestone) or insoluble rocks. Locally, at tectonically disturbed places, there was surface settlement and formation of dissolution collapses. Over the period, 1992–1998, four dissolution collapses formed approximately 800 m downstream, near the regulating pool, and later (2003–2005) in different parts of the dam site (Fig. 7.7).

Immediately after the first problems were detected (1986), maintenance grouting work started. A massive grouting technique was applied. Composition of the mixture was as follows: Cement weight = 465 kg, Sand/Cement ratio ≈ 2:1, so sand weight is 930 kg, Water/Cement ratio = 1:1, so water is 415 L, Bentonite/Cement Ratio = 4%, so bentonite weight is 18.6 kg. The total dry weight of the mix was 1413.6 kg/m³. The quantity of different grout mixes consumed between 1986 and 2015 reached 95657.43 tons (Adamo et al., 2015).

Still the seepage losses due to the dissolution of gypsum and anhydrite beds under the foundation of the dam could not be stopped. This caused great concern about the

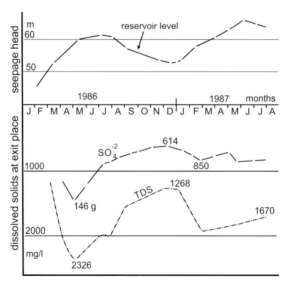

**Figure 7.6.** Mosul Dam, Iraq. Relationship between reservoir levels and dissolved solids in the seepage water (Anagnosti, 1987).

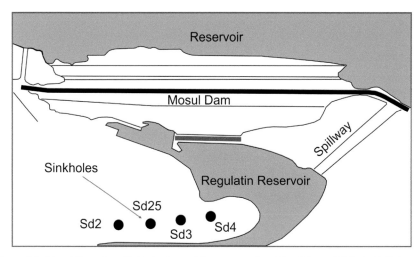

**Figure 7.7.** Mosul Dam, Iraq. Sinkholes located downstream from Mosul Dam—Right bank (Adamo et al., 2015).

possibility of the dam failure. The capital of Iraq, Baghdad, 638 km downstream of Mosul Dam, and the very large populated area in between, is endangered. Due to the adverse geological properties and intensive dissolution in the foundations, Mosul Dam has been declared to be one of the most dangerous dams in the world.

**Joumine Dam** (Tunisia) is a 57 m high embankment structure constructed in 1983. The dam site consists of Cretaceous limestone underlain by Triassic formations. A gypsum bed was detected at the site. During the first impounding seepage began at 0.5 m³/s and when the reservoir reached its maximum level, increased to about 1.0 m³/s. After a few additional stages of grouting, this was reduced to 90 L/s (2013) but with an increasing amount of sulphates in the seepage water (Sari, 2013).

**Kamskaya Dam** on the Kama River (Russia) is 25 m high and 2.5 km long (Fig. 7.8).

The dam was constructed in 1954. At the dam site and reservoir area there is a complex lithological sequence: argilites, sandstone, gypsum, dolomites and anhydrite (Gorbunova et al. (1991); Lykoshin et al. (1992), and Maximovich (2006)). During the impoundment up to a height of 22 m, seepage began to be detected below the dam foundation. Abrupt increase of sulfates in the seepage water indicated that gypsum dissolution was increasing significantly. As a consequence of fluctuations in reservoir water levels new collapse sinkholes were created in its vicinity. Before dam construction two (natural) collapses had been recorded during a period of 50 years. After the reservoir started to operate, there were 11 new collapses in six years. According to Pechorkin and Pechorkin (1979) due to the intensive solution 40 to 50 m³ of gypsum has been dissolved from each linear meter of the reservoir banks.

To improve the resistance and density of the grout curtain in the gypsiferous rock under the Kama dam, a chemical gel-forming grout mix with penetration capability close to that of water was introduced. This grout mix (an oxalo-aluminosilicate) consists of two components—sodium silicate with a density of 1.19 g/cm³ and a

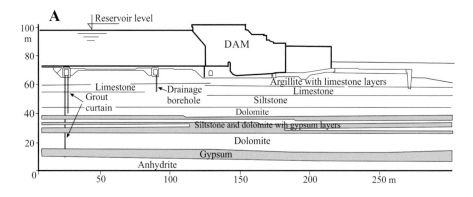

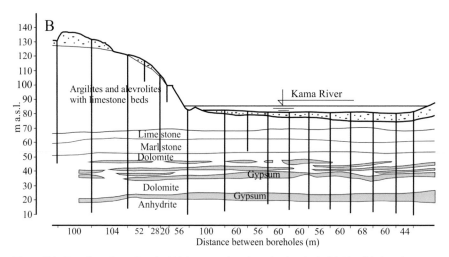

**Figure 7.8.** Kamskaya Dam, Russia. (A) Cross-section along the river bed; (B) Simplified geology cross-section along the dam axis. According Maximovich, 2006.

hardener (Maximovich, 2006). After application of this chemical, the seepage through the Kama curtain was reduced and the stability of the dam increased.

**Bratsk Reservoir** in the Angara River valley (Russia) is retained by a concrete gravity dam 124.5 m high and 4417 m long. A part of the reservoir is situated on a soluble Lower/Middle Cambrian formation (dolomite, limestone, anhydrite-dolomite and gypsum). Under natural conditions, the karst forms at the surface included large collapse sinkholes and shafts (up to 60 m in diameter and 25 m in depth), blind creeks, and caves and karst channels underground (Kozyreva and Trzhtsinsky, 2004). During reservoir filling (1963–1966) about 200 new sinkholes per km$^2$ were induced every year, ranging 2–30 m in diameter and 5–28 m in depth. Genesis of these sinkholes was a consequence of reactivation of paleokarst features and the fluctuations in reservoir water levels. Some of them occurred at distances as great as six km from the river channel. Close to the reservoir fresh caverns in gypsum are now up to 30 m deep and about 2 m in diameter. There have been a number of landslides due to solution. Based on detailed investigations and cartographic volumetric modeling it is concluded that the reservoir shore line will retreat as much as 10 m per year in some sectors. In many areas of gypsum – anhydrite outcrop new collapses of 20–38 m depth have been created. As a consequence, a number of buildings have been damaged and much land has become unsuitable for use.

**San Juan Reservoir** (Huesca Province, Spain) with a capacity of 850,000 m³, was built in 1999 on gypsiferous mantled pediment deposits overlying Tertiary dispersive clay sediments (Gutierrez at al., 2003). A failure occurred during the first test filling in 2001. There was a huge breach in the earth dam due to the use of old alluvial deposits containing considerable amounts of gypsum in the dam foundations. There was a flood discharge of about 300,000 m³ in a short period. Highly soluble gypsum grains and fillings provoked intensive piping and, finally, the dam failure.

**La Loteta Reservoir** (Spain) with storage capacity of 105 hm³ was created by construction of a 34 m high and 1.5 km long embankment dam in the Ebro River valley. The bedrock is an 850 m thick formation including anhydrite, halite and glauberite. The foundation and abutments of the dam include a horizontally bedded, laterally extensive gypsum unit 11 m thick. The water-retaining structure includes a vertical clay core, a horizontal clay blanket, a cut-off wall, and grout curtains 675 m and 255 m long in the left and right abutments (Gutierrez et al., 2015). A cement-flyash-bentonite cut-off wall was built along a continuous single-slot cutter trench using the panel wall technique: it is 1567 m long, 0.8 m wide with an average depth of 23 m. Grout curtains were constructed in both abutments. Figure 7.9 presents the general layout of the curtains in both abutments, and partial sketches showing the distribution of borehole rows in different sectors. Sections with higher grout consumption are indicated. Locally, in these sections, grout mix consumption up to 1496 kg/m was recorded. Such high take indicates the existence of a cavernous zone in the lower part of the gypsum unit, directly beneath the body of the dam.

When the reservoir reached its maximum level (2011), the total leakage was only 46 l/s, i.e., negligible. In spite of additional grouting in the left abutment after impoundment, in 2013 leakage had risen to 55.5 l/s. This, still small but is increasing, volume of seepage suggests the possibility that there is progressive dissolution along the leakage route(s).

**Caspe Dam** (Spain) is an earth-filled structure with a crest length of 400 m and 51 m in height. The dam and reservoir are located on sub-horizontal beds of marls and gypsum (in joints) which contain smaller amounts of anhydrite, halite, glauberite and other minerals with complex reactions with water. The grout curtain, 25 m deep, was connected to the clay core. Distance between grouting holes is 3 m. During the first filling of the reservoir (1987), seepage was recorded at three places. It increased up to 200–250 l/s. Besides dissolution, the transformation of glauberite to gypsum with formation of mirabilite led to a volume reduction of 28%, creating cavities. With additional remedial works the seepage is under control "although injections are required to maintain the grout curtain" (Mancebo Piquera et al., 2012).

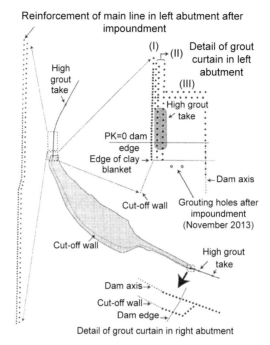

**Figure 7.9.** La Loteta Dam, Spain. General layout of the grout curtain (Gutierrez et al., 2015).

A number of reservoirs partly on evaporites also suffer from significant water pollution problems. Nearly two-thirds of the **Marun Reservoir** in Iran is on the Gachsaran Formation (Miocene, gypsum, anhydrite, marlstone, siltstone and shale). The main problem with the Gachsaran beds is that their erosion and weathering causes high rates of siltation in the reservoir and also has some effects on the water quality (Barjasteh, 2012). The same author analyzed seepage from the **Jarreh Reservoir** on the Rud-e Zard River, which rests on many beds of the Gachsaran evaporites.

**Gotvand Dam** (Iran) is a 178 m high rock-fill structure located in the Bakhtiary Fm, a conglomerate with thin gypsum veins. Approximately 4 km from dam the left bank of the reservoir begins a three kilometer section that is in direct contact with the approximately 120 m thick Gachsaran Formation, i.e., in direct contact with salt intercalated with gypsum and layers of plastic clay and marl. Each salt bed is 10–15 m thick (Fig. 7.10).

Locally, massive and compact salt diapiric structures penetrate through the upper beds of Gachsaran Formation (Fig. 7.11). Twelve of 16 investigation boreholes intercepted cavities occupying up to 16% of borehole depth (Milanović, 2004–2010). Water in boreholes drilled in the salt contained on average: Na – 4,337 (meq/l); Cl – 5,000 (meq/l); SO4 – 31.7 (meq/l); EC – 184,080 $\mu$S/cm$^2$ and TDS – 337,832 (mg/L): i.e., these are essentially saturated solutions. pH ranged between 5.4 and 7.1.

Due to the very rapid dissolution of the salt there appear to be three crucial problems: leakage from reservoir, stability of reservoir banks and water pollution.

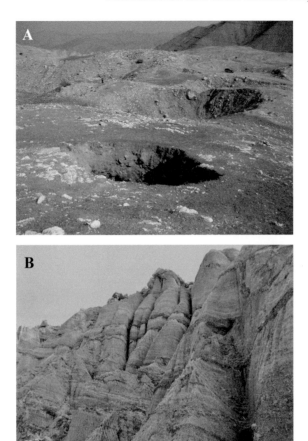

**Figure 7.10.** Gotvand Reservoir, Iran. (A) Karstified salt at left bank before impounding. (B) Collapses in gypsum beds.

**Figure 7.11.** Gotvand Reservoir bank. A small salt diapiric structure.

Because this water has to be used for irrigation, its very large chloride content in water is problematic.

In flood water conditions dissolution has been very rapid. The stepped forms seen in Fig. 7.12 are a consequence of salt dissolution during one flood with rapid river flow. According to experiments (physical models) by the Gotvand staff, the estimated dissolution rate is approximately 10 cm/day on exposed salt cliff faces. Based on this rough dissolution estimate it was concluded that 11.5 million ton of salt would be dissolved during the first year of reservoir operation. According to detailed hydrogeological analysis, off-site leakage through the evaporite strata is unlikely, however.

To minimize the intensity of direct contact between the reservoir water and the halite outcrops cover with different materials were tested: geomembrane, polyurethane, asphaltic materials, concrete blocks and clay blanket. Finally, clay blankets and rip-rap protection were applied as covers. Large collapses were filled with clay slurry (Mahjoob et al., 2014).

During the first year after filling the dissolution rate was higher than expected from the numerical models. However, in the following year the solution rate decreased. A 30 meter deep layer of dense, supersaturated water has been formed on the reservoir bottom. This water is distinctly separate from the overlying water where TDS is to the same as that of the river in its previous natural conditions (Fig. 7.13).

The upper 70 m of the reservoir water are used for irrigation. Three additional low elevation intakes (at 90, 110 and 158 m a.s.l.) have been constructed to drain away the oversaturated water, which can be used in the oil industry.

**Figure 7.12.** Gotvand Reservoir, Iran. Effects of salt dissolution on cliffs in the reservoir bank following a few hours of high flood (upper fresh scars) and a horizontal corrosion notch (the cavern-like feature) formed at the waterline over approximately 10 days.

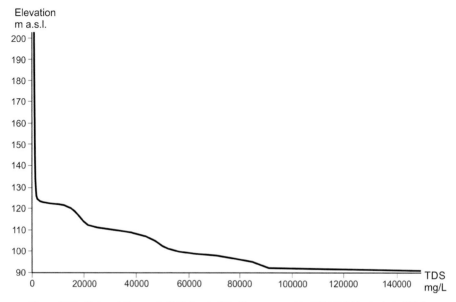

**Figure 7.13.** Gotvand Reservoir. Salinity stratification measured in 2014 (Mahjoob et al., 2014).

**Khordad 15 Reservoir** (Iran). The reservoir is for water supply and irrigation. The dam itself is built on granite. However, the floor of much of the large reservoir behind it consists of karstified limestone, partly covered with alluvium and evaporites. During site investigations, a few collapses and a number of cavities were noted and some investigated with tracer tests. After filling the reservoir (1993), water pollution (TDS, due to the evaporites) increased tremendously. To improve the water quality, reservoir water was mixed with water from the Dez River (transported through a tunnel) and with water from deep wells.

**Huoshipo Reservoir** (China) is constructed on a karstified limestone with 48 gypsum interbeds or partings (Wuzhou, 1988). The height of this earth-filled dam is 23 m and the length is 200 m. Leakage of 237 l/s occurred in the initial stage of reservoir impounding and a number of fresh collapses opened in the reservoir bottom; evidently laminar flow filtration had been replaced by turbulent flow. Because the reservoir capacity is only 4.7 million cubic meters, the reported leakage was declared to be too high. With additional sealing work (grout curtain, plastic membranes, clay cover) seepage was reduced to ≈ 80 l/s; however, the dissolution has not been eliminated.

The **McMillan Dam** on the Pecos River in New Mexico (USA) was constructed in 1893. Gypsum was known only in the abutments. During investigation and construction no caverns were observed, but during the first 12 years of operation they were quickly created in the left abutment and the reservoir dried up. According James and Lupton (1978) during the period from 1893 to 1942 dissolution of the gypsum created approximately 50 million meter cubic of karst channels and caverns.

**Horsetooth Reservoir and Carter Lake Reservoir** (Colorado, USA, 1940) are situated in a formation of interbedded siltstone, shale, limestone and karstified gypsum. In period, 1980–1990, sinkholes formed in the reservoir and the seepage increased dramatically. In 2000, a sinkhole formed in the upstream toe of the dam (Johnson, 2004).

**Cedar Ridge Dam Site** (Texas) is an example of relocation following adequate field investigations of the original site. Gypsum was exposed in outcrop at only one place, a little upstream of the proposed location for the dam. Drilling discovered a gypsiferous sequence, thick 30–45 m, consisting of eight gypsum beds of one-three meter thickness interbedded with shale. This led to a new dam site being adopted about eight km upstream. At the new location the uppermost gypsum is at least 23 m beneath the dam site (Johnson and Wilkerson, 2013).

**Kavsak Dam Site** (Turkey) is located in a Permian karstified limestone (Fig. 7.14).

The concrete face dam rises 88 m from its foundations (73 m from the river channel level). A number of cavities are developed along fault zones (tectonic breccia) or subvertical joints. According to borehole data, all caverns were naturally plugged with clayey/sandy material including some gravel and limestone blocks. An additional problem is the presence of gypsum (30 m thick in some boreholes) situated beneath the dam foundations, close to the basal contour of the grout curtain. More than 30 years of intensive investigations and analyses were undertaken before dam construction started. It was completed in 2013.

**Figure 7.14.** Kavsak Dam site, Turkey.

The **Foum el-Gherza Dam** in Algeria is a 65 m high concrete arch dam situated in a 200 m thick Cretaceous limestone lying over marly shale. Reservoir volume is 47 million m³. Limestone formations at the dam site are in direct contact with marine gypsum. The gypsum formations are seen in outcrop at a few places in the catchment area.

The dam was constructed in 1952 and, after the first filling a number of new small seepages were noted in both banks downstream of it (Hocini and Moulla, 2005). Maximum water losses of 20.7 Mm³ from the reservoir were recorded during the 1981/82 season. Collapses at the surface were also reported. A program of tracer tests and monitoring of conductivity and temperature in piezometric boreholes was begun. A sharp increase in conductivity from 1000 and 1450 μS at a depth of 30 m was found in Piezometer S26 (Fig. 7.15). The abrupt increase in conductivity and the temperature gradient indicates that there is intrusion of groundwater at depths of more than 45 m.

Since the first filling was completed in 2002, the seepage rate has been mostly less than 150 L/s. In a separate investigation conductivity was found to have increased to close to 2000 μS, however. This indicates the minor influence of nearby gypsum formations.

**Khersan III Reservoir** (Iran) is divided from downstream Einakak River by the ridge composed of marlstone, gypsum, gypsiferous marlstone and marly limestone (Gachsaran formation). This formation is intensively tectonised. Distance between reservoir space and spring zone in valley is about 800 m and the difference between reservoir level and spring zone in the valley is 52 m. Due to huge dissolution capacity of gypsum the large shaft was created by water dropping only from pipeline during approximately 10 years (Fig. 7.16).

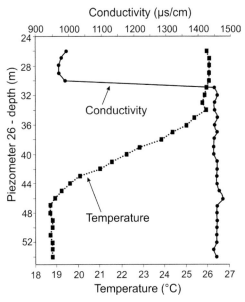

**Figure 7.15.** Foum El Gerza Dam, Algeria. Schematic layout and temperature/conductivity graph in piezometer 26, by depth. Benfetta et al., 2017.

**Figure 7.16.** Karst shaft created in gypsum by dropping water only.

Large area of evaporates is covered with very pervious overburden. Some outcrops of gypsum in the reservoir area will be directly exposed to solution process. Once the solution process starts karstification of gypsum and gypsiferous marl has to progressively increase. Construction of cut-off wall structure have priority with regard to other waterproofing measures.

CHAPTER 8

# UNDERGROUND DAMS AND RESERVOIRS

## 8.1 UNDERGROUND DAMS – NEW CHALLENGES

To make the best use of underground water resources that cannot or have not been tapped by building structures on the surface, unconventional installations such as underground dams are the only solution in a number of cases. Each cut-off structure (grout curtain, cut-off wall) and karst channel plug serves as an underground dam. However, the role of these structures is usually to prevent seepage from a surface reservoir or to protect the dam structure against destructive uplift forces. The term 'underground dam' in this chapter means a structure built underground to prevent rapid loss of water through the karst voids, to increase the groundwater level behind the structure, and to use this artificially stored water in a controlled manner for different purposes (water supply, irrigation, industry or power production). In a pioneering volume *Geology and Engineering* (Legget, 1939) presented the instance of an underground dam, 4.5 m deep and 273 m long that was constructed in the alluvial sediments of the Dry River (Harrisburg, U.S.A.) to retain the dry season underflow there.

In natural conditions very large quantities of the groundwater in karst regions are beyond any control from the surface because the basal flows are too deep and little known, without the possibility to be used optimally. Due to the large volumes of flow through high capacity solution conduits, the retention capacity of a karstified rock mass is generally poor, in contrast to its extremely high transmission capacity. To control the groundwater regime in such cases artificial underground storage is the only possibility, if the geological conditions will allow it. Construction of underground dams and reservoirs in karst is relatively new in dam engineering and engineering karstology. Experiments, design and construction of underground dams started in the second half of the 20th century.

In the Chinese karst from the late 1950s until the beginning of 21st century many underground dams and reservoirs with storage capacities from several hundred thousand to several million cubic meters were constructed. The volume of the largest one is 141 million cubic meters of water.

The first idea to construct underground dam and reservoir behind the large karst springs particularly in Dinaric karst area was presented by Perić, 1963.

In the *Handbook of Karst Hydrogeology*, Burdon and Papakis (1963) concluded that "More and more attention is being paid to the underground storage of groundwater, whose importance has increased as sites for additional surface storage become more and more difficult to find".

Pavlin (1973) analyzes the possibility of subsurface barriers for construction the underground storages in the coastal karst spring zones and their protection against sea-water intrusion.

A particularly challenging and complex underground dam project in the French karst was the submarine Port-Miou project (1968/1976). The Minafuku underground dams in the Miyako Islands in Japan were designed and constructed in the period, 1974–1992. Small underground dams in Indonesia (Bribin Cave) and Iraq (Gali Basera) are noted by Stevanović (2015).

One of the world's largest projects is a 300 m high underground dam behind Ombla Spring on the coast of Croatia. Site investigations and the design phase of the Ombla Dam project are already finished. A huge underground reservoir will be created.

Depending on the geological structure and its hydrogeological properties, there are a few general options for underground storage:

- the plugging of individual karst conduits must be at the geologically correct location;
- building an underground dam (positive grout cut-off or cut-off wall) in the rock behind a large karst spring can be very efficient;
- construction of an underground dam or plugging a karst channel to achieve water storage in a conventional reservoir on the surface is also effective where it is practicable;
- building a surface dam in front of a large karst spring to obtain coupled surface and underground reservoir storage; and
- underground karst storage physically coupled to a surface reservoir by storage in karstified banks.

## 8.2   GENERAL GEOLOGICAL PROPERTIES PREREQUISITE FOR UNDERGROUND DAM INSTALLATION

Every karst aquifer will have distinctive hydrogeological features, so that a model concept for underground damming must consider a significant number of variables: the lithology, geological structure, the base of karstification, the location of base flow conduits, the number of discharge points, their discharge capacity, hydraulic characteristics such as volumes of sedimentation, possible connections with adjacent karst aquifers, the physical accessibility of large karst channels, possible influence on upstream areas (e.g., inadvertent back-flooding), the influence on local seismic balances, on water quality, on the ecosystem, plus many secondary influences.

The purpose of storing the water is also an important matter. For instance, if the purpose is hydro-power production there are specific requirements: a sufficient quantity of stored water, the capacity of the aquifer to release large volumes of water in a short time, and its capability of achieving and maintaining a high water head. In the cases of need for irrigation or for drinking water, the physical requirements can be much more relaxed because the need to release large quantities of water quickly is usually low.

If the natural discharge point is at sea level or a little below it, there may be the problem of salt water intrusion. If the purpose is for drinking water supply, this can be very serious. To prevent sea water intrusion into the water-tapping structure an underground dam (grout curtain or cutoff wall) is one of the most effective solutions.

As is the case with surface dams and reservoirs, building underground dams has advantages and shortcomings, Milanović, 1988. The main advantages are:

- By using artificial underground storage, settlements and arable land on the surface are not subject to deliberate inundation. The problems of expropriation, relocation of infrastructure, residential facilities and historical monuments, etc. does not exist.
- Water from underground reservoirs is generally the same as spring water from the host aquifer, which usually implies reliable high water quality.
- Water temperature is constant.
- Water stratification, common in surface reservoirs, does not exist in underground storage in most instances.
- Losses due to evaporation are negligible, without practical significance.
- Sedimentation processes normally do not jeopardize underground reservoirs.
- By proper design and operation, turbidity can be eliminated or kept to a minimum.
- Questions of dam stability and risks of failure scarcely exist in practice. Underground dams (grout curtain or cut-off wall) are geotechnically part of the surrounding rock, not a separate body as is case with surface dams.

The main shortcomings are:

- Selection of proper underground dam site locations depends on understanding the complex lithological, structural and hydrogeological properties, including the nature and extent of the karst evolution. The precise determination of the necessary variables, particularly the water tightness of the underground storage space, needs complicated investigation and monitoring.
- Storage volumes are comparatively limited and require comprehensive investigations and analysis to define them.
- Effect of underground storage on upstream depressions (karst poljes, large sinkholes) can provoke flooding of arable lands.
- In the case of shallow overburden above the maximum storage level, collapses at the surface can be a serious problem.
- Construction of watertight cut-offs (underground dams) may require complicated technology because tolerance for seepage losses is much lower than in the case of surface dams.

To construct a successful underground dam in a natural karst channel, good speleological information is crucial. Selection of the narrowest channel locations with good geotechnical properties (lack of significant fracture and matrix porosity, adequate mechanical strength) for the plug foundation is a key prerequisite. The plug location has to be as far as possible behind the channel outlet. In this case the piezometric surface/

water table is shifted deeper into the rock mass. For proper operation, an underground dam has to be equipped with appropriate control structures like any conventional dam on the surface: a water intake, a bottom outlet and an overflow installation. For the overflow, the location which obtains the highest controllable water level in the underground reservoir should be selected.

## 8.3  SELECTED CASE STUDIES

**Port-Miou and Bestouan** (near the port of Cassis in southern France) is one of the first damming projects of an underground karst river below sea level (Potie et al., 2005). Two submarine karst springs are discharge points of a sub-horizontal solution channel that has been explored upstream by divers for a distance of about 2 km. Spring discharge varies between 2.6 m³/s and 45 m³/s. The deepest point in the channel is 147 m b.s.l. but at 800 m from the outlet the roof lies only 10 to 20 m below sea level. The size of the channel varies but is never less than 100 m². Much of it is about 15 m in height over deposits of fine-grained sediments known to be several meters thick. The fineness of the sediment reflects the very low water velocities in the system, even during floods.

The main purpose of the underground dam is to reduce the concentration of chlorides in the spring water. The concept is based on the density difference between fresh water and sea water. A specially designed dam structure was constructed at a distance of 530 m from the outlet (Fig. 8.1). The work was done in two phases between 1968 and 1978. The Phase One structure consisted of an upstream dam built on the conduit floor and a downstream dam suspended from the roof. Upstream of the dam the salinity decreased from 18–20 g/L to 3–4 g/L at a depth of 20 m b.s.l. In Phase Two the initial project was modified by completely closing off the gallery, constructing several 'spillway pipes' through the base of the dam for flood control, and excavating a free-flow spillway for the fresh water in the rock above the gallery. The content of chlorides was reduced by a further two-thirds.

As noted above, many underground dams have been constructed in China. According to Lu (1986) more than three thousand underground rivers have been cataloged in southern China. Underground dam sites and reservoir areas were selected carefully, based on massive and meticulous geological and speleological investigations

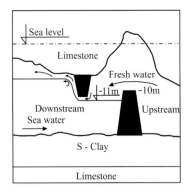
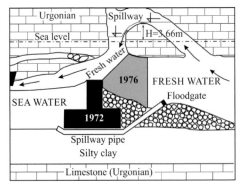

**Figure 8.1.** Port-Miou, France. The underground submarine dam. Left: design of Phase One; Right: Phase Two (Potie et al., 2005).

and hydrological measurements, coupled with various geophysical techniques (Cao et al., 2016). Underground rivers or caves protected for scientific, ecological, or tourist reasons are not used for damming.

Among the hundreds of underground dams and reservoirs in China, mostly for irrigation (Yuan, 1990), the most prominent are: **Yuhong** (Hunan Province), a concrete plug 10.5 m high, 7 m wide and 3 m thick; **Jijiao, Yidong, Yuzha** (Guangxi Province); and **Shuanglung underground dam** (Chekiang Province). In some cases, plugging underground channels created man-made reservoirs that are partially underground and partially on the surface: **Fengfa** (Guizhou Province), **Neiwan** (Hunan Province), **Wanger** (Guizhou Province), and **Fu Liulang** (Guangxi Province). Examples of construction for power production are **Beilou** (24 m water head, Guangxi Province) and **Yuhong** (10.5 m high, 7 m wide and 3.0 m thick arch dam, Hunan Province).

Four important types of underground reservoirs selected from the numerous examples presented by Lu (1986) are shown in Fig. 8.2.

Figure 8.2(A) shows a simple dam structure close to the conduit outlet. The dam is equipped with a control valve for water transfer to the irrigation canal. In Sketch (B) the dam is constructed as a full barrier (a plug) in the main channel to raise the water level up to a natural overflow outlet where it can be used for gravity irrigation. (C) show two dams constructed at two different levels in the cave channel, with a pump in each of the ponds created that raises the water to the surface for irrigation. In cross-section (D) the underground dam (a complete plug) was constructed to achieve the head required for power production in an underground hydropower plant.

Lu (1986) shows a number of concepts for creating joint subsurface and surface reservoirs. Two examples are presented in Fig. 8.3. In the upper sketch the dam is constructed on the surface to prevent stream flow into a large cave downstream. The surface reservoir is coupled with a large underground system upstream, creating one compact water body. The lower cross-section shows a case where a karst channel has been fully plugged to compel the ground water in a large system of caverns and conduits to flow through an artificial tunnel into a surface reservoir.

Another Chinese author, Yuan (1990) has analyzed the problems that must be considered in the construction of underground dams in karst. To begin with, detailed analysis of the possible impact of such a disturbance of the natural underground water regime in both upstream and downstream directions is of primary importance. Analysis should include, first of all, the problems of water supply and flood control. For instance, the **Wanger** underground dam in Guizhou province (Fig. 8.4) is 8 m long, 9 m high, and 6 m wide, with a reservoir capacity of one thousand cubic meters. However, its backwater is 3000 m long and impacts (floods) a series of karst depressions upstream. To prevent the floods a special surface spillway had to be built.

The **Jijiao** underground dam was constructed in 1958 to block a subterranean river with an annual mean discharge of 3–4 $m^3$/s and minimum discharge of 1.5–2 $m^3$/s. In flood periods discharge increases to 14–16 $m^3$/s and the water table rises 19–34 m. The underground dam (a full plug) is constructed in the downstream side of a karst window (Fig. 8.4). After damming, the mean groundwater level rose 19 m in the dry season and 30 m in the rainy season. An artificial tunnel 49 m long drains water from the underground reservoir for use in irrigation. However, in periods of

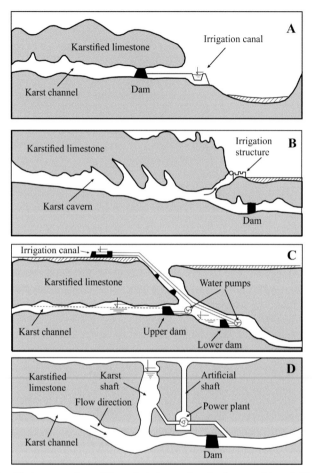

**Figure 8.2.** Simplified cross-sections of four selected examples of underground dams in China (Lu, 1986).

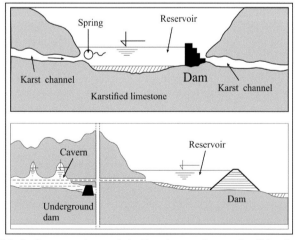

**Figure 8.3.** Surface reservoirs coupled with underground reservoirs, China (Lu, 1986).

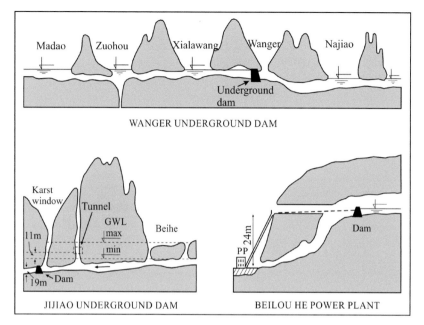

**Figure 8.4.** Examples of underground dams in south China, Yuan (1983, 1990).

high water level in the underground reservoir every year the Beihe depression, 2 km upstream, becomes flooded (Yuan, 1983).

The **Beilou** subterranean river (Fig. 8.4) is impounded by an underground dam with a crest overflow outlet. Minimum discharge is 0.15 m³/s. The water is transported by a tunnel and a penstock with a head of 24 m to a power plant.

A few more underground projects are presented in Yuan (1990).

For the **Yuzhai** underground reservoir in Guizhou province a dam 5.97 high, 9 m long, 4 m wide at the base and 1.8 at the crest was constructed. The length of the underground reservoir it creates is 1200 m, with a volume of 96 thousand cubic meters.

The **Yuhong** subterranean river (Hunan province) is developed along the contact between limestone and impervious beds. An arch dam structure in its channel is 10.5 m high, 7 m wide and 3 m thick. It raises the level of the underground reservoir 34 m in the dry season and 115 m in flood. The water is used for power production.

In 2004 the **Siziguan** underground dam in western Hubei province was completed (Yuan, 2012, pers. comm.). The dam was built in a canyon-like cave passage on the Zongjian River and is 192 m high and 16 m wide. The total capacity of the reservoir is 141 million m³, with a major portion being in a surface gorge upstream of the cave (pers. information, Yuan D). Currently, the catchment area of 71 km² is being expanded by excavating a tunnel 6300 m in length into an adjoining basin.

**The Liulangdong** underground reservoir was one of the first of the larger underground reservoirs to be constructed in China (late 1950s). The reservoir was created by construction of a 15 m high dam. The average flow is 23.8 m³/s (max > 100 m³/s). The water is used for power production, with a head of 109 m.

**Suduku** underground dam (Yunnan, China) is a 12 m high concrete structure constructed to block a karst channel 12 m below the surface. The channel is 1.8 m high and 2 m wide. The captured groundwater flow is 1.3 $m^3$/s during the dry season.

The **Fenfa** underground reservoir in Guizhou Province, China, is an engineering example of the common problems faced when plugging individual karst channels (Mingzhang, 2006). Construction of an underground dam at the exit of a large karst channel was initially organized by local residents (Fig. 8.5). Due to insufficient understanding of the hydrogeology and groundwater flow system, leakage occurred shortly after the structure was completed. To solve the problem, a second and third dam was built further and further up the karst channel. Finally, after detailed hydrogeological investigations, the faults along which seepage losses were occurring were discovered and the problem was cleared up. A fourth underground dam site was selected 55 m upstream of the cave entrance. As consequence of plugging the cave river the water level was raised by 26 m. Two artificial reservoirs at the surface (Hongmei and Powoxi) with combined storage capacity of 2.2 x $10^5$ $m^3$ were formed. The water is used for irrigation.

**Pijiazhai Spring** in Yunnan Province has been tapped by construction of a grout curtain and a 'U-form' dam to make a small reservoir to concentrate the flow, see Fig. 8.6 (Cao et al., 2016). The head on the karst conduit was increased by 4.4 m to make it possible to supply a farm irrigation channel by gravity.

An unusual and interesting solution is the example of the **Fu Liulang** underground dam (Fig. 8.7). By plugging a steeply descending cave channel, the groundwater flow was backed up to the surface to create a reservoir there.

**Wulichong Underground Dam** (Yunnan, China) is constructed inside highly karstified Triassic limestone with two underground rivers. The dam was completed between 1991 and 1996. It consists of a grout curtain 1,333 m in length with a maximum depth of 260 m, impounding a surface storage area of 262,000 $m^2$ (Fig. 8.8).

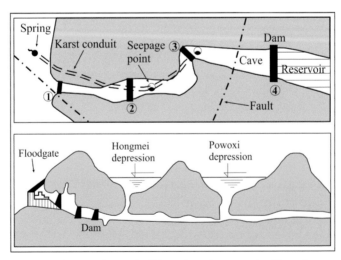

**Figure 8.5.** The Fenfa underground reservoir, China, schematic presentation. Upper–Layout, and Lower–cross-section (Mingzhang, 2006), from Cao et al., 2016.

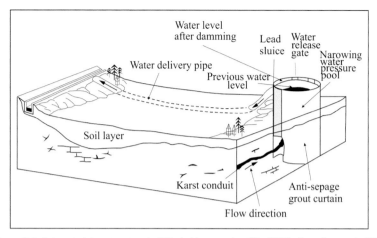

**Figure 8.6.** Integrated dam structure in Pijiazhai karst spring (Yu, 2008), from Cao et al., 2016.

**Figure 8.7.** Fu Liulang Reservoir, Guangxi Province, China. Surface reservoir established by the construction of an underground dam. Upper left: simplified cross-section with location of the underground dam (sketch based on explanation at the site).

The section with karstified and highly brecciated rock along the curtain route had an area of 3200 m². The investigation program included 10,343 m of boreholes, 2783 m of adits and a variety of geophysical methods. The vertical separation between the three grouting galleries was 50–70 m. The grout curtain is the suspension type, consisting of upper, middle and lower layers. The grouting boreholes were arranged in a single row, double rows and multiple rows, with the holes spaced at 2 m and rows spaced at 1.5 m (Zhang and Wu, 2000). In a dissected karstified rock mass between

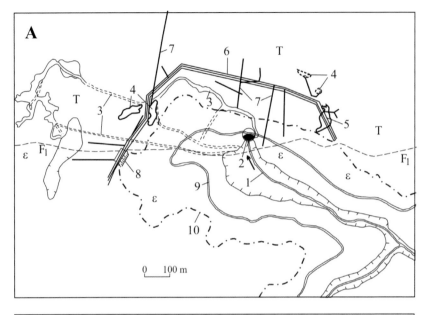

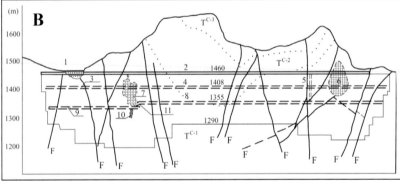

**Figure 8.8.** Wulichong underground dam, China. A. Layout: (1) River; (2) ponor; (3) underground river; (4) cavern; (5) brecciated and karstified rock; (6) grouting galleries; (7) adits; (8) dam; (9) road; (10) reservoir level. (T) Middle Triassic limestone; (ε) Upper Cambrian formations. B. Cross-section: (1) Auxiliary dam; (2) upper grouting gallery; (3) access adit; (4) middle grouting gallery; (5) ventilation shaft; (6) brecciated and karstified rock; (7) cutoff wall; (8) lower grouting gallery; (9) access inclined adit; (10) concrete plug; (11) access inclined adit (from Zhang and Wu, 2000).

two large caverns, a cut-off wall 100.4 m high, 50–30 m long, 2.5–2.0 m wide was built. 14.775 m³ of rock mass was excavated for its construction and replaced with 15.152 m³ of concrete. The entire rock mass behind the wall (28–35 m) was consolidated. A concrete plug was built to block a cavern containing an underground stream. The plug was 33.46 m high, 13.9 m wide, and 2 to 10 m thick.

In the most critical middle layer a seven row grout curtain was constructed. It consists of three vertical and two inclined rows of grouting holes in the downstream section, plus two rows of vertical boreholes extended down from the upper layer. The upstream and downstream rows were grouted first and the vertical curtain rows second. In this manner a grout curtain about 10 m in thickness was created in the middle layer of the curtain. Maximum grouting pressure for this middle layer was 40 bars and 35 bars for the upper layer. A stiff grout mass (water:cement ratio of 1:1) was used. The reservoir was successfully impounded and this project operates at full capacity.

On Miyakojima Island, Japan, **Minafuku,** an experimental underground dam in karstified limestone was completed in 1979 (Yoshikawa and Shokohifard, 1993). In the following years up to 2001, two more underground dams (**Sunagawa** and **Fukuzato**), were constructed. Fukuzato Dam has one main dam and two auxiliary dams. The island consists of highly porous Quaternary Ryukyu coral limestone on a basement of Tertiary mudstone. The effective porosity of limestone is about 10%. Hydrogeological compartments created by a number of sub-vertical, near-parallel faults, was the basis for conceiving the pattern of the underground dam and reservoirs (Fig. 8.9).

The dams (cut-off walls) were constructed from the surface down through the porous limestone, to be tied into the impervious mudstone below. The maximum depth of the Sunagawa underground dam is 50 m, length 1.677 m, and the volume of the underground reservoir is 9.5 million m$^3$. The dam crest is at 31 m above sea level. The crest elevation of the Fukuzato main dam is 46 m, depth 27 m and length 1,790 m. Reservoir volume is 10.5 million m$^3$ (Satoshi et al., 2003). To prevent lateral seepage losses from this reservoir two auxiliary underground dams had to be constructed: Auxiliary 1 is 21 m deep, and 786 m long; Auxiliary 2 is 6 m deep and 332 m long. Water from the underground reservoirs is extracted by 147 pumping wells. Due to the spatially varying karst porosity, the capacity of the wells may differ. Frequently, a well with a capacity of 100 m$^3$/day is only a few meters from another with a capacity of 2000 m$^3$/day. Locally, the hydraulic conductivity of the limestone rock mass is seriously disturbed by presence of clay in voids in the coral.

The **Ombla Power Plant** (Croatia) is one of the largest projects of underground damming and storage. The idea of underground storage there was born from hydrogeological analyses during the geological investigations (1969–1971) for the Trebišnjica Hydrosystem Project (Milanović, 1977).

Both of the crucial structures, dam and reservoir, are located underground in karstified strata. The large Ombla Spring is at sea level on the Adriatic coast near the town of Dubrovnik. The spring outlet is on a reverse fault in the Mesozoic carbonate complex (karstified limestone and dolomite), which is overthrust on autochthonous Eocene flysch sediments. The flysch acts as a hydrogeological barrier, now eroded down to the sea level (Figs. 8.10 and 8.11).

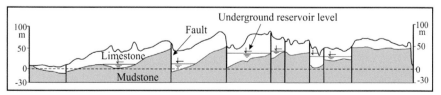

**Figure 8.9.** Japan, Miyakojima underground dams. Simplified geological cross-section (Satoshi et al., 2003).

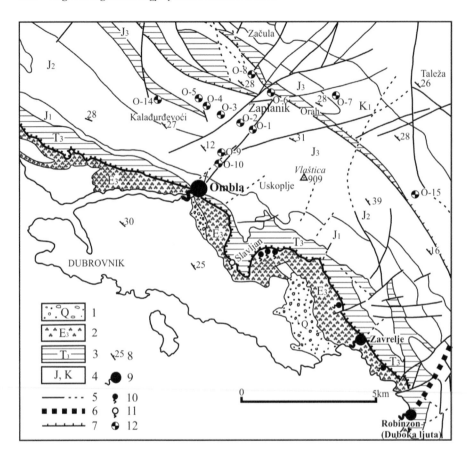

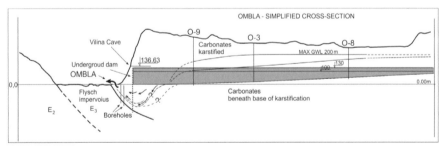

**Figure 8.10.** Ombla Spring. Simplified geological map. Above – Part of the Ombla Spring catchment area: (1) Quaternary sediments, (2) Eocene flysch, (3) Triassic dolomite, (4) Jurassic and Cretaceous limestone, (5) fault, (6) large fault zone, (7) overthrust, (8) dip direction, (9) large spring, (10) small spring, (11) temporary spring, (12) borehole. Beneath – Simplified cross-section with approximate position of the underground dam.

The flysch barrier becomes higher on both sides of the spring, forming a shallow 'V' profile (Fig. 8.11). The average discharge of Ombla Spring is $Q = 24.4$ m³/s. The recorded minimum discharge rate was 2.3 m³/s, while its maximum rate has reached 112.5 m³/s.

**Figure 8.11.** Ombla Spring, showing the approximate outline of the crest of the underground dam.

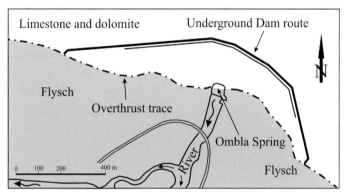

**Figure 8.12.** Ombla underground dam route, layout.

The spring zone was investigated in detail by means of 19 deep piezometric boreholes drilled from the surface, plus 1100 m of galleries and 30 boreholes drilled from those galleries. The geophysical methods used included gravimetry, thermometry, different electrical methods, cross-hole seismicity, reflection, refraction, borehole radar, etc. Speleologists and divers explored about 3 km of karst channels.

The principal groundwater circulation occurs through deep siphonal conduits (Fig. 8.13).

The deepest part of the siphon zone was discovered by drilling, thermometry and borehole radar application to be at a depth of about 150 m below sea level and about 200 m behind the chief spring outlet. The siphon was explored by speleologist-divers to a depth of 54 m below sea level. This part of the siphon channel was deep inside the karst massif, about 500 m behind the spring. The vertical position of the flysch and of lateral overflow springs limits the height of the underground dam to about 100 m above sea level (maximum 130 m). Due to this height requirement and the alignment of the overburden, the arch-form impermeable dam had to be located at least 200 m behind the spring outlet.

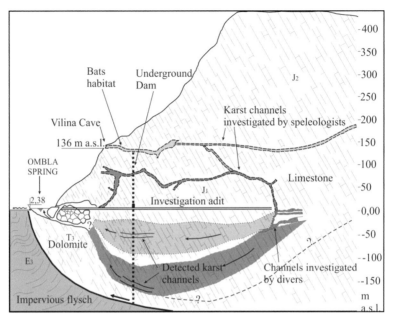

**Figure 8.13.** Ombla underground dam. Positions of deep siphonal flows and the cave system investigated by speleologists.

Fifteen deep piezometric boreholes were drilled to study the underground storage space. Estimation of the storage volume was based on the analysis of: (1) water level fluctuations, (2) spring discharge, and (3) geophysical investigations. According to the data from geoelectrical soundings, it was possible to define the boundaries (vertical and horizontal) of a highly karstified rock mass behind the spring with a larger storage capacity than in the surrounding, less-karstified rocks. These geoelectrical contour lines represent the *base of karstification.*

The underground storage space is intensively karstified, mainly large conduits and caverns. Because the aim of water storage was power production, an important property of storage space had to be of a capacity to discharge the water rapidly. As a consequence of evolution of the Ombla karst aquifer, the different levels of the karst channels are hydrogeologically (physically) and hydraulically well connected. In Fig. 8.14C these levels are schematically marked as I, II and III. The base flows are at levels I and II. Piezometers P1 and P2 are in direct hydraulic connection with them. Level III channels are not in direct hydraulic connection with the base flow zone except during very high ground water levels.

The excellent hydraulic inter-connections within the Ombla karst aquifer, particularly in the base flow zone, have been confirmed many times. As noted, an essential hydraulic requirement for power production is the capacity of the aquifer to release huge amounts of water in short periods of time, almost instantaneously. Such capacity is not important for water supply or irrigation purposes but is crucial for power production. Effective porosity of the hydrogeologically active part of aquifer (the rock

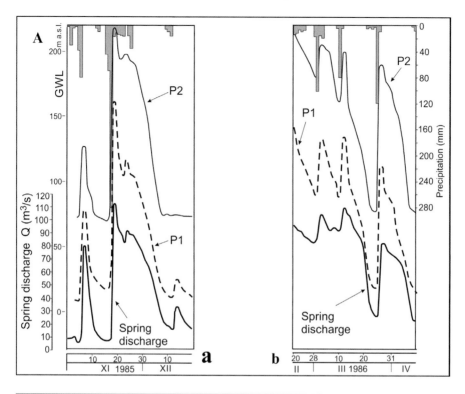

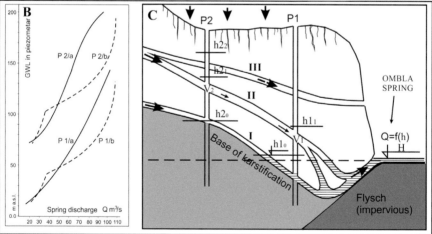

**Figure 8.14.** Ombla Spring. Hydraulic properties of the Ombla aquifer. (A, **a** and **b**) simultaneous graphs of spring discharge, precipitation and water level fluctuation in two piezometers. (B) the relationship between spring discharge and groundwater levels. (C) Simplified sketch of the Ombla cross-section including position of piezometers (not to scale).

mass above the base of karstification) ranges between 1.5 and 2.5%, depending on the depth. During the dry season there is base flow only. In that period seepage from the nearby Gorica reservoir (1–2 m$^3$/s) is the key source of water. At the end of the dry season the water stored in the 315 x 10$^6$ m$^3$ of the rock mass that can contribute to the spring yields a discharge of only about 1–2 m$^3$/s. During the wet season, the catchment area varies in space and time (600–700 km$^2$): the differences can be as great as approximately 15%. After rain begins, saturation of the main conduits is rapid. When the flow comes under pressure the spring discharge (Q) can be defined as:

$$Q = \emptyset \sqrt{2g\ (hl_0 - H)}$$

Q – spring discharge,

$\emptyset$ – product of mean conduit aperture and discharge coefficient $c$ that depends
   on conduit roughness,

H1$_0$ – ground water elevation in piezometer

H – elevation of discharge point

   This equation assumes that outside influence (external inflow into karst conduits) does not exist or is negligible.

   During periods of physical saturation, the upper conduits gradually become included in the system under hydrostatic pressure (phreatic). When water level in the piezometer nearest the spring (P1) reaches an elevation 40–45 m and P2 reaches elevations around 100 m new conduit(s) are activated (dashed lines in Fig. 8.14, graphs B (P1/b and P2/b). In Fig. 8.14(C) this part of the system is schematically presented as 'Level II'. If after heavy rain the submerged system of karst conduits becomes saturated extremely quickly, flows under pressure are seriously disturbed by external hydraulic influences, and the graphs (solid lines) take different shapes. In Fig. 8.14B the curves P1/a and P2/a represent water level increases after extreme precipitation (280 mm/24 h). The upper section of Curve P2/a changes direction due to the ground water level reaching an upper (much karstified) zone. While the form of the graph still reflects a flow system under pressure, the upper part of graph, however, indicates free flow conditions in the highest parts of the aquifer. Inclusion of new conduits into the system is not presented on the graph.

   The underground dam consists of an upper section above 0.00 m (sea level) that is 135 m high in the middle, and a section below 0.00 m down to a depth of 200 m (locally more) in the middle. Total height of middle of the dam is 315–350 m, and its total surface area 280,000 m$^2$. Lateral parts of the dam have to be tied into the flysch impervious formations. The dam consists of a positive cut-off, plus plugging of karst channels and construction of cut-off wall(s). Geotechnically it would be one of the most complex and challenging structures ever undertaken, particularly its permanently submerged part with very large underground flow.

### Perućac-Potential Underground Dam and Reservoir

The Perućac karst spring is located on the right bank of the Drina River, Serbia. A large karst aquifer discharges through this spring only. The average discharge is 1.2 m$^3$/s; the recorded minimum discharge was 0.4 m$^3$/s, and the maximum 9.0 m$^3$/s. The spring outlet is situated on the contact between impervious Lower Triassic sandstone and karstified Middle and Upper Triassic limestone. The contact has a 'V' form, with the spring at its lowest point. Water balance analysis indicates a significant retardation storage capacity (Milanović and Milanović, 2001).

Investigations included detailed geological mapping of the entire area to define watersheds, including important parts of adjacent catchments (1:10.000); detailed mapping of the spring area (1:1000); tracer tests to define watersheds; hydrological measurements; investigation/piezometric boreholes; geophysical investigations including borehole logging, and excavation of an investigation adit.

A very strong correlation between ground water fluctuations in the piezometers and the spring discharge was established. This supported the conclusion that this aquifer is a hydrogeological system under pressure. It is one of the important parameters of underground storage, particularly in the case when it is to be used for power production.

A few different types of retaining structure were considered, including the possibility of constructing a dam in front of the spring outlet. The accepted solution is an arch-like structure (grouted cut-off) located in the karstified limestone and tied into the impervious Triassic sandstone underneath (Fig. 8.15). The upper part of the curtain is a one row grout curtain and the deeper section is a three row grout curtain. All karst channels along this route, including base flow(s) sources must be plugged. At spring level the height of underground dam crest is 86 m. The deepest point along the lowest contour line is approximately 170 m deeper than the crest (Fig. 8.16).

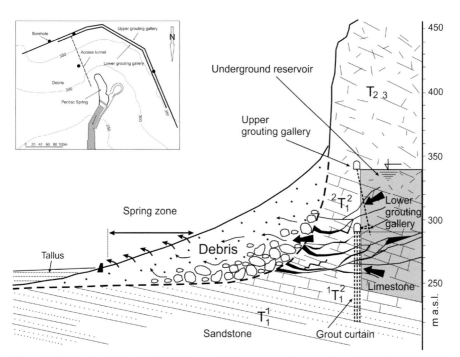

**Figure 8.15.** Perućac Spring, Serbia. Underground dam concept. Cross-section and simplified sketch of layout (Milanović and Milanović, 2001).

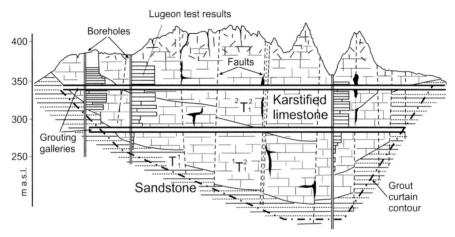

**Figure 8.16.** Perućac Spring, Serbia. Concept for the proposed underground dam (Milanović and Milanović, 2001).

The storage volume at the highest underground crest level selected is estimated to be about $3 \times 10^3$/m. Analysis of all available data confirms that the underground storage space at this proposed level is watertight.

# CHAPTER 9

# MODELING IN KARST

## 9.1 MODELING AND ITS LIMITATION IN ENGINEERING KARSTOLOGY

Modeling of underground flow in karstified rock is an extremely complex hydrogeological task. The rules of groundwater flow in karst aquifers are completely different from those in the intergranular (porous media) environment. The foremost characteristic of a karst aquifer is hydrogeological inhomogeneity. Flow occurs through solution conduits of different sizes and in enlarged joints that are surrounded by the lower effective porosity of the rock matrix (a double- or triple-porosity situation). Relationships between a given input into a karst system and its output can differ. Rains of the same intensity will not always have the same discharge as a consequence. Transformation of precipitation into underground flow after a long dry period (when the epikarst zone is dry and the vadose or aeration zone is deep) is not the same as it will be during a lengthy rainy period. After a dry period, precipitation of 30–40 mm/24 h will probably have a negligible effect on karst aquifer storage. In contrast, during the wet period of the year 50 to 80% of precipitation typically becomes part of the storage.

The surface extent of catchment areas depends on the location of the underground watersheds. The latter depend on the aquifer saturation, especially the possibility of upper, normally dry, conduits being incorporated into the flow net. The existence of zones of bifurcation (divergence of flow) is an additional uncertainty. Bifurcation zones play a role in many underground karst watersheds, and are changeable in space and time.

The Elemental Hydrogeological Volume (EHV) does not exist in karst. The EVH concept depends on a properly selected scale for the volume of the aquifer that is to be analyzed. From the hydraulic point of view, the rules of flow in karst aquifers are closer to those for flow through the pipes under pressure or with a free surface than for filtration through a porous medium as formulated using Darcy's Law. Over the past 50 years many authors have tried to describe karst flow models by applying different approaches. Generally, based on two different approaches, models in karst may be physical or mathematical.

There are three types of physical models:

a. A simulation of a given karst hydrogeological system (e.g., a karst polje or karst solution channel system), constructed following the principles of simulation models built in hydraulic laboratories.
b. 2D visualization (horizontal or vertical) of a karst system based on all available geomorphological and hydrogeological data and measurements (e.g., Figs. 2.32 and 2.37, Chapter 2). Every hydrogeological or geotechnical cross-section is a 2D model. Its accuracy depends on the amount of investigation and its interpretation.
c. 3D visualization of all available (measured) geomorphology, geology and hydrogeology data about a karst aquifer, including all changes in the aquifer as consequences of its saturation and emptying.

The variety of mathematical models is enormous. However, not one of the existing models is able to answer the two crucial questions in dam engineering karstology – (i) how to determine the exact spatial location and hydraulic properties of each karst conduit(s) at a given point, e.g., along a grout curtain route and (ii) how to define the exact flow capacity of the karst channels.

In the 1960s and 70s a number of karst aquifer models based on the recession sector of the discharge hydrograph and on tracer tests were developed ('black box' approach). Some of prominent authors are: Schoeller (1962); Plotnikov (1959); Castany (1963); Forkasiewicz and Paloc (1965); Mijatović (1968); Mangin (1975); and Drogue (1980). Other authors have presented karst systems in the form of physical models, e.g., Smart (1983); Lauritzen (2000); Milanović, S. (2010).

Mangin (1969) built a physical model of the ebbing and flowing (intermittent) spring at Fontestorbe, France. It worked very well, Ford and Williams (2007).

Mathematical models are based on laws of probability and statistics (probabilistic models) or deterministic (analytical and numerical). These models are described in more detail by Krešić (2010), and Mikszewski and Krešić (2015). According to Krešić (2010) "A very important factor to limit wider use of stochastic models in hydrogeology is the lack of recorded data. Since these models are based on statistical and probabilistic calculations, very short time series do not allow for meaningful derivation of model parameters".

From an engineering point of view models of seepage from reservoirs or below and around the dam sites and models for optimal operation of reservoirs are the most important, particularly in the case of reservoirs built for hydropower production. In some cases the operational models incorporate natural surface reservoirs (karst poljes or other depressions), man-made reservoirs and complex karst aquifers mutually connected to karst conduits and artificial tunnels.

In the hydrogeological-hydraulic approach, the karst aquifer model is simplified as a system of conduits connected to ponors (inlets), estavelles (inlet/outlet) and spring(s) – (outlets). The distribution and discharge capacity of individual conduits is not known. Saturation of the aquifer has as a consequence the piezometric relationships observed by piezometers in boreholes. The quality of the groundwater level data depends crucially on obtaining a good hydraulic contact between the borehole and the base karst flow or base flow zone. The hydrogeological-hydraulic karst system is partially under pressure, partially with free flow surface. Ponors can be under pressure or have free surface inflow, also. The activity of estavelles, when they discharge, is always under pressure conditions (Fig. 2.34). Saturation of the aquifer occurs concentrated via the ponors and estavelles and also via the general epikarst when there is precipitation over the entire catchment area.

Theoretically the 'transparent box' model can be an appropriate basis for seepage analysis and waterproofing structural works. However, due to the complexity in most karstified rocks, even after numerous investigations in practice it is impossible to incorporate into the analysis the hundreds of hydraulic relations that mathematically define the flows in each karst joint and channel and the hydraulic relationships of flow transformations along these joints, channels and in the numerous larger caverns that may be present.

A few examples of different approaches in karst modeling are presented below.

## 9.2  PHYSICAL MODELS OF KARST AQUIFER

To create a physical model (3D) of a karst aquifer is a complex and time-consuming task. Many authors from the 1970s to the present time have tried, however. Each of these models is based on a quantity of exact (measured) data but, because of the inaccessible nature of much of an aquifer, the database must be considered to be very incomplete. Development of investigation and software techniques that allow collection of large amounts of high quality data are improving the situation and the number of researchers interested in the subject has increased significantly since 2000; e.g., Kovacs (2003); Springer (2004); Butscher and Huggenberger (2007); Jeannin et al. (2007); Filipponi et al. (2009); Borghi et al. (2010); Milanović (2010).

The case study of a 3D model that is briefly presented here is that of the Beljanica karst aquifer in eastern Serbia (Milanović, S. 2010). The aquifer is developed in a limestone area of about 300 km². The limestone is surrounded by impervious rocks. The karst aquifer is thus entirely 'dammed' by hydrogeological barriers and its discharge points (springs) have developed at the contact aquifer/barrier.

Krupaja Spring is a deep siphon spring at the contact between the limestone and sandstone, investigated by divers to the depth of 105 m (which is not the full depth). Discharge varies between 0.3 and 9 m³. Mlava Spring is a deep siphon spring investigated by divers to a depth of 73 m (again, not the full depth); its discharge ranges 0.3 m³–16 m³/s.

The model is based on nine different layers of input data: topographic contour information, geological, hydrogeological, geomorphological (plotting dolines and sinkholes), map of groundwater levels, map of the distribution of forest, shrubs and bare karst, map of the distribution of known cave stream channels, tectonic map, contour map of the base of karstification and a plot of the hypsometric location of the sinkholes, caves and karst springs (Milanović, S. 2010). The input data were collected from 69 caves, 15 sinkholes, 1,682 dolines, sevev major springs and about 70 small springs. More than 6,000 data points were entered into the data base. The ArcGIS program was used for the 3D model, together with some additional associated special software (Fig. 9.1).

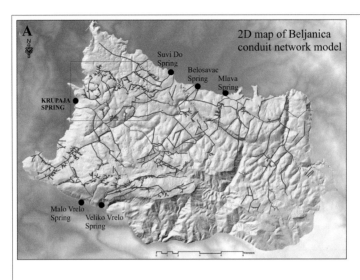

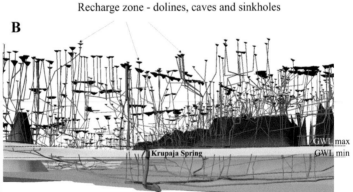

**Figure 9.1.** Physical models of the Beljanica karst aquifer. (A) 2D map of estimated underground flows, and (B) 3D model of the Krupaja karst aquifer. Brown areas above the GWL max indicate rocks not prone to karstification. Milanović, S. (2010).

Based on this 3D model, the volume of the limestone Beljanica aquifer above the base of karstification is 482 km³. The thickness of the saturated zone is more than 200 m, total length of karst channels is 647 km, and the estimated average groundwater potential is 4 m³/s annually.

## 9.3 MODELING DISTURBANCES OF NATURAL FLOW REGIMES DUE TO CONSTRUCTION OF RESERVOIRS

A simplified (conceptual) physically based model of water regimes in karst areas has been used to estimate the disturbance of the natural flow regime caused by construction of the Trebišnjica Hydropower System in eastern Herzegovina. The model was applied to determine the effects of a partial transfer (diversion) of water from the catchment of the Bregava Spring into the Bileća Reservoir in the catchment of Trebišnjica Spring. Two impacts seemed possible: disturbance of the Bregava Spring discharge, and change of the flood regimes in the Dabarsko and Fatničko poljes. To transfer the water of the two tunnels are already constructed: one 3.4 km in length links Dabarsko Polje to Fatničko Polje and the second (15.84 km) links Fatničko Polje to Bileća Reservoir (Fig. 9.2). Hydrogeologically, it is an extremely complex

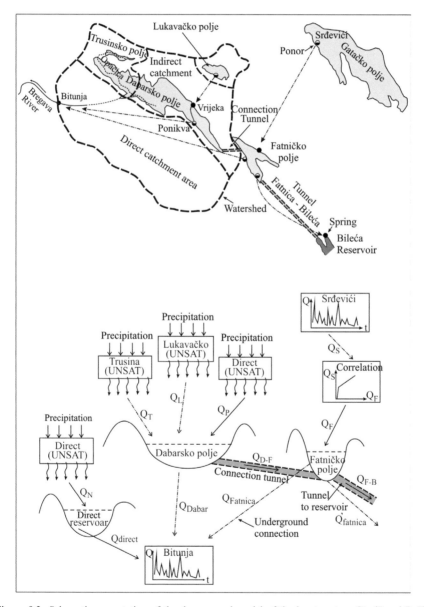

**Figure 9.2.** Schematic presentation of the decomposed model of the karst system, Stanić and Dašić (2005).

system. Under natural conditions the water flows only through underground conduits between these large poljes. The watersheds between the catchment areas can vary in both space and time, and so cannot be defined exactly. The groundwater flow regimes at a bifurcation (groundwater divergence) area in Fatničko Polje are particularly dynamic. The capacities of the ponors and estavelles depend on the extent of aquifer saturation downstream of them. The complexity of the model that was needed was further increased because the catchment area of Bregava Spring consists of two hydrogeologically separated units (aquifers).

Hydraulically, the flow regime was described as a system of reservoirs and pipes with flow under pressure and with free surfaces (Stanić and Dašić, 2005).Vertical water movement through the epikarst and vadose zone is simulated by applying the UNSAT model. The model for water balance in the horizontal plane is based on quasi-steady hydraulic simulation. This part of the karst complex has been modeled as a system of reservoirs (nodes) tied together by permanent active flow passages, dewatering passages from poljes and tunnels equipped by water gates. The source catchment area for each node was defined, including recharge from the vadose zone. Vertical water movements in the vadose zone have been modeled by a separate subprogram – UNSAT that transforms precipitation into runoff (Fig. 9.3).

From the hydrogeological properties, three physical subsystems have been recognized: the direct Bregava catchment, the indirect Bregava catchment, and part of the Trebišnjica Spring(s) catchment. Direct and indirect catchments are separated by long and deep hydrogeological barrier beneath the Dabarsko Polje. Direct hydrogeological connections between those two catchments (and aquifers) is not possible. All the water that saturated indirect (northern) aquifer discharges along the northern border of polje and after the short surface flow sinks into the ponors along the south polje border to become part of aquifer that discharges at the Bregava springs.

The vadose zone was modeled by UNSAT model, and separately, mostly horizontal water movement was modeled by net of nodes including connections between them.

A period of 10 years was simulated by the model. Simulations were made for (i) natural conditions; (ii) for the period after excavation of the tunnel between Dabarsko and Fatničko poljes, and (iii) for the period after excavation of the tunnel between Fatničko Polje and Bileća Reservoir.

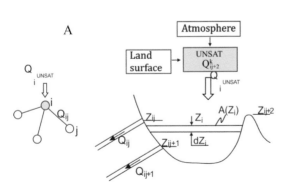
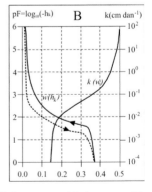

**Figure 9.3.** (A) Schematically presented portion of the system, and (B) Common graphical patterns of the vadose zone saturation $w(h_g)$ and hydraulic conductivity$(w)$, Stanić and Dašić (2005).

Accordingly the model output, the impact of the Dabarsko–Fatničko Polje tunnel on the discharge regime of Bregava Spring and the flood regime in both poljes is negligible. Some small effects (15–20% reduction of discharge) were obtained for periods when spring discharge was above 20 m³/s. When spring discharge is less than 10 m³/s there are no consequences of the water transfer. In contrast, however, the model shows that the Fatničko Polje – Bileća Reservoir tunnel has a considerable impact on the flood regimes in both poljes. By proper controls of flow in both tunnels and in the downstream reservoir, flooding in both poljes can be reduced to a minimum.

## 9.4  MODELING THE CONSEQUENCES OF DAM CONSTRUCTION ON PROCESSES OF KARSTIFICATION

From the very beginning, i.e., from the construction of the first dams on karstified rocks, the question of the possible influence of the high water heads created in the reservoirs on the intensity of karstification has arisen. Coupled with it is also the question of the effects of the high water pressure on karstification and degradation processes through grout curtains or beneath the bottoms of suspended grout curtains. Under laminar flow conditions carbonate dissolution is slow and widening of the aperture along a discontinuity requires a lot of time. However, when an aperture increases to a size that allows turbulent flow, the solution rate increases abruptly. Is this process fast enough to provoke failure due to increased seepage losses during the life of a dam and its reservoir? There are no analyses of these solution effects based on precise (monitoring) data for dams situated in the carbonate rocks reported in the literature at the present time.

Significant questions arise if a grout curtain is not perfectly tight. It is well known that grout curtains in karstified rocks are rarely absolutely impervious. Almost always some joints will remain unplugged. Due to this fact, dissolution and grout curtain degradation are to be expected as possible risks.

An attempt to analyze this problem with a mathematical model has been developed by Dreybrodt et al. (2005). The model (750 m wide and 350 m deep) consists of a two-dimensional cross-section of soluble rock below the dam foundation (Fig. 9.4). The basic parameters used in the model consist of a set of network parameters, parameters of the dam site and the chemical solubility parameters of the limestone. The rock is dissected by a network of discontinuities (joints and bedding planes) into blocks measuring 7.5 x 7.5 x 1 m. The aperture width of these uniformly distributed fractures is 0.02 cm. Reservoir depth (hydraulic head) is 100 m and the tail water level is 0.0 m. At places where the bedrock is impermeable (in the region of the grouting curtain for example) fractures are omitted from the network. Among the number of possibilities analyzed by the authors the scenario presented below assumes that the width of the dam foundation (W) is 262 m and the depth of the grout curtain (G) is 97 meters.

In Fig. 9.5, frames **a** and **b** show the modeled effects of dissolution after 31 years, **c** and **d** after 71 years, and **e** and **f** after 88 years. The maximum aperture widths $a_{max}$ (cm) and times are shown in the lower right corners. According to this model, during the first period (**a** and **b**) a channel is developed vertically down the widened fracture (Fig. 9.4) to the lower limit of the grout curtain. After 71 years (**c** and **d**), horizontal karst channels have extended outwards by solutional enlargement in fractures of the

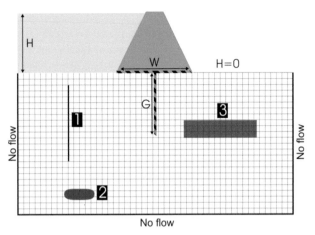

**Figure 9.4.** The model of a dam with a fractured rock foundation. (1) A widened fracture; (2) a cave/conduit; (3) An area of the rock having greater solubility than in the remainder of the network. Dreybrodt et al. (2005).

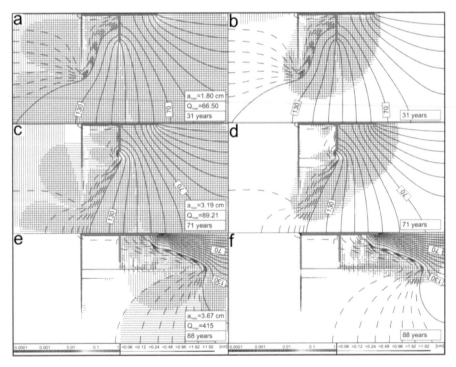

**Figure 9.5.** Evolution of fracture aperture widths, pressure heads, flow rates and dissolution rates. Dreybrodt et al., 2005.

standard aperture toward the bottom of grout curtain, and after 88 years an upper horizontal channel thus created has become the main recipient for groundwater flow under the curtain and into the patch of more soluble rock (Fig. 9.4;3) downflow of it.

According to the solubility values adopted in this model, after 91 or more years the discontinuities will have become large enough for flow to become turbulent.

A few more scenarios for the grout curtain made with the same model (Romanov et al., 2005) were: (A) width of the dam is 262 m and grouting depth 97 m; and (B) width of grouting section is 82 m and depth is 187 m. Depth of reservoir water is 150 m and the initial standard joint aperture remains 0.02 cm. To estimate widening rates of each joint over time the fluid dynamics of laminar and turbulent flow and the dissolution rates for dissolved calcium carbonate were applied. For scenario (A) the grout curtain is impervious and fractures of aperture 0.02 mm extend down to a depth of 187.5 m. For scenario (B) a widened fracture is located at a depth of 22.5 m and in (C) there is an ungrouted fracture at a depth of 97.5 m. In scenario (D) there is an ungrouted fracture at a depth of 165 m, close to the bottom of the grout curtain. According to the model, in scenario (A) after 28 years a vertical karst channel parallel to the grout curtain has developed down to the bottom of curtain. At the bottom of this new conduit the hydraulic head is almost equal to the reservoir head (150 m). Figure 9.6 shows graphs of evolution of total leakage beneath the dam for scenarios A, B and C.

The authors applied the same model for dams constructed on or in gypsum. The only difference was in the change of dissolution constants, those for gypsum being applied. Model results indicate that dissolution is approximately five times faster if the dam foundations consist of gypsum rather than of limestone; Dreybrodt et al. (2001).

According to these results, the processes of karstification will be highly accelerated under or around dams. As a consequence, the expected seepage losses through any fractures enlarged by solution increase unusually rapidly, implying that during the design lifetime of the dam the seepage can endanger the physical integrity of the dam and the water tightness of the reservoir. These findings, at a more or less theoretical

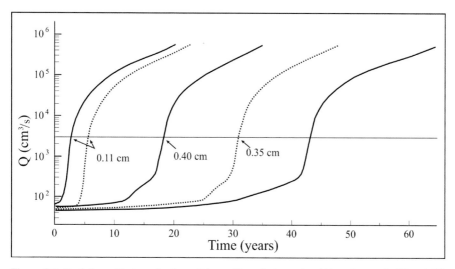

**Figure 9.6.** Evolution of leakage for 1 m of dam with perfect grouting (A), and scenario (B) and (C) depict aperture for an unsealed fracture of 0.1 mm; the dotted lines stand for aperture width of 0.2 mm. The numbers on the curves give the aperture of the unsealed fracture at breakthrough from laminar to turbulent flow (Romanov et al., 2005).

level, lead us to recognize the necessity of careful monitoring during the operation of a reservoir. Careful monitoring is important to confirm the application and validity of these and similar models and the likely necessity for regrouting during the life of the reservoir.

# CHAPTER 10

# LEAKAGE FROM RESERVOIRS—A COMMON PROBLEM IN KARST

## 10.1 THE LEAKAGE – KEY PROBLEM IN ENGINEERING KARSTOLOGY

During the operation of dams and reservoirs situated in karstified rocks a number of different destructive processes may endanger their integrity. The most important is the process of rejuvenation in which old, naturally plugged (infilled) karst features such as conduits and caverns are re-activated and flushed partly clear of their detritus by increasing pressure due to the reservoir filling, and later due to fluctuation in reservoir level. Piping followed by rapid erosion can play a key destructive role. Both processes are provoked by the unnaturally rapid and drastic changes in the hydrostatic and hydrodynamic conditions within the rock mass (huge pressure increases, turbulent flow, increased dissolution capacity) and the many secondary destructive processes and events (induced earthquakes, air-hammer and water-hammer effects).

The ultimate consequence of all these processes, sometimes mutually coupled, is leakage from the reservoir during its first filling and later during its operation. In many cases leakage is coupled with collapses in reservoir bottoms and in the vicinity of the reservoir and its dam.

In approximately 80% of dams and reservoirs in karst, unexpected and/or significant seepage problems have occurred during the first filling or later operation. In many cases the results of prevention or remedial works during construction were not effective enough—seepage was not completely eliminated. Reported leakages range between a few 10 of liters per second up to a few 10s of cubic meters per second.

In many cases the problem is too complicated for the available sealing technology. Despite extensive investigation programs and long-lasting sealing treatments, leakage problems could not be overcome and some dams have suffered from intensive leakage for many years during their operation. It is the chief problem for dam designers in karst. By its nature, karst is to complex to be detected and plugged every single cavern or conduit.

## 10.1.1 Dams with High Leakage

Listed below are some of well known dams in karst with high leakage after first filling or during operation: **Hales Bar,** USA, 50 m³/s; **Vrtac**, Montenegro, 25 m³/s; **Taoqupo**, China, 27,8 m³/s; **Dalongdong**, China, 15 m³/s; **Keban,** Turkey, 26 m³/s; **Zaixiangkou**, China, 20 m³/s; **Maotiao 4th Cascade**, China, 20 m³/s; **Višegrad**, BiH, 1.5 m³/s increased to 14,5 m³/s; **Iliki**, Greece, 13 m³/s; **Camarassa**, Spain, 11.2 m³/s; **Lar,** Iran, 10,8 m³/ s; **Salakovac**, BiH, > 10 m³/s; **Ataturk**, Turkey, > 10 m³/s; **Center Hill Dam**, USA, ~ 10 m³/s; **Mavrovo**, FYUR Macedonia, 9,5 m³/s; **Marun**, Iran, 9,5 m³/s; **Great Falls**, USA, 9,5–12,7 m³/s; **Canelles**, Spain, 8–10 m³/s; **Slano**, Montenegro, 8 m³/s; **Seymareh**, Iran, 7 m³/s; **Dokan**, Iraq, 6 m³/s; **Karun I (Shahid Abbaspoor Dam),** Iran, 1–6 m³/s; **Buško Blato**, BiH, 3–5 m³/s; **Montejaque**, Spain, 4 m³/s; **Fodda**, Morocco, 3–5 m³/s; **Contreas**, Spain, 3–4 m³/s; **Gorica**, BiH, 3–4 m³/s; **Tang Ab,** Iran, 3.5 m³/s; **Kowsar (Tange Duk),** Iran ~ 3 m³/s; **Hutovo**, BiH, 3 m³/s; **Karun III,** Iran, 2–3 m³/s; **Samanalawewa**, Sri Lanka, 2,8 m³/s (max 7 m³/s); **Špilje Dam**, FYUR Macedonia, 2 m³/s; **Liverovići,** Montenegro, 1–2 m³/s; **Piva (Mratinje)**, Montenegro, 1–2 m³/s, **Shuicaozi**, China, 1.8 m³/s; **El Cajon**, Honduras, 1.65 m³/s; **Krupac**, Montenegro, 1.4 m³/s; **Peruća Dam,** Croatia, 1.2 m³/s; **Bin el Ouidan,** Morroco, 1 m³/s; **Beni Haroun**, Algeria, about 1 m³/s; **Guanting**, China, 1 m³/s; **Francisco Zarko** (Mexico), 1.0 m³/s; **Kruščica (Sklope)**, Croatia, 0.8 m³/s; **Charmine** France, 0.8 m³/s; **Foum El Gherza Dam,** Algeria, 0.66 m³/s; **Karun IV**, Iran, 0.6 m³/s; **La Bolera**, Spain, 0.6 m³/s; **Salman Farsi**, Iran, 0.5 m³/s; **Times Ford Dam**, U.S.A., 0.5 m³/s; **Perdikas,** Greece, abandoned; **Mornos,** Greece, 0.5 m³/s; **Mujib**, Jordan, 0.3 m³/s. Seepage is reported in the case of following dams: **Lone Pine**, U.S.A.; **Onac** (Turkey); **Apa**, Turkey; **Chevizli**, Turkey; **May**, Turkey; **Schoroh** (Switzerland); **Yingtaoao**, China; **Farim Sahra Dam,** Iran; **Bitvaz Dam,** Iran; **Bayar Dam** (Jordan); and many others.

As noted earlier, some of these dams are abandoned ones (Hales Bar, U.S.; Montejaque, Spain, Vrtac and Liverovići, Montenegro; Perdikas, Greece), or operate with unacceptably heavy leakage (Lar, Iran; Samanalawewa, Sri Lanka; Iliki, Greece; Salakovac BiH; Ataturk, Turkey). Progressive erosion and dissolution are the most common processes that occur during reservoir operation in karstified rocks. However, in a number of cases remedial works were successful or partially successful.

## 10.2  CASE STUDIES OF LEAKAGE AS A CONSEQUENCE OF PROGRESSIVE EROSION AND DISSOLUTION

**Mosul Dam** (Iraq) is a well known example where there is intensive gypsum and anhydrite dissolution and groundwater seepage through a 1000 m long section of the dam foundation. Gypsum and anhydrite layers are inter-bedded between limestone and clayey/marly strata. All formations are greatly disturbed by a number of sub-vertical discontinuities. In spite of the grout curtain (4,454 m + 1,506 m), during a first partial impounding (1986) the leakage through the dam site area rapidly increased up to 1,400 l/s. The dissolution rate ranged from 42 to 80 t/day (Guzina et al., 1991). To replace the volume of dissolved materials, extensive repair grouting was applied almost immediately and has been almost continuous since then (more than 25 years). During this period a few large sinkholes have formed short distances downstream on the right bank. These long-lasting remedial works have been without success. Dissolution and leakages are still active and the Mosul Dam is declared to be a high-risk structure (see 7.4).

**Višegrad Dam** (Bosnia and Herzegovina) is located on the Drina River. The foundations are limestone and dolomite rocks karstified by solution and partially by hydrothermal action (hypogene karstification). After excavation of the surface of the foundation (20,000 m²) discharge of hydrothermal water with total outflow of 360 l/s was found at 140 points along the exposed joint network. The deepest karst features detected were 283 m below the river bed. During the first filling of the reservoir (1986), seepage of 1.4 m³/s deep beneath the dam foundation was observed. This process provoked intensive washout of clay from caverns and joints. In spite of intensive regrouting the leakage rate gradually increased to 6.5 m³/s (1996), 9.4 m³/s (2003) and 13.92 m³/s (2008). During this period an enormous amount of clay from joints and caverns has been washed out. Active karst channels are situated at depths more than 130 m below the dam foundation (Fig. 10.1); see Chapter 13 for more information.

In the case of **Gorica Dam** (Herzegovina) only part of the designed grout curtain was constructed. Intentionally, a grout curtain in the right bank (about 50% of the complete curtain as designed) was not installed. From the first filling of the reservoir in 1965 until 2003 the seepage beneath and around the dam increased from 1.4 m³/s to 4.4 m³/s (Fig. 10.2). Progressive erosion of infilling sediments was intensive, particularly during a strong earthquake in 1979 with its epicenter only about 60–70 km distant in the Adriatic Sea. For more than 24 hours afterwards muddy water with a lot of bubbles and fragments of plastic clay was discharged close to the downstream perimeter of the dam. Some of the compacted clay clasts squeezed out of the karstified joints weighed as much as 1–1.5 kg. However, in spite of this intensive and progressing erosion the stability of dam is not thought to be endangered; see Chapter 13.

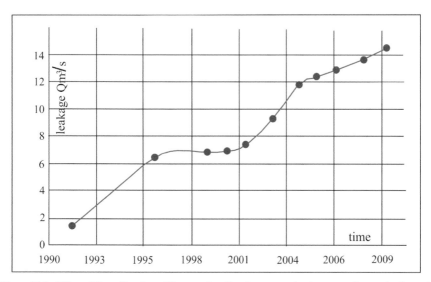

**Figure 10.1.** Višegrad Dam, Bosnia and Herzegovina. Graph: seepage due to progressive erosion beneath the grout curtain. Milanović, S. 2015.

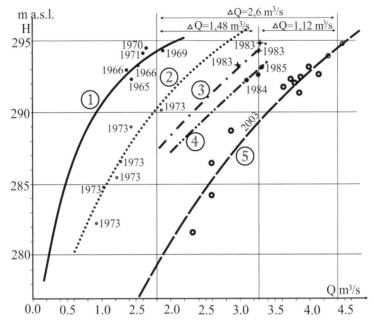

**Figure 10.2.** Gorica Dam, Bosnia and Herzegovina. Graph showing seepage increasing due to erosion of clay in karstified joints below the dam (Uljarević et al., 2003).

In the case of the **Great Falls Reservoir** (Tennessee, U.S.A.) relatively small seepage was recorded at first, but this increased rapidly from 0.47 m³/s in 1926 to more than 6.6 m³/s in 1939 and to 12 m³/s in 1945 when the water level in the reservoir had fallen by 8.1 m because of the leakage.

**Keban Dam** (Turkey) is made of two parts: rock-filled and concrete gravity sections. The dam is situated in heavily karstified marble and limestone. During its construction about 30 caverns were treated, including 'Crab Cavity' with a volume of 105,000 m³. During the reservoir filling in 1976, when the water level reached an elevation 10 m below the design maximum a large karst shaft opened near the dam and losses from the reservoir increased sharply to 26 m³/s (Fig. 6.27). This water discharges into Keban Creek, 2.5 km south of the dam site. The karst shaft led into the huge Petek Cavern. After the cavern was filled with about 605,000 m³ of limestone blocks, gravel, sand and clay, the flow rate in Keban Creek decreased to 8 to 9 m³/s.

**Lar Dam** (Iran) is an earth-filled structure located in a very complex geological setting: lava, tuff, lake deposits over intensively karstified Jurassic and Cretaceous limestone. In natural conditions the groundwater table was at a depth of approximately 200 m. Immediately after impounding began (1980) seepage started at two springs, as well as a few new collapses appearing in the reservoir floor (Fig. 10.3A and B). The measured discharge was 9.85 m³/s at the highest water level that has been achieved in the reservoir during the past 30 years (24 m lower than the design level). A large part of the seepage percolates underground along the karstified right bank (Fig. 10.4B and C). During the past 30 years massive remedial works have been undertaken, including intensive grouting and the filling of a large cavern (volume –90,000 m³),

**Figure 10.3.** Lar Dam, Iran. (A) large collapse in the reservoir floor; (B) one of the outlets of Haraz Spring; (C) and (D) Open karstified joints in the right bank of the reservoir.

however without success. The bottom of the cavern is approximately 150 m beneath the level of the dam foundations, but underground seepage flows are much deeper, roughly deeper than 300 m. The Lar Reservoir has never completely filled up, and the quantity of leakage is now almost constant.

**Hammam Grouz** (Algeria) is constructed for water supply. The limestone formation in the dam foundation and right bank is karstified by hydrothermal water (probably a hypogene karstification). To achieve the necessary water tightness in the reservoir floor a protective impervious clay blanket (two layers totaling 1.8 m in thickness) was laid down and a reinforced shotcrete blanket (thick 0.17 m) covered the limestone banks. In 2003, after 17 years of successful operation, seepage from the reservoir was detected. The seepage gradually increased and the reservoir dried up completely in 2007 (Fig. 10.4). The leakage point was treated by filling it with a channel plug covered by a reinforced slab and a compacted clay blanket. After many years of remedial works the seepage problem still remains, however.

In the case of **Samanalawewa Reservoir** in Sri Lanka, seepage was recorded during the first filling in spite of the installation of a grout curtain 1,300 m long, and 100 m deep along the right bank (Laksiri et al., 2005). A principal leak was noticed in an extremely tectonized zone in the right bank 300 m downstream of the toe of the dam (Fig. 10.5). Within a few hours of an initial leak of 75 l/s, the discharge dramatically increased to more than 7 m³/s (1992) for a short period, then fell quickly again, finally stabilizing at a more or less constant value between 1.8 m³/s and 2.2 m³/s. An attempt to halt the leakage without emptying the reservoir was made by laying down a 'wet clay blanket'. This was done by using bottom-opening barges to

**Figure 10.4.** Hammam Grouz, Algeria. (A) Upstream face of the dam; (B) Collapse at the perimeter of reservoir floor.

**Figure 10.5.** Samanalawewa Dam, Sri Lanka. Seepage spring 300 m downstream from the dam site (right bank).

dump graded earth material along the 700 m suspected seepage zone. The aim was to make an impervious blanket but, in spite of dumping of about 500,000 m$^3$ of earth, the final result was not successful.

**Huoshipo Reservoir** (China) is constructed on a karstified limestone formation that includes 48 beds of gypsum (Wuzhou, 1988). The height of this earth-filled dam is 23 m. Leakage of 237 L/s occurred in the initial stage of reservoir impounding when a number of sinkholes opened up in the reservoir floor. Because the reservoir capacity is only 4.7 million cubic meters this loss was declared to be too high. By sealing works (grout curtain, plastic membranes, clay cover) seepage was reduced to ≈ 80 L/s; however, the dissolution has not been eliminated.

Dams and reservoirs situated in evaporites are particularly sensitive (Chapter 7). In general two different kinds of processes and problems are common:

1. Disturbance of the dam stability, endangering the integrity of the reservoir and destruction of watertight seals, are considerable problems: sometimes there is the potential of outright failure (Mosul Dam, Iraq); and
2. Deterioration of water quality due to large increases in the concentration of sulfates and chlorides in the reservoir water (Khordad 15 and Gotvand Reservoir in Iran).

## 10.3　COLLAPSES (SINKHOLES) AS CONSEQUENCES OF RESERVOIR OPERATION

Collapses are the most frequent and most visible destructive landforms seen in the reservoir floors or nearby. Many of them are sources of reservoir seepage losses, but not all of them.

The origin of collapses generated by bringing a reservoir into operation is quite different from the formation of the majority of natural sinkholes (dolines). Reservoir collapses tend to occur suddenly and instantaneously ('catastrophic collapses'). Most natural sinkholes are the consequence of slow solution processes. Abrupt collapses of natural cave roofs are relatively rare phenomena.

Karstified bedrock covered with unconsolidated deposits is the key prerequisite for the majority of such catastrophic collapses. Almost always the collapses occur in the unconsolidated sediments (alluvium, alluvial fans, colluviums, etc.) as consequences of the mechanical action of water: piping that is caused by the frequent fluctuations of reservoir water levels; progressive erosion; the hydraulic pressure of the impoundment; major increases in dissolution if evaporites are present. In some cases collapse sinkholes, 'alluvial ponors', had developed under natural conditions but became buried beneath thick deposits of later sediments. Existing investigation techniques are still not good enough to detect the top of a karst solution channel at the contact between the karstified rock and unconsolidated cover sediments. Collapses in the floors are reported in many reservoirs in karst (Fig. 10.6). Some of them occurred during the first filling of reservoir: May and Keban, Turkey; Hutovo, Herzegovina; Slano, Montenegro; Lar, Iran; Kamskaya, Russia; Haditha and Mosul, Iraq; Anchor Dam, U.S.A., Wyoming. Alternatively, over time hazardous collapses can occur after many years of successful reservoir operation: Horsetooth Reservoir, U.S.A., Colorado,

**Figure 10.6.** Cover sediment collapses that occurred during operation of reservoirs on karst rocks. (A) Salamanawewa Reservoir, Sri Lanka; (B) Bileća Reservoir, Herzegovina (C) Hutovo Reservoir, Herzegovina; (D) Mavrovo Reservoir, FYUR Macedonia.

after 50 years; Mavrovo Reservoir, FYUR Macedonia, after 25 years; Hamam Grouz, Algeria, after 17; Bileća Reservoir after 18 years, Salamanawewa, Sri Lanka, after more than 15 years of operation, and La Loteta, Spain, after 2 years.

## 10.4  ACCEPTABLE LEAKAGE

The matter of achieving watertight conditions in a reservoir in karst has both technical and economic components. Reservoirs in karst should not necessarily be absolutely watertight. If filtration occurs through karstified foundations, the crucial question should be the possible negative influence on dam stability and grout curtain integrity. Additional work to increase the impermeability after the first impounding of a reservoir is also a question of technology and economy. In many cases it is technically too complex, long lasting and expensive to try to prevent small seepages. If the expected results justify the money invested, then remedial works are reasonable.

The term *acceptable (tolerated) leakage* from reservoirs itself is problematic and almost impossible to define. Because of this, there is not a strict and widely applied definition of acceptable leakage in karst dam engineering. Usually where leakage is around 3% of the mean discharge of the source river/catchment, it will be questioned and become a matter for consideration. Acceptable leakage depends on many parameters including the mean river inflow, the storage volume and the purpose of the reservoir. At many cases *acceptable leakage* is forcibly accepted amount of water due to technical, economical or political reasons. A significant number of dams operate successfully in karst with seepage between 100 and 500 L/s (or more). In some cases leakage of a few m$^3$/s is declared as acceptable. For instance, in the case of Poliphiton Reservoir in Greece with a storage capacity of 1900 million m$^3$, leakage of 6 m$^3$/s (10% of the mean river flow) is declared as acceptable. Losses of 1,2 m$^3$/s from Peruča Reservoir (Croatia) are considered acceptable against 50 m$^3$/s of mean discharge. Quite different is the example of Boqaata Reservoir in Lebanon, a water supply facility with a storage capacity of just 7 million m$^3$. Its maximum acceptable leakage is only 35 L/s.

Where the seepage loss discharges into a different catchment area, it can be considered as a loss to its own catchment but may, however, be of benefit to the new one. If this water can be used for downstream power production, irrigation, or to guarantee ecologically necessary flow, the merit of undertaking expensive and problematic measures to halt it is open to question. In many of these cases, in fact, proposed remedial works have been declared as unacceptable and too expensive for the expected benefit.

During the construction of the Gorica Dam (Herzegovina) the grout curtain designed for the left bank was not built, being intentionally ignored. The by-pass leakage of 3 m$^3$/s that resulted from this decision was considered acceptable because of its role in guaranteeing water flow to sustain the ecological system in the urban area downstream. As consequence of seepage losses from Hutovo Reservoir (Herzegovina), the capacity of water supply wells supplying the town of Metković in Croatia has more than doubled!

In many cases, particularly in shallow reservoirs with large surface areas, leakage is not the only loss. Evaporation from reservoir surface is also important: however, this parameter is not easy to be measured precisely. Leakage from many reservoirs in karst has been declared as acceptable, essentially because the cost of remedial works is too high and success of any waterproofing too doubtful: Karun 3 (Iran) 250–300 L/s; Sklope Dam (Croatia) 0.5–0.8 m$^3$; Camarasa (Spain) ~ 2.6 m$^3$/s; Salman Farsi (Iran) 0.5 m$^3$/s; Hutovo Reservoir (Herzegovina) 1.2 m$^3$/s; Piva (Montenegro) ~ 1 m$^3$/s; Špilje (FYUR Macedonia) 2 m$^3$/s.

# CHAPTER 11

# ENGINEERING AND THE ENVIRONMENTAL IMPACT OF DAMS AND RESERVOIRS IN KARST

## 11.1 INTRODUCTION

Many karst regions in the world are rich in water resources but due to the irregularity of their distribution spatially and temporally, people living in karst areas have had to cope with of the extremes of misfortune, both floods and droughts. The only strategy for socio-economic development of these regions is to optimize water resources development. Dams and reservoirs are the only strong civil structures able to maintain karst water on the surface for long periods and in the service of man.

Building these structures changes the natural surface and groundwater regimes. Instead of flowing through natural underground conduits large amounts of water are stored in reservoirs and transferred by tunnels and canals towards regional base levels. Such modifications of natural groundwater regimes can have considerable detrimental impacts on hydrogeological, hydrological, climatological, ecological, archeological and socio-economic features in a region, including impact on historical monuments and other national rarities. Starvation of karst aquifers is one of the undesirable consequences. Transfer of water from one catchment to another provokes different impacts also.

Protection in karst is particularly difficult due to changes that are unexpected and because the source of a problem is often at some distance from the impacted area, and is likely to be out of sight underground. The majority of impacts can be foreseen and analyzed during feasibility studies and eliminated or mitigated by appropriate design. One of the crucial problems is that the standard criteria for determining the appropriate measures for environmental protection, as well as the regulatory procedures that are applicable for dams and reservoirs in non-karstic regions, are generally not suitable for karst terrains.

Successful protection solutions require serious and comprehensive investigations and the close cooperation of a wide spectrum of scientists and engineers. Determining and applying the optimal strategy for dam and reservoir construction in karst areas is a key requirement for socio-economic development at local and regional scales.

## 11.2 IMPACT ON REGIONAL REGIME OF SURFACE AND GROUNDWATER

The major aims of constructing dams and reservoirs in karst terrain are to minimize naturally negative conditions and to maximize positive socio-economic development. In most instances, the impact of dam construction in karst is positive: flood control, irrigation, food production, water supply, hydropower production, fish farming, recreation, reduction of deforestation, infrastructure development and many secondary benefits.

The Tennessee Valley Authority (Tennessee River, USA), founded in 1933, is one of the largest dam-reservoir operators of flood control, navigation, power production and irrigation installations. A large proportion of its projects have been constructed in karst areas. In natural conditions the water regime was unfavorable for agriculture and life. Thirty percent of the population in the Tennessee Valley was affected by malaria. To construct 29 dams and reservoirs more than 15,000 families had to be displaced. Setting aside the electricity generation, the flood control and general re-organization of the water regime have been a great benefit for this very large region. More than 1000 kilometers of navigation channels have been constructed as part of this project, also. One of the important positive environmental impacts has been control of the surface water regime, resulting in the elimination of malaria.

In the past two centuries alone a great many people have died along the Yangtze River, China, due to catastrophic floods. In 1840 about 156,000 persons lost their lives during a flood; in 1931, 145,000; in 1954 about 33,000, and during a more recent flood in 1998 over 1,500 people died. Millions of hectares of arable land have been destroyed or rendered temporarily out of use. A large number of villages completely disappeared. During the flood of 1954, 18 million people had to escape from the region. Wuhan, a city of eight million people, was covered with flood water for three months. With the construction of the Three Gorges Dam (181 m high, 2335 m long) the frequency of major floods is reduced to a minimum and, after the project is finalized, large ships will be able to sail from Shanghai 2,400 km upstream. Because of requirements for the project about 1.24 million of people have been relocated.

Construction of several large multipurpose projects in the Dinaric karst (Croatia, Bosnia and Herzegovina and Montenegro) started in the early 1950s. The most prominent example is the Trebišnjica Multipurpose Hydrosystem in eastern Herzegovina. This is one of the most karstified regions in the world, and also with the highest mean annual precipitation in Europe – 1800 mm/a. Water is the most important natural resource of the region, in fact. However, of the 367 m³/s of mean annual precipitation, only about 145 m³/s is discharged by surface runoff. Underground karstic flow, droughts and floods are the chief natural characteristics. To change and control the surface and groundwater regimes is the only means of regional economic development. This had to be done by construction of an integrated hydrosystem. It is one of the most complex projects undertaken in any karst region in the world. Due to the geological complexity of the karst, intensive investigations took place over a period of 50 years. Over 600 exploration boreholes were drilled, mostly for groundwater level monitoring. Geophysical investigations covered an area of about 450 km², including all dam sites and tunnel routes. There were more than 130 tracer tests, using 6000 kg of Na-fluorescein, the radioactive tracers Bromine-82, Iodine-131, Chromium-51, Tritium ($^3$H), post-activated isotopes, Lycopodium spores, smoke and gas tracers.

The system has seven dams, six artificial reservoirs, six tunnels and four channels providing for multi-purpose use of the water resources from 1000 m a.s.l. down to the sea l (Figs. 11.1 and 11.2). This complex hydrosystem enables optimal management of the hydro-potential, by storing water during the wet period of the year and controlling utilization throughout the year. After completion, the average annual output of the Trebišnjica Hydrosystem will be 856.2 GWh. In addition to power generation, the system enables drainage of temporarily flooded karst poljes and irrigation of about 240 km² of arable land. The largest European sinking river, Trebišnjica, is blanketed

**Figure 11.1.** The Layout of the Trebišnjica Multipurpose Hydrosystem, Herzegovina. (1) Karst polje elevation; (2) Permanent flow; (3) Canal; (4) Tunnel; (5) Dam (gravity and arch); (6) Power plant (operational); (7) Power plant under construction.

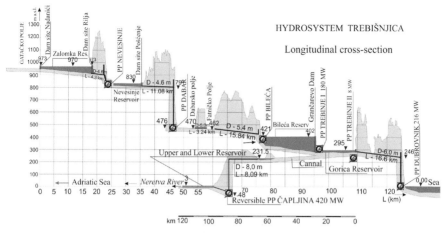

**Figure 11.2.** Long section of the Trebišnjica Multipurpose Hydrosystem, Herzegovina.

with 2 million m² of shotcrete to become a permanent surface channel. Construction of such a complex hydrosystem in one of the most karstified areas of the world faced unique technical, ecological, social and political challenges. In addition, the problem of trans-boundary aquifers has become a serious political problem. However, the best strategy for water resources development in eastern Herzegovina is to keep a balance between the need for development and the preservation of a very sensitive environment.

## 11.3 IMPACTS ON SUBTERRANEAN AND TERRESTRIAL FAUNA AND ARCHEOLOGICAL MATERIAL

Subterranean terrestrial and aquatic species are found in the karst underground all over the world. As a consequence of disturbing the natural underground and surface water regimes, very distinct impacts on a wide spectrum of biodiversities is to be expected. On the one side are the needs of the supreme species, human beings, and on other side is nature, the number and fragility of the diverse underground species. Often, building any structure in karst, particularly dewatering of temporarily flooded karst depressions and construction of dams and reservoirs, has resulted in negative influences on caverns inhabited by endemic species.

For example, in a survey of 82 caves in the USA, 1042 of the 1475 subterranean species are known from fewer than five sites, and 316 species are known from a single cave (Roth, 1999). In the Dinaric karst area, 470 aquatic and 860 terrestrial cave species have been found. It is one of the richest karst regions with subterranean fauna (Sket, 2003).

In the case of the Ombla underground hydropower project (Croatia), in accordance with an EU Habitats Directive, some upper parts of the underground dam structure were redesigned and adjusted to protect the natural bat habitat and to conserve other wild species. The underground dam Ombla is designed to raise the water level within karstified limestone behind Ombla Spring to create an underground reservoir (Fig. 11.3). The underground dam (grout curtain) will stretch from 280 m below sea level to 130 m above it. Vilina (Fairy) Cave is just above the top of the upper grout

**Figure 11.3.** Sekerpinar Spring, Turkey. A shoal of fish deep in the karst channel.

curtain. The upper section of the cave at an elevation of 136 m is an important bat habitat. Any adverse impact on the bat fauna and any reduction in the quality of the habitat by physical and visual disturbance is prohibited, particularly in the areas where the grout curtain will be constructed. During operation of the Ombla underground power plant any flood effects on Vilina Cave, including on humidity and airflow conditions, has to be strictly controlled. Monitoring bats in the land around the cave must be under strict control over a monitoring with a minimum radius of 10 km.

Sekerpinari Spring in Turkey (elevation 800 m a.s.l.) is a karst spring used for bottling water. Discharge varies from 1 to 15 m³s. The water discharges from two karst channels about 20 m apart. The Spring is the habitat of a particular species of unusually big cave fish (length up to 40 cm). After the main channel was fitted with pumps, the shoal of several thousand fish changed their local habitat and colonized the lesser (second) channel (siphon section) at a depth of about 30 m (Fig. 11.3).

A well-known cave-dwelling endemic blind aquatic species, the cave salamander *Proteus anguinus* (known as the 'human fish' ) is frequently found in the Dinaric karst. In the Popovo Polje, Trebinje Town and Gorica Reservoir area *Proteus* was found at more than 40 localities. It is interesting that the large and deep siphon springs along the Adriatic Sea coast are not the common *Proteus* habitat. Rarely, *Proteus* can be transported by turbulent karst flow from upstream parts of the aquifer and detected at these springs.

*Triplophysa shilinensis* Chen and Yang, discovered in Weiboyi Cave (1991) is one of the 11 species of cave fish and the fourth of the *Nemacheilinae* family found in the karst caves of China (Xie and Li, 2001). In the karst of Jianxi the other blind fish spp. is *Anophthalmus* (Lu, 1986). The cave fish *Typhilichthys subterraneus* and *Amblyopsis rosae* and cave salamanders *Typhlotrition spelaeus* and *Eurycea lucifugia* are part of the Ozarks, US, cave aquatic fauna (Palmer and Palmer, 2009).

In the case of the fish *Paraphoxinus ghetaldi* that inhabits siphons of estavelles in Popovi Polje (BiH) problems arise during dry periods of the year and in the intermittent lakes of the karst poljes during the flood season. After floods in the polje were eliminated, fishing at the openings of estavelles, which for centuries was an important tradition and food source for the local people, was largely destroyed.

The unique tubeworm *Mariphugia cavatica* (Fig. 11.4) that colonize temporarily flooded karst channels is seriously endangered after dewatering of temporarily flooded karst poljes.

Large colony of theses worms was discovered in the karst channel beneath 10 m of alluvium in Popovo Polje (Fig. 11.5). After construction of an impervious blanket over the alluvial deposits to prevent seepage from the reservoir floor, the survival of Mariphugia is questionable.

Seymareh concrete arch dam (in Iran) is situated in karstified Asmari limestone. During excavation of the head race tunnel, cavities inhabited by reddish fishes (*Iranocypris*, 3–6 cm long) and other species described as 'shrimps' have been found (Fig. 11.6, pers. information by K. Solgi, 2010). By construction of the grout curtain a part of their habitat will be endangered. However, it is expected that these species will adapt to the new aquatic conditions.

Outlets of submarine springs along the sea coast are frequently used for commercial oyster and mollusc farms. In many cases, the submarine springs are supplied from large swallow holes (ponors) situated far inland. If outflow of fresh water is considerably reduced due to building control structures at the ponors within reservoir, the operation of these farms can be endangered.

According the EU Habitats Directive (May 1992) almost all cave-dwelling (terrestrial and aquatic) species, particularly endemic ones, need special protection. EU Member States are required to take measures to maintain or restore natural habitats of cave species to a favorable conservation status and to introduce robust protection for those habitats and species. In karst regions where every cave and many saturated (phreatic) karst channels are habitats for numerous species (many of them are endemic)

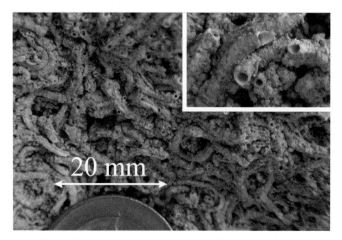

**Figure 11.4.** Popovo Polje, Herzegovina. *Mariphugia cavatica.*

**Figure 11.5.** Hutovo Reservoir, eastern Herzegovina. Karst channel below the man-made reservoir is the habitat of *Mariphugia cavatica.*

**Figure 11.6.** Seymareh Dam (Iran). Cavefish, *Iranocypris,* from the underground at the dam site. Photo, courtesy K. Solgi (2010).

strict requirements for protection can be source of conflicts between needs for regional development and preservation of nature and, in extreme cases, can be a source of transboundary conflicts.

Many caves contain important archaeological artifacts: prehistoric human and animal bones; carvings; paintings; sometimes ruins of complete settlements (e.g.,

Shapour Cave in Iran). In some cases reservoirs may flood caves. These potential impacts are often reason enough to abandon a new project. For example, construction of the Lower Gordon Dam in southwest Tasmania would have flooded a large karst area containing caves of great archaeological importance. The project was abandoned for legal and environmental reasons in 1983 (Kiernian, 1988).

The speleologically important caves at Cavuto (Italy) are interrupted by construction 30 m high dam and Letino Reservoir.

## 11.4  SUBMERGENCE OF LARGE KARST SPRINGS BY RESERVOIRS

As consequences of dam construction, some large karst springs in different karst regions have been submerged. In the complex hydrogeological conditions in karst some of the key questions for the final decision about the feasibility of constructing a dam are related to: possible leakage from reservoir through the submerged spring; protection of the spring water for water supply and bottling; how would the submergence affect the local seismicity; what would be the interference of submergence on the spring discharge; and, will the surrounding depressions and caves be endangered by floods (environmental impact).

Some important examples are:

Due to construction the Oymapinar Dam (185 m high) the high-discharge Dumanli Spring (Turkey) was submerged by a head of 120 m of water (Fig. 11.7). Spring discharge varies between 20 and 100 $m^2$/s. A discharge rate of 35.6 $m^3$/s was

**Figure 11.7.**  Dumanli Spring, Turkey.

measured at the very end of the dry period in October 1978 (Günay and Karanjac, 1980). Risks due to possible negative consequences of submerging the spring was the crucial question for the final decision. On the basis of detailed geological investigations and analysis a positive decision was made and, after many years of reservoir operation, no water losses from the reservoir have been observed.

By construction of the 123 m high Grančarevo Dam, the Trebišnjica Spring is submerged 75 m in the Bileća Reservoir (Fig. 11.8). In natural conditions, spring discharge varies from 0.4 to 219 m³/s. A part of the aeration–zone in the immediate vicinity of the spring became a zone of saturation.

The submergence of the spring zone affects the dynamics of emptying of the karst aquifer (Milanović, 1986). In the area around piezometric borehole (PB-1) at the distance of 12 km from the spring, lowering of the water table from 500 to 450 m requires twice as a long a period as before submergence of the spring.

From long-time monitoring in natural conditions, 150 hours were needed to lower the water table from 500 m to 450 m a.s.l. Since the Trebišnjica Springs were inundated by the reservoir about 300 hours were needed to decrease the water table by the same amount. Due to the spring submergence the duration of flooding in the upper karst polje has been prolonged.

TREBISNJICA SPRING - CROSS-SECTION

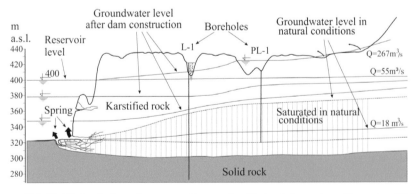

**Figure 11.8.** Trebišnjica Spring submergence, Herzegovina. Photos, left–opening of the spring; Right–the circle of mounded water at the reservoir surface during high discharge from the submerged spring. Simplified cross-section through the spring background with different reservoir and groundwater levels.

The large spring of Neraidha (Greece, Macedonia) discharges about 10 m³/s. After construction of Poliphiton Reservoir the Neraidha Spring was flooded by 40 m of water. Calculations based on detailed geological and hydrogeological investigations, including hydraulic calculations, suggested possible water losses of 6 m³/s from the reservoir, corresponding to 10% of the mean river runoff. Such a risk was considered acceptable but, in the meantime, a full-scale test after filling of the reservoir in 1974 indicated that no losses occurred.

After construction of the 220 m high Piva Dam (Montenegro) the Pivsko Oko karst spring was flooded by 70 m of water. Spring discharge varies from 1.4 to 154.2 m³/s. No seepage problems have occurred.

Three of the main karst springs supplying the Rama River (BiH) are submerged by 'Rama' Reservoir. The depth of the water varies between 40 and 60 m. Once again, there have been no known seepage problems.

Yarg Spring (Iran) is submerged by 27 m of water after construction of the Salman Farsi Dam. Its average discharge is 0.7 m³/m. At present the spring is submerged, but not to the designed maximum head; no seepage has been recorded.

The Oko Spring (Herzegovina) is submerged by 17 m of reservoir water by the construction of the Gorica Dam (Fig. 11.9). The spring, a rising karst solution channel, was tapped betwen 1899 and 2007 for water supply for the town of Trebinje. Discharge varied between 0.8–20 m³/s. After the deep siphon karst channel was discovered a new intake structure was constructed. The pumps were installed directly into the natural channel more than 25 m deeper than the outlet. According to divers' explorations the channel continues much deeper and it further parts are still unknown.

Because the Oko Spring is used for water supply the problem was mixing of water from reservoir with aquifer water creates possible problems of pollution.

A consequence of construction of the Daryan Dam on the Sirvan River, Iran, was the submergence of the Bel Spring. The dam is 230 m in height (190 from the river level). Discharge of Bel Spring ranges from $Q_{min} = 0.2$ m³/s to $Q_{max} = 6$ m³/s. The spring is used for water supply by the local community and for two water bottling factories. After impounding, Bel Spring was submerged by more than 100 m of water. During the feasibility study it was important to predict the possible effects of submergence on

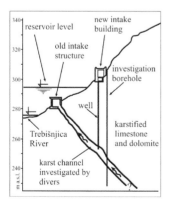 

**Figure 11.9.** The Oko Spring Tapping structure, eastern Herzegovina, after the old control house was submerged by a reservoir.

the karst aquifer, and whether it could be used for the same purposes as before. The most important questions to be answered were (Milanović, 2007):

- Is this requirement technically feasible?
- Is it possible that there will be water losses from the reservoir into the adjacent catchment via the spring?
- What will be the effects of this interference of the groundwater regime on the aquifer and the discharge of its spring?

Very sophisticated investigations were organized. A deep siphon channel was investigated by divers to a depth of approximately 50 m below the spring outlet. The solution that was reached is shown in Fig. 11.10 (Solgi, Mahab Ghodss, 2015, pers. comm.). The main spring outlet is closed by a massive concrete plug. All natural channels are filled with concrete.

The experience in these cases of submerging karst springs (Oymapinar, Turkey – 120 m of water; Bileća Reservoir, Herzegovina – 75 m; Neraidha Spring, Greece – 40 m; Piva, Montenegro – 70 m; Rama, Bosnia – 60 m) is that after many years of operating the reservoirs, there have been no negative effects of water losses from them.

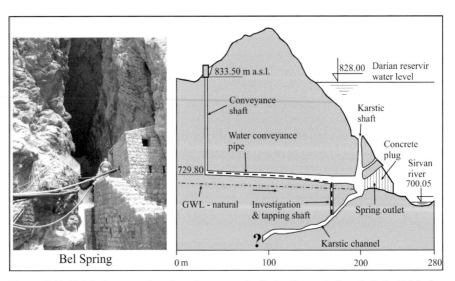

**Figure 11.10.** Bel Spring protection after submergence by Darian Reservoir, Iran. Left–the Bel Spring outlet cavern (Milanović, 2007); Right–The solution that was executed, Cross-section.

## 11.5  RESERVOIR BANK STORAGE

Due to the cavernous nature of karst porosity, water can readily saturate any karst in the banks of reservoirs. When the reservoir level rises above the level of the water table in any adjoining karst aquifer, reservoir and aquifer becomes hydrogeologically and hydraulically coupled. Parts of the vadose zone are reactivated and become a part of the active underground storage. In some karst reservoirs the amount of the water stored in the banks may be large enough to contribute significantly to the total storage in the reservoir. If the reservoir is used for hydropower production this additional amount may be very beneficial. However, this water is almost never included in estimates of the reservoir capacity.

In spite of the low average bulk porosity in a karst the storage capacity in banks of large reservoirs can be huge. Underlying that storage is the base of karstification. There is no significant bank storage below this base. The lateral boundaries of the storage formation depend on its structural and lithologic characteristics or on variations in the base level of karstification. Due to the very good hydrogeological and hydraulic connections, the reservoir and karst banks are a unique, physically coupled storage capacity. As the reservoir fluctuates the saturation level of the karst aquifer changes simultaneously. A number of hydrographs in karstified reservoir banks show active bank storage extending 5 to 15 km out from the reservoir. Linkage of the underground storage with the reservoir consists of systems of mutually connected conduits with large flow capacity.

The effect of bank storage coupled with surface reservoir has been analyzed by several authors:- Coffin, 1970; U.S. Bureau of Reclamation, 1974; Isailović, 1976; Öziş et al., 1981; Günay et al., 1986.

In the cases of some well-investigated reservoirs in karstified rocks the physically coupled nature of the surface and underground storage has been clearly confirmed. The Bileća Reservoir, located entirely in karstified carbonate rocks, is one of the best examples. Its storage capacity is estimated to be 1.27 Billion cu m. A large area along the right bank and far behind the spring zone has been investigated by geo-electrical sounding and profiling, and measured by a number of piezometers. Figure 11.11 shows annual fluctuations of the water table in piezometers and cross-sections approximately perpendicular to the reservoir bank. As the reservoir fluctuates, the saturated depth of underground storage in the middle sector ranges from 350 up to 400 m a.s.l. in elevation, i.e., 40 to 50 m in a year with average humid.

From Fig. 11.11 – A and B, it is apparent that the water level in Piezometer B-4 and B-2, at a distance of more than 4 km from the reservoir, has a direct, swiftly acting, hydraulic connection to it. Bank storage levels are closely dependent upon the reservoir levels but also under influence of the other local hydrogeological control. Recharge from precipitation is also important. During the wet winter season active underground storage can rise much higher than in the reservoir. In some piezometers GWL is 20 to 25 m higher than the reservoir level for periods up to some weeks in the duration. Decrease of the groundwater levels is much slower.

In this part of the rock mass the minimum water level in piezometers is above the elevation of 360 m while in vicinity of Piezometer M-4 it decreases below the elevation of 355 m. The bank storage when reservoir is full, saturate a huge volume of the right bank, Fig. 11.11C.

Simultaneous graphs of GWL in piezometers and reservoir fluctuations (Fig. 11.11A) show the perfect hydraulic connection between the reservoir and underground storage along the principal direction of groundwater flow, from piezometer H-1 (Fig. 11.12) toward Trebišnjica Spring and that the bank storage depends on the reservoir level as well as on influence of the more variable tributary underground inflows.

After the dam construction and reservoir impounding the storage function is reactivated in a portion of the aeration (vadose) zone. This change has been most

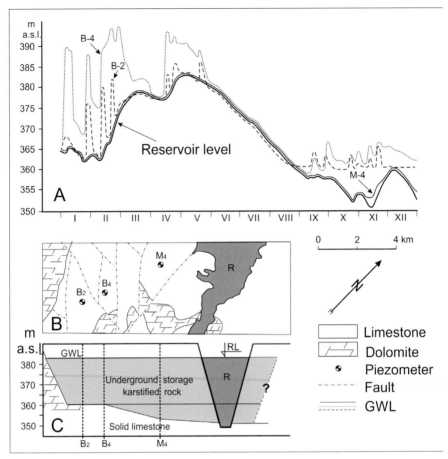

**Figure 11.11.** Bileca Reservoir, an example of physically coupled reservoir (surface water) and groundwater stored in cavernous karst.

prominent in parts of the vadose zone closest to the Trebišnjica Spring. However, the influence of the reservoir is clear at a distance of 12 km. Due to the effects of bank storage the dynamics of the emptying of the karst aquifer is smaller than they were under to the natural conditions (Fig. 11.12, graph H-1), i.e., the rate of draining of the bank storage component of the aquifer is slower than it was in the prior natural regime (Milanović, 1986). If it is assumed that the average depth of groundwater held in the rock mass is 50 m at maximum and the effective porosity is 1%, the active bank storage can amount to 13 to 15% of the surface reservoir storage capacity.

Bank storage along the Oymapinar Reservoir in Turkey was analyzed by Günay et al. (1986). The reservoir is situated in a deep canyon 5 km in length and has a storage capacity of $310 \times 10^6$ m$^3$. The reservoir is in hydraulic connection with aquifers in karstified carbonate rocks on both sides. In this study net inflow into the reservoir and the bank storage has been calculated using the difference between the inflow and reservoir volume for a given water level. The data used for calculation were collected on a daily basis during the authors' field work. According to their analysis, for the

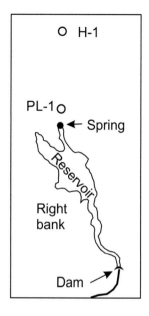
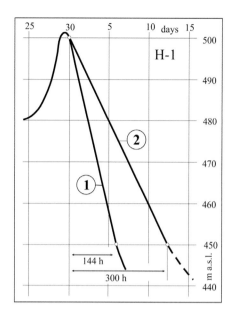

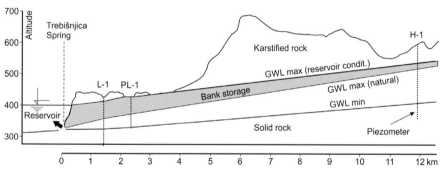

**Figure 11.12.** Bank storage in the vadose section of the karst aquifer behind Trebišnjica Spring and the Bileca Reservoir.

initial reservoir volume of 86.7 x 10⁶ m³, bank storage volume was about 42 x 10⁶ m³. When impounding was completed, the bank storage increased to roughly 122 x 10⁶ m³, which was 35–40% of the volume of the surface reservoir.

The Oymapinar bank storage capacity was analyzed by Öziş et al., 1981. Based on the recession graph analysis the bank storage was estimated to be more than 10% of the surface storage capacity. The upper possible limit was not estimated.

The Beitu underground stream system was submerged by dam construction in a canyon on the Longjiang River (Guanxi, China). Natural discharge of the Beitu underground stream was 0.14 m³/s. As a consequence of the submergence the groundwater level in a number of karst shafts (karst windows) increased. At Dengui karst window eight km from the reservoir, the pumping capacity more than doubled (Yuan, 1983).

The problem of deciding an optimal policy for water use from coupled surface-underground storage was also analyzed for general cases by Isailović (1976). One of the components considered was aquifer recharge under karst specific conditions. In view that underground storage in karst can significantly affect reservoir performances, more attention should be paid to investigation of the complex flow mechanism between the two components of the reservoir.

## 11.6  KARST-TRIGGERED SEISMICITY

Relationships between precipitation and earthquakes were first analyzed by Imamura (1937). He concluded when heavy precipitation occurs after a long dry period, seismic activity increased. The same conclusion was reached by Cvijanović (1980).

A similar relationship has been established for the time when large reservoirs are being impounded – 'induced (triggered) seismicity'. From the very beginning the concept of reservoir-triggered seismicity has been controversial. The first documented case was that of the Hoover Dam (Lake Mead, USA). Today more than 60 cases of reservoir-triggered seismicity are frequently cited in the literature. Earthquakes recorded in karstified rocks during reservoir filling or karst conduit plugging, or during extremely heavy rain indicate that, by its nature, karst may play an important role in generating these shocks. Native people in many karst regions have reported ground shaking and underground shocks after sudden and heavy rains. Powerful air flow from many piezometric boreholes has been reported during rapid saturation of karst aquifers as well.

In some karst areas local seismic activity occurs during intensive rainfall when there is abrupt filling of the aquifer and rapidly rising groundwater levels. During dry periods of the year large parts of many aquifers become drained and so can store enormous volumes of air in karst channels and caverns. After the first heavy rains, much of this air may become trapped and subject to increasingly strong pressure. The explosive release of the air after a critical pressure level is reached may cause local earth shaking and loud noise. Such events have been observed in the field and also reproduced in physical models. Under strong pressure from rising groundwater, 'pillows of air' trapped in siphoning segments of karst channels suddenly escape, creating strong explosions that are recorded by nearby seismological stations. The environmental impact of the karst-induced seismicity is generally local. However, there can be psychological stress for the local people.

One of earliest documented examples of induced seismicity due to the rapid saturation of karstified rock masses was registered in 1837 at the Timavo Spring near Trieste, Italy. Earth shaking as consequence of the quickly rising water table following heavy precipitation was reported by local inhabitants. Seismicity in karst thus can be triggered mostly by heavy rain or by the rapid filling of a reservoir. Rain-triggered seismicity has been observed in Italy, Herzegovina, China, and recently in Germany, France and Switzerland. Seismic activities are always associated with the rainy season when there is intensive filling of karst aquifers and rapid rising of groundwater levels. Induced seismic activity in other karst regions due to rapid saturation of karst aquifers has been reported by a number of authors, e.g., Petrović, 1965; Roksandić, 1970; Stojić, 1980; Ai, 1982; Milanović, 1984, 2004; Roth, 1994; Chen et al., 1995; Lu and Duan, 1997; Miller, 2008.

In Fatničko Polje in the Dinaric karst, local inhabitants living close to the large overflow Obod Spring have noticed the ground shaking with periodic strong seismic shocks (explosive releases), 15 to 30 hours before there is discharge at the spring itself ($Q_{max} \sim 60$ m³/s). From time immemorial these events have been well known in the local community. The principal source of concentrated recharge is in Cerničko Polje ($Q \sim 50$ m³/s) which is directly connected with Obod but 13 km to the north and about 350 m higher in elevation. Obod Spring is functioning as an overflow when regional base flow to the Trebišnjica springs is obstructed because the downstream parts of the karst aquifer are back-flooded by the Bileca reservoir plus locally derived surcharge in the bank storage.

In 1964 there was an experimental plugging of the Obod outlet channel (Fig. 11.13). A massive concrete plug, 10 m high and 3 m wide (in average) was built inside the spring entrance (Petrović, 1965). The plug was designed to withstand a pressure of 40 bars. The proposed duration of the experiment was one hydrological year. The pressure behind the plug was controlled by manometers. However, the first rain after the plug was installed was 230 mm/24 h, an unexpectedly large amount. After a few hours underground torrents completely saturated all conduits. Piezometric levels in

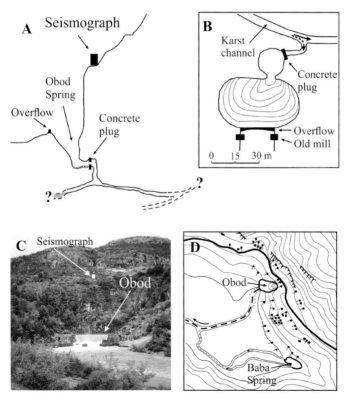

**Figure 11.13.** Obod Spring, Herzegovina; an example of seismicity triggered by rapid saturation of a karst conduit system, (A) Simplified cross-section of Obod Spring; (B) Plan of the outlet channel and position of the concrete plug; (C) Obod Spring during high discharge; (D) The neigborhood of Obod Spring, showing new springs created as a consequence of the plugging.

the rock behind the spring increased abruptly. More than 45 new springs appeared in vertical cliffs above the plugged outlet and in the wider surroundings of the spring. Some of them were 100 m higher and burst out inside houses in the village. The total discharge of newly formed springs was about 11 $m^3/s$. Water pressure behind the concrete plug increased up to 10.6 bars. Five strong, and a greater number of smaller, quakes were noticed. A number houses at distances of 250–300 m were damaged. The local road settled by 50 cm along a 30 m section. One hour after part of the concrete plug had been deliberately destroyed, the uppermost springs dried up and all other springs ceased after six hours. This experiment established that the Obod karst aquifer had high transmissivity but negligible storage capacity!

To further analyze the seismic activities in the Obod area a temporary seismic station was installed above the spring (Z-component only) in 1975. Results registered by the seismograph confirmed that there were ground vibrations and shocks after heavy rains, one or two days before the spring itself began to discharge (Milanović, 2002).

During the rapid filling of the 'Bileća' Reservoir (in extremely karstified rocks) some vibrations and shocks were registered by seismograph that cannot be explained by normal seismic activity. These vibrations were interpreted as explosive releases of air trapped in karst channels (karst-specific seismicity).

During impounding of the Salanfe Reservoir (Switzerland, 1953) a causal relationship between impounding, seepage from the reservoir and seismic activities in the surrounding area was also established.

At many examples in different karst regions (China, Switzerland, Herzegovina) has been registered that karst induced seismicity correlating with fast saturation of karst aquifer, particularly at beginning stage of the abrupt increasing of large karst springs discharge.

According Yuan, 1991, 14 earthquakes induced by reservoir impoundment have occurred in China, among which 11 are located in karst areas. Earthquakes occurred mostly when the initial stress of rocks approaches the critical value for giving rise to an earthquake, if an earthquake can be triggered off. The earthquake of M = 5.1 occurred in 1973 in the area of Danjiangkou Reservoir. The height of the dam is 97 m and the reservoir is situated in karstified limestone. In the case of the Wujiangdu Dam, 165 m high in Guizhou Province the magnitude of the triggered earthquake was M = 1.

Triggered earthquakes are not consequence on reservoirs depth only. Some of them are triggered by impounding of medium or small-size reservoirs. Qianjin Dam (Hubei Province) is 50 m high. The reservoir is located at an active fault in limestone. During impoundment in 1972, when reservoir level reached its maximum, the earthquake M = 3.0 occurred. Thirty meters deep, the Fengcun Reservoir, in Shanxi Province, is situated in karstified limestone. After impounding 364 earthquakes were recorded during April-June 1984. The magnitude of highest recorded was M = 2.9. The earthquake that occurred in small Dengjiaqiao Reservoir, Hubei Province (13 m deep), M = 2.2, was probably triggered by collapsing the karst cavern roof, Yuan, 1991.

Lu and Duan (1997) have suggested that there may be three different types of induced seismicity caused by reservoir water storage in karstic areas: A – Loading fault creation type (due to reservoir weight); B – Pneumatolytic type (explosions due to uprising boils of water from below the reservoir); and C – Cave damaged type (explosive release of air compressed in the caverns).

# CHAPTER 12

# TRANS-BOUNDARY PROBLEMS AND GROUNDWATER PROTECTION

## 12.1 A SIMPLIFIED MODEL OF TRANS-BOUNDARY AQUIFERS IN KARST

An aquifer is a hydrogeological formation or group of formations created naturally during the geological evolution of an area that can, in many cases, cross political borders. Aquifers are permanent, borders are artificial and changeable.

It is now well established in this volume that the hydrogeological properties of karst differ significantly from those of non-karstified rocks. The term 'trans-boundary aquifer' in karst does not mean only zones located near state boundaries or narrow borderland zones. In many cases trans-boundary problems are a consequence of water sinking into ponors located far from a boundary, i.e., close to watersheds with other catchments inside a state, at distances of 10, 20, 30 km or more from the border. As a result, the entire catchment area can be declared as potentially problematic for groundwater management within borderland areas. In general the trans-boundary problem in karst appears in a case where a political border passes between much or all of the recharge zone inside a catchment and the areas of discharge along the erosion base level (sea coast, deep valleys, etc.).

Two simplified models of trans-boundary aquifers (better to write 'trans-boundary concentrated flows') are presented in Figs. 12.1 and 12.2. A number of expensive investigations need to be done to define the connections between sinking and discharging points with the greatest precision. Knowing the location of the principal solution channels and monitoring the groundwater regime and water quality in a borderland can be particularly complicated. The groundwater regime in the dry period of the year (when there is much free surface flow) may be quite different from that of the wet period when, after periods of heavy precipitation, karst aquifers are fully saturated and water within the channels is under hydraulic pressure. The groundwater

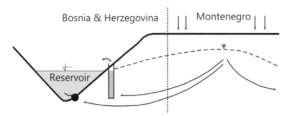

**Figure 12.1.** Illustrating trans-boundary concentrated underground flows crossing a state border.

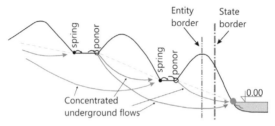

**Figure 12.2.** Trans-boundary underground flows crossing internal and international (state) borders.

regime is very dynamic and frequently changes its behavior in space and time. Within some parts of an aquifer, bifurcation of the flow during floods can further complicate the hydrogeological behavior.

## 12.2  TRANS-BOUNDARY PROBLEMS COMMON IN KARST

Internationally significant trans-boundary aquifers exist in number of countries that have extensive karst areas: between Slovenia and Italy; Slovenia and Croatia; Croatia and Bosnia-Herzegovina; Bosnia-Herzegovina and Montenegro; Montenegro and Albania; FYUR Macedonia and Greece; FYUR Macedonia and Albania; Serbia and Bulgaria; Iran and Iraq; China and Myanmar; China and Vietnam; Switzerland and France, and many other regions with soluble rocks. Some of trans-boundary aquifers cover huge areas. The Chinese sector of the aquifer between China and Myanmar (the Nu River Valley) extends over an area of 35,500 km$^2$; the Upper Zuojiang Valley Aquifer (China, Guangxi/Vietnam) is 32,000 km$^2$ in area and the Chinese portion of the Beilung River karst aquifer (also China/Vietnam) is 30,000 km$^2$ (Zaisheng, 2006).

Trans-boundary aquifers also exist inside a country in areas divided by politically lower level boundaries (between federal administrative areas, provinces, cantons, municipalities, etc.). In China alone 15 large hydrogeological units cross the boundaries between several provinces. For a pilot study Zaisheng (2006) selected two karst aquifers of the eastern Erdos Basin that cross boundaries between the provinces of Shaanxi and Inner Mongolia and Chezhoushan karst aquifer that crosses the Tianjin/ Hebei provincial border. Trans-boundary problems at these lower levels are frequent and extremely complicated in the Dinaric karst.

With increasing demands for water resources these aquifers can lead to conflicts between the political agencies on opposite sides of the border. The Internationally Shared Aquifer Resources Management (ISRAM), Framework Document (UNESCO,

2001) distinguishes the following aspects of trans-boundary aquifers: hydrogeological, legal, socio-economical, institutional and environmental. In many of these cases the trans-boundary problem is coupled with, and intensified by, construction of dams and reservoirs that will change the natural hydrological regime. The major aims of these changes are to minimize negative and to maximize positive socio-economic and environmental impacts.

All the complexity of trans-boundary problem(s) can be demonstrated in the history of the Dinaric karst, particularly after the disintegration of Yugoslavia created a number of new boundaries, some of them contested. A number of natural lakes and man-made reservoirs cross the political borders or have direct underground connections with aquifers on the other side; e.g., Skadar Lake (Montenegro/Albania); Piva Reservoir (Montenegro/Herzegovina); Bileća Reservoir (Herzegovina/Montenergo); Špilje Reservoir (FYUR Macedonia/Albania); Lake Ohrid (FYUR Macedonia/Albania); Lake Prespa (FYUR Macedonia/Greece/Albania); Lake Dojran (FYUR Macedonia/Greece); Buško Blato Reservoir (Bosnia and Herzegovina/Croatia); Peruća Reservoir (Croatia/Bosnia and Herzegovina).

The Dinaric karst area was selected for the case study here due to the extremely complex hydrogeological characteristics of the karst aquifers that are crossed by boundaries and the availability of a large amount of good quality data collected during more than 50 years of investigations and monitoring. Because they are divided by several state boundaries the Dinaric karst aquifers are a potential source of conflicts. A regional project known as the Dinaric Karst Trans-boundary Aquifer System (DIKTAS) "is the first ever attempt to globally introduce integrated management principles in a trans-boundary karst freshwater aquifer system of such magnitude"…. "the DIKTAS project provides a technical-political interface that will ensure wide participation and joint commitment to sustainable and equitable use and protection of transboundary groundwater of the Dinaric Karst" (Kukurić, 2015). The Dinaric Karst region extends from north-eastern Italy through Slovenia, Croatia, Bosnia and Herzegovina, Serbia, Montenegro, Albania, FYUR Macedonia and into Greece.

## 12.3 THE QUESTIONS THAT ARISE

One of the best examples of a project under development in karst is the large Trebišnjica Hydrosystem in eastern Herzegovina. It participates in the DIKTAS project. The entire region was part of one state – Yugoslavia. As a consequence of rational adaptations of the natural water regime for optimal use, socio-economic development has improved significantly in the region. From the very beginning the potential impacts of future reservoirs on downstream springs were carefully analyzed. Some of key topics were:

- The influence of reservoir operation on downstream spring discharge and water quality (reservoir and spring are located in different political entities).
- Starvation of aquifers due to water is stored at surface, in reservoirs, instead to follow the natural underground routes.
- Impacts of water re-routing as consequences of reservoir construction.

Before construction started a total of 120 springs that could potentially be affected by Hydro-system construction were cataloged, and 26 of them were equipped with discharge gauging stations. A long-term investigation has been undertaken to study the effects of hydrological and hydrogeological changes on living conditions of a commercial oyster and mollusc farm on the Adriatic Sea coast.

A number of the impacts were readily predictable and measures to mitigate them were suggested during the design stages. However, during the construction and later during operation of the first structures (dams and reservoirs), some unpredicted negative impacts were detected. Environmental aspects of these impacts have become one of the top priorities during the System operation and the building of further new structures.

Before new state boundaries were established, the focused infiltration points (sinks) and springs were located entirely within one country. Boundaries between republics had no role or posed no serious obstacle. The question of defining exclusive rights to underground water was not very important. Water needs for power production and water supply were analyzed as part of integral projects based on their hydrogeological and hydrological properties. Only the natural boundaries that defined the physical limits of watersheds were recognized. Projects were focused on designing and building multipurpose utilities to obtain the greatest potential water development for the entire region.

Immediately after the new international boundaries were established (boundaries between republics became state boundaries, Fig. 12.3) a number of very important questions appeared that needed to be solved. Newly established regulations for each country were not harmonized. Criteria for determining environmental protection, as well as standard regulatory procedures that are applicable for non-karst regions, are generally not suitable for karst terrains. For example, the exceptional velocities of groundwater flow in karst, and their range, are a basic problem. In some cases (dry periods) the velocity of a labeled tracer is 1.13 cm/s, but for the same route when the aquifer is saturated (wet period) velocity increases to 7.53 cm/s or more.

One of the largest set of springs in the entire region are the Trebišnjica Springs, which discharge between 2 and > 300 m³/s. The majority of its catchment area is presently situated on one side of the border (in Bosnia and Herzegovina) but a part of it is on the other side (in Montenegro). Due to construction of the 123 m high dam and reservoir in Herzegovina, these springs were subjected to a hydraulic head of 75 m. The newly established state border locally crosses a small part of the reservoir. The average discharge of the spring has been estimated at 80 m³/s. This part of the aquifer can be defined as a trans-boundary aquifer. The problem of trans-boundary flows did not exist when it was contained within one state. Immediately after the new state borders were established, however, questions of exclusive rights to the spring waters were raised. Groundwater governed by rights in one state now crosses the border and discharges at springs in the other state. Request for rights to a part of this discharging water for power production because part of its catchment is located on the other side of the border is the subject of very tough negotiations.

At the western border of the region, along the Neretva River valley, the largest springs are Buna (2.95–380 m³/s), Bunica (0.75–207 m³/s) and Bregava (0.45–58.7 m³/s). Buna and Bunica springs and their entire catchment area are located in just one country (Bosnia and Herzegovina); however, the two springs are in one political sub-unit (entity) but almost their entire catchment area (approximately 90%) is located in another sub-unit. By constructing a large hydropower system, a part of the spring waters would

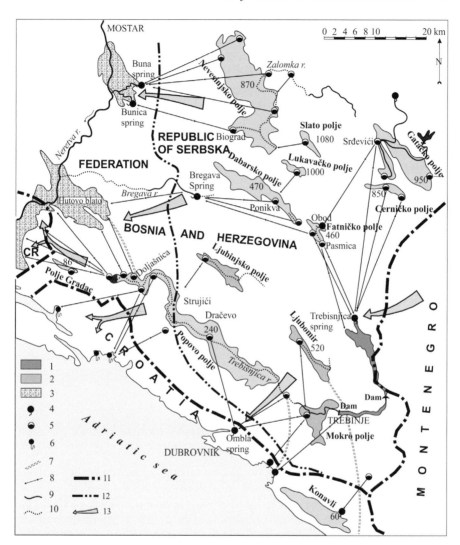

**Figure 12.3.** SE Dinarides. Boundaries between Bosnia and Herzegovina, Croatia and Montenegro. (1) Reservoir; (2) Karst Polje; (3) Alluvial deposits; (4) Large spring; (5) Large ponor; (6) Submarine spring; (7) Fault zone; (8) Underground connections; (9) River flow; (10) Temporary river; (11) State boundary; (12) Entity boundary; (13) Mine directions of underground flows.

be re-routed into an already operational part of the hydro-system. Bunica Spring (el. 36 m above sea level) is directly connected to Biograd ponor at 800 m asl and 20 km distant. The groundwater flow crosses the internal political border at depths of 600–700 m. The entire Zalomka River sinks into the Biograd ponor (about 110 m³/s at maximum). During summer time this river does not exist, its bed becoming completely dry. However, the minimum (base) flow of Bunica Spring also depends on it's local aquifer; it is not entirely dependent on the water sinking in Biograd ponor.

The consequences of water re-routing on the regime of the Buna Spring is negligible in periods of minimum as well as in maximum discharge. The discharge of Bunica Spring, however, is greatly disturbed but only during periods of high discharge. For discharges less than 4 m³/s, there is no impact by any man-made structure within the catchment area. This is confirmed by 40 years of hydrogeological/hydrological investigations and monitoring, and by some different mathematical models.

A part of the project in the area named 'Upper Horizons' is focused on improving water supply, food production and environmental conditions in the Nevesinjsko (800–900 m a.s.l.) and Dabarsko (400 m a.s.l.) poljes and to improve water supplies along much of the Bregava River and in Hutovo Blato Nature Park (Fig. 12.4).

In the solution that has been designed, geotechnical measures are proposed to reduce the permability of the porous and karstified Bregava River bed, with the consequence that in dry periods the flow through the town of Stolac would increase by three to seven times of that in its natural conditions. A few kilometers downstream from this urban area all water sinks in the river bed and discharges in the Hutovo Blato Nature Park (Fig. 12.5).

There is no doubt that construction of dams and reservoirs at higher levels upstream of Stolac are designed in a manner that will increase the river flow through it in the minimal flow season and will have a positive influence on critical minimal inflow to the Nature Park. Graphs presented in Fig. 12.6 confirm the positive changes in the Nature Park as a consequence of dams in the upper parts of the catchment.

This example shows that in most instances construction of complex hydropower systems in karst regions can have positive impacts at the regional scale. The critical question in this particular example is how to harmonize the optimal criteria and how to monitor their efficiency on the other side of the border. Trust between both parties, a professional approach and agreed means of verification are the only ways to provide effective protection of trans-boundary aquifers.

## 12.4  GROUNDWATER PROTECTION ZONING IN KARST – A CONCEPTUAL MODEL

Protection zones for public water supply in karst cannot be based on the same criteria as in non-karstic aquifers. The important difference between karst aquifers and those in intergranular or weakly permeable fractured rocks is the much shorter contact of the contaminant with the rock matrix in karst because of the rapid flow through the conduits. In highly developed karst groundwater velocities are mostly > 2 cm/s and can be 20 to 50 cm/s, which leads to very short residence times. In the case of heavy rains that fully saturate an aquifer, karst water can travel distances as great as 80 km in only 6–8 days. Discharges of springs then may be 10 to 40 times higher than in dry periods of the year. Dilution of any pollutants is very high, i.e., their concentrations are low. The opposite conditions occur in dry periods when there is likely to be slow, free-surface base flow only. Groundwater velocities and the dilution of pollutants are low. For the same distance (about 80 km) the water needs 30 to 50 days to pass through the system. For many pollutants this can be enough time for them to decompose—but not all, particularly not for many kinds of manufactured chemical pollutants.

In general, in comparison with non-karst rock, the self-purification capability of karstified aquifers is very limited. If the distance between sinking and discharge points is

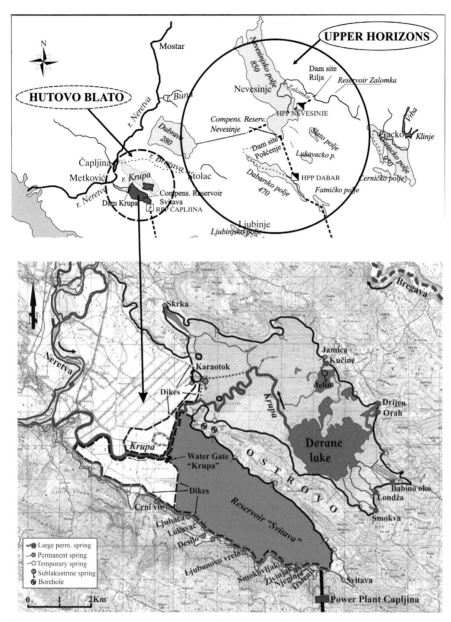

**Figure 12.4.** Eastern Herzegovina. Location of high karst poljes (Upper figure), with dams under construction, and Hutovo Blato Nature Park (Lower figure), elevation ~ 3 m a.s.l.

short (10 km or less) the consequences of pollutants in the groundwater can be hazardous, even tragic.

However, results of some field experiments have challenged this generalization. The goal of a large experiment organized in 1974 in the Pivka underground river (Slovenia)

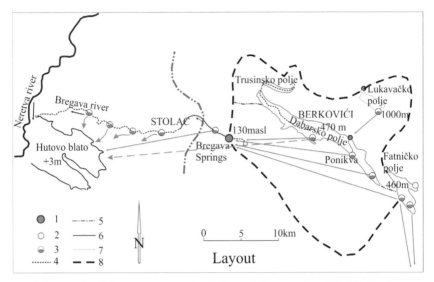

1. Large spring; 2. Large temporary spring; 3. Ponor; 4. Tempoary flow; 5. Political border;
6. Permanent flow; 7. Underground connection; 8. Watershed.

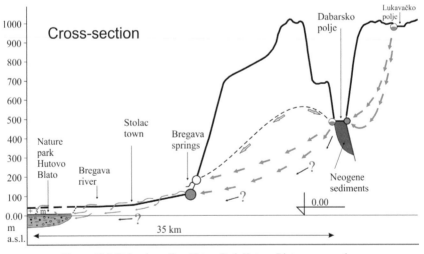

**Figure 12.5.** Dabarsko polje – Nature Park Hutovo Blato, cross-section.

was to determine how rapid and effective were natural processes of self-purification along
its 6.5 km underground flow path. The source of contamination was a sewage system.
Samples were taken at four locations every two hours for 10 days (Preka and Preka, 1976).
To pass a section of three km of contaminated water (Q ranging between 0.69–3.0 m³/s)
required 96 hours. Comparing results at upstream and downstream sampling points the
computed degree of self-purification was 62%. A few other experiments in the Dinaric
karst have confirmed the results of the Pivka experiment.

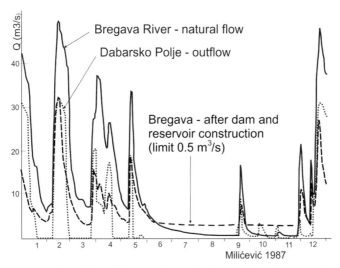

**Figure 12.6.** Bregava River hydrograph, in natural conditions (solid line), and as expected after the water regime is changed by construction of dams in upper parts of the catchment (Milićević, 1987, adapted).

But in spite of these particular field results, underground karst water flow must be presumed to be too fast for efficient self-purification, as the following data suggest. In the period 1925–1973, there were 75 epidemics recorded in Dinaric karst region due to potable water at karst springs or in karst wells being contaminated with pathogenic bacteria. About 6,600 individual cases of sickness were registered; however, it was estimated that the true number was more than 30,000 persons with serious health problems (Pokrajčić, 1976). Epidemics usually occurred 13 to 20 days after rains. In many of these cases the water purification equipment (particularly chlorinators) did not work properly.

One recent well-known example is the Walkerton disaster, Canada (Fig. 12.7). Groundwater contaminated by pathogenic bacteria from a cattle farm flowed for a few hundred meters through solutionally enlarged fractures in dolomite to municipal Well No. 6. This is one of the small cluster of water supply wells. In May 2000, some 2300 people become ill and seven died due to this contamination. It is significant that the other water supply wells very close by were not contaminated. "The low effective porosity of about 0.05% indicates susceptibility to rapid transport of bacteria through the aquifer, and this was undoubtedly an important factor in the tragic contamination of the water supply at Walkerton" (Worthington et al., 2012).

To minimize the public health risks the criteria for determining the zones of sanitary protection in karst should be based on the specific hydrogeological particularities of each karst aquifer. During the past few decades a few different approaches have been proposed: Milanović, 1990, 1995; Biondić et al., 1998; Zwahlen (ed.), 2004; Ravbar and Goldscheider, 2007; Worthington, 2011; Živanović, 2015.

In 1989 some very detailed investigations of an area of 3100 km$^2$ were undertaken to define zones of sanitary protection for springs along 100 km of the Dubrovnik (Croatia) littoral (Milanović, 1990). From a total of 72 springs 33 are tapped of which 7 provide municipal water supplies. There are some extremely rapid flows; e.g., only 16 hours to travel 20 km underground. Based on detailed geological mapping, more

**Figure 12.7.** Walkerton disaster. Dr. Stephen Worthington explaining the hydrogeological reasons for the disaster.

than 100 tracer tests, monitoring of all tapped springs and listing of all important ponors, the criteria for determining the zonation for sanitary groundwater protection were proposed.

In general the zoning concepts were based on the following parameters: hydrogeological maps, groundwater velocity, groundwater residence time and distance between possible sources of pollution and water intake structures. However, an important question arises—Is it possible to apply a general rule for zoning (i.e., same rules for all hydrogeological formations) or do karst aquifers need distinct rules designed for the specific hydrogeological and hydrological properties of karst rock? It is certain that in karst a standard groundwater velocity cannot be applied automatically. Optimal zoning in karst depends on many other factors as well: the saturation state of the aquifer, the time necessary to fully replace a given volume of water, the locations of ponors and their swallowing capacities, protective measures in sinking zones, their discharge capacity, and similar other concerns. In many cases the zone of immediate protection, requiring very strong protection and restrictions, will cross state borders and encompass large areas on both sides of them. Because of this, the criteria proposed should be applied with flexibility and should allow for changes to be made when local hydrogeological conditions require it.

Four zones of sanitary protection are proposed:

Zone 1. The intake area (sinkpoint). Area to be protected by a fence at least 50 m from the intake structure in the direction of the inflow. A detailed hydrogeological map (1:1000) of the surrounding area is used to define the exact location of the fence.

Zone 2. A local (near) protection area should be defined with a hydrogeological map scaled at 1:5000. This zone includes those parts of the catchment area where groundwater flow will need less than 24 hours to reach the tapping structure in the case of a fully saturated aquifer (rainy period of the year). Each ponor in this area should be protected and under strict control. Industry and waste disposal facilities are prohibited in this zone.

Zone 3. An intermediate protection area should be defined based on a hydrogeological map at 1:25000. At times when the aquifer is fully saturated ground water from sink points (ponors) or epikarst water percolating from the surface needs two up to 10 days (depending on aquifer saturation) to flow to the tapping structure. Some ponors in Zone 3 that have more direct connection to the intake structure should be marked as Zone 2 enclaves inside Zone 3.

Zone 4. A general protection area that includes all of the area between the Zone 3 limits and the regional watershed. This zone is optional and should be considered based on the hydrogeological characteristics of the entire catchment area. All water from Zone 4 catchment areas flows towards discharge areas at the main base level but not directly. In dry period water needs 10 or more days to reach final discharge point 4.

In general any facility constructed on karst (industry, farms, highways, dams, gas stations, waste disposal sites, etc.) requires special protection treatment. One of the crucial trans-boundary problems in karst is the likelihood that there can be very rapid transport of polluted water across political boundaries.

The presence of a man-made reservoir on a groundwater route, between drainage points and intake structures, can have a positive role because the dilution of any pollutants in the reservoir water will be high, i.e., the concentration of pollutants is reduced. A reservoir has also a considerable retardation capacity, i.e., it reduces average velocities and retains the water for long periods. In extremely dangerous situations a reservoir can be required to play the role of a 'trap' to capture and hold the polluted water. However, this can be a risky practice regarding water quality in the reservoir over the longer term.

# CHAPTER 13

# A CATALOG OF SELECTED DAMS AND RESERVOIRS IN KARST

## 13.1 A LIST OF DAMS KNOWN TO BE CONSTRUCTED IN KARSTIFIED ROCKS

A number of case histories show the problems encountered at old dams in karstified rocks because of inadequate investigation and weak anti-seepage measures or due to the inadequate anti-seepage technology at the time. These dams were constructed mostly at end of the 19th and the first part of the 20th century. Because of that, many dams and reservoirs have suffered unacceptable leakage for long periods in spite of frequent remedial measures: **McMillan Dam** (USA, 1893); **Miller Dam** (USA, 1900); **Hales Bar Dam** (USA, 1905–1913); **Shenandoah Dam** (Virginia); **Montejaque Dam** (1924); **Volhovskaya** (1926, the first dam constructed in karstified rock in Russia, 10 m high); **Wolf Creek Dam** (USA constructed in 1938–1952); **Center Hull Dam** (USA, constructed 1938–1948); **Freeman Dam** (Kentucky, 1960s); **Grand Rapids** (Manitoba, Canada, 1960s); **Biver Dam** (USA, 1966); **Clearwater Dam** (USA, 1948); Stewartville Dam (Canada, 1948); **Walter F. George Dam** (USA, 1963); **Mississinewa Dam** (USA, 1967); **Camarassa Reservoir** (Spain, 1927–1931); **Kentucky Dam** (USA); **Salanfe Reservoir** (Switzerland, 1952); **Maria Cristina and Sichar dams** (Spain); **Ira Price Dam** (USA); **Jinlong Reservoir** (China); **Anchor Dam** (USA); **Civitella Liciana** (Italy); **Cheurfa Dam, Ouizert Dam, Tichy Haf Dam** (Algeria); **Fremont Reservoir** (USA); **North Dike** (USA, Florida); and **Cuber Dam** (Spain).

The list of dams and reservoirs with very high leakage is presented at Chapter 10.

Well-known case studies related to dams and reservoirs situated in the karstified rock include: **Adam Beck Upper Reservoir** (Canada), Niagara Escarpment, built on top of 10–40 m of glacial deposits on dolomite and limestone; **Apa** (Turkey); **Armagan Dam** (Turkey); **Ataturk Dam** (179 m high fill dam (Turkey); **Atbashinskaya** (Russia), rock-filled, 80 m high; **Beni Haroun Dam** (Algeria) RCC, 118 m high, for irrigation and water supply; **Biver Creek River Dam** (Alabama); **Boljunčica Reservoir**

(Croatia); **Boqaata,** 68 m high rollcrete dam (Lebanon); **Cautillo Reservoir** (Cuba); **Cevisli** (Turkey); **Center Hill Dam** (U.S.A.), 80 m high concrete-earth embankment founded directly on karstified limestone; **Chabrouh Dam** (Lebanon); **Charmine Dam** (France); **Charvakskaya** (Russia), rock/earth filled, 168 m high; **Cheboksarsky Reservoir** (Russia); **Chichik Dam** (Uzbekistan) 168 high earth-filled structure situated on karstified Lower Carboniferous limestone **Chicoasen Dam** (Mexico), 146.7 m high rock-earth filled structure; **Clarence Cannon Dam** (Missouri); **Claytor Dam** (Virginia, U.S.A.) concrete gravity structure, 41.8 m high and 1,142 m long; **Darwin Dam** (Tasmania), 21 m high gravel fill embankment structure; **Diyawini Dam** (Sri Lanka) 38 m high including two dikes 7 and 8 m high; **Dongmenhe Dam** (China); **Doosti Dam** (Turkmenistan/Iran), 78 m high rock-filled structure; **East Fork Dam** (Kentucky, USA); **El Kansera Dam** (Morocco) 51 + 6 m high; **Foum El-Gherza Dam** (Algeria) 73 m high concrete arch structure; **Gergebilska Dam** (Russia, Dagestan); **Green River Dam** (Kentucky, USA) 43 m high, rock-fill dam; **Guangting** (Hebei, China) 45 m high dam; **Hoa Binh Dam** (Vietnam), earth-filled, 128 m high; **Ilarion Reservoir – Elati (**Greece) 130 m high earth-filled dam with specific seepage problem in the reservoir, 10 km upstream – Elati Basin; **Jinlong Reservoir (**Guangxi, China); **J. Percy Priest** (USA) earth/concrete dam, 35.4 m situated high on thin bedded limestone, sinkholes; **Kahovskaya** (Russia), earth-filled, 16 m high; **Kasseb Dam** (Tunisia), arch dam, 70 m high); **King Talal** (Tunisia), 100 m high, karstified dolomites and caverns in sandstones (Jordan); **Kopili Dam** (India); **Kozjak Dam** (FYUR Macedonia), concrete arch dam, 110 m high, situated in karstified marbles, dolomitic marbles and schists; **Krupac Reservoir** (Montenegro); **La Bolera Reservoir** (Spain); **Lazići Dam,** 125 m high, earth filled dam (Serbia); **Logan Martin Dam,** 30 m high (Alabama, USA); **Lone Pine Reservoir**, 31 m high dam (USA); **Madden Dam** (Panama); **May** (Turkey); **Maifedoun**, 4 km long and 250 m deep curtain (Lebanon); **Megenin** (Libya); **Messochora** (Greece, Achileos River) 150 m high concrete-face rock-filled dam; **Miatlinskaya** (Russia), rock/earth filled, 90 m high; **Mississinewa Dam** (Indiana, cut-off wall, 2004); **Mornos Reservoir** (Greece), 135 high rock-filled dam; **Mujib Dam** (Jordan), 67 m high, composite RCC/CCR; **Niukouyu Reservoir (**Beijing, China); **Nolin River Dam** (Kentucky, USA), 40 m high, rock-fill; **Onac Reservoirs** (Turkey); **Okukubi Dam** (Japan); 40.9 m high gravity dam in Okinawa, **Ourkiss Dam** (Algeria), 35 m high structure; **Paderborn Dam,** 8 m earth-filled dam (Germany); **Panix Dam** (Switzerland); **Patoka Dam** (Indiana, USA), 46 m high, embankment dam; **Pavlovskaya** (Russia), earth-filled, 56 m high; **Patagh Dam** (Iran); **Patoka Dam** (Indiana, sinkholes in reservoir); **Pengshui** (China), gravity dam on the Wujiang River, 120 m in height; **Perdikas Resevoir,** 40 m high earth-filled dam (Greece); **Plyiavinskaya,** earth-filled, 40 m high (Russia); **Pueblo Viejo Dam,** (Guatemala) 130 m high dam; **Qaraoun** and (Lebanon); **Quinson** (France) 45 m high arch dam; **Ričica Reservoir** (Croatia); **Rizhkaya** (Russia), earth-filled, 18 m high; **Rough River Dam** (Kentucky); **Salacola Dam** (Georgia); **Scrivener Dam** (Canberra, Australia), 33 m high; **Shaorskaya** (Russia), earth-filled, 13 m high; **Shibazi Reservoir** (Hunan, China); **Tang Ab Dam** (Iran) embankment, 51.5 m high; **Volzskaya** (Russia), concrete gravity, 40 m high; **Vudu Reservoir** (Sichuan, China); **Wadi Gattara** (Libya) earth-filled, 52.3 m high, Miocene karstified limestone; **Wala Dam (**Jordan);

**Yumaguzinsk Reservoir** (Russia), karstified limestone and dolomite, problems with karstification and ecology; **Zubcovskaya** (Russia), earth-filled, 15 m high.

Structural failure of a dam is rare in karst, however. One of the very few examples of a dam that failed completely was that on the **East Fork of Pond River** – Kentucky. The 15 m high earthen embankment structure was completed in 1978. During its first filling, a rainstorm of 35 mm quickly raised the water up to the spillway level. Failure occurred as a consequence of intensive piping along karst solution features.

The **Quail Creek Dike** failure (Utah, 1989) is another failure, as explained in Chapter 7.

## 13.2  LIST OF DAMS WITH PROBLEMS DUE TO LARGE CAVERNS OR HYPOGENE KARSTIFICATION

Large caverns have been detected below the dam foundations in many cases: **Keban Dam** (Turkey), huge caverns below the dam body and in the left abutment foundation; **Ourkiss Dam** (Algeria), a number of large and empty, interconnected caverns at a depth of 45 m to 52 m below the foundation surface; **Grabovica Dam** (Bosnia and Herzegovina), a long and wide cavern in the dam foundations, filled with clayey-sandy material, locally about 40 m deep; **Seymareh Dam** (Iran), caverns on both sides of the dam close to the foundations; **Diyawini Dam** (Sri Lanka), large cavities discovered by excavation and boreholes in the left bank; **Karun IV** (Iran), a cavern beneath the left dam foundation; **Salman Farsi Dam** (Iran), a large cavern system in the right bank; **Sklope Dam** (Croatia), huge caverns in both banks close to the surface; a large cavern below the **Lar Dam** foundation (Iran); **Wudongde** dam site (China), a large cavern filled with clay, sand and blocks in the right bank; and further examples at a number of other dam sites and along grout curtains.

At some dam sites and reservoirs the leakage is a consequence of hydrothermal water (hypogene karst features) or the combined influence of hydrothermal and meteoric waters: **Salman Farsi dam site** (Iran), **Francisco Morozan-El Cajon dam site** (Honduras); **Višegrad dam site** (Bosnia and Herzegovina); **Hamam Grouz** (Algeria); **Salanfe** (Switzerland), **Samanalawewa** (Sri Lanka), **Berke** (Turkey), **Chicik Dam** (Uzbekistan) and probably **Wudongde** (Yunnan, China).

## 13.3  DAMS WITH SUCCESSFUL WATERPROOFING DURING CONSTRUCTION

If dam sites and reservoir areas are selected on the basis of very detailed geological and hydrogeological analysis the risk is low but, however, can never be absolutely eliminated. Of course, proper adaptations and modifications during the dam construction and execution of the waterproofing structures are crucial. In many examples of leakage during the first filling and later during reservoir operation have been acceptably low or negligible. Some examples are: **Castillon, Genissiat, Greoux, Sainte Croix, Ceirac, La Rouviere, Quinson and Conqueyrc** (France); **Ekbatan** (Iran); **Panix** and **Punta Del Gall** (Switzerland); **Rama** and **Grabovica** (Bosnia and Herzegovina); Štikada (Croatia); **Quad Fodda** (Algeria); **Altapina** (Turkey); **Nebaana** (Tunisia); **Hsinankiang** and **Chinwuo** (China); **Santa Guistina** (Italy);

**Bin al Ouidance** (Morroco); **Beniner (**Spain); **La Angostura** (Mexico); **Pueblo Viejo (**Guatemala); **Rama Dam** (Bosnia and Herzegovina); **Doosti**, border between Iran and Turkmenistan; **Storglomvatn (**Norway); and **Zhai Xiang Kou** (Gueizhou Plateau, China).

The most prominent examples of successful watertight sealing work during construction include:

**Berke Dam** (Turkey, Ceyhan River), a concrete arch structure, 201 m in height, situated in tectonized and locally karstified Mesozoic formations. There were three springs with discharges of different temperatures at the dam site: 13–16°C, 19–28°C and 33°C. The grout curtain was made from galleries at five different levels between 346 m and 150 m asl. The bottom of the lower curtain limit is 50 m below sea level. Karst channels (aperture 20 to 200 cm) are developed along bedding planes. In heavily karstified zones, a triple-row grouting array with a cross pattern of holes was used. The deepest holes were 225 m. The grouting hole pattern was: 12–6–3 and 1.5 m. Spacing between quinary holes was 0.75 m. The grouting pressure ranged from 30 (upper sections) to 60 bars (lower sections). The average grout mix consumption was: in slightly karstified limestone – 54.6 kg/m; in transitional zones – 88.8 kg/m; and in heavily karstified limestone 518 kg/m. Maximum grout mix consumption varied between 500 and 700 tonne/m. It is one of most complicated grout curtains in the world (Altug and Saticoglu, 2001).

**Boqaata Dam,** is a 71 m high RCC structure in Lebanon. Because it is a water supply project and the reservoir volume is very small (7 x $10^6$ m$^3$, 12 mill. dynamic volume) the maximum acceptable leakage was set at only 35 l/s. The water table is 35 m beneath the surface of the dam foundations.

The bottom of the reservoir consists of two areas of karstified limestone, plus igneous rocks, thick deposits of tuff and an alluvial cover of variable thickness. The tuff varies between impermeable and very permeable over short distances. A number of karst shafts and sub-vertical solutional discontinuities were observed inside the reservoir as well as in the near vicinity (Fig. 13.1).

**Figure 13.1.** Boqaata Dam site area, Lebanon. Karstification developed along sub-vertical discontinuities.

Two different waterproofing solutions were considered: a grout curtain and surface treatment. The dam foundation was in intensively karstified limestone. Twenty-four caverns of different sizes were found during excavation of the foundations (Fig. 13.2). All caverns were filled with concrete and the dam was successfully constructed.

**Figure 13.2.** Boqaata Dam site, Lebanon. Arrows show largest caverns discovered in the foundations. Courtesy Gicome Co, Beirut.

**Doosti Dam** is constructed on the Harirood River at the border of Turkmenistan and Iran. The height of the earth-filled dam is 78 m. Average annual flow is 15 m³/s and reservoir storage capacity is 1,250 million cubic meters. The dam site area consists of steeply dipping rocks including sandstone, limestone, sandy limestone, marlstone and chalky limestone. Karstification occurs in the sandstone due to a soluble carbonate matrix as well as in the limestone, with cavities well developed along the joints and bedding planes.

A two row grout curtain was constructed beneath the dam body (mostly karstified formations) and a one row curtain in the right bank (mostly impervious formations). Row distance was 1.5 m and final spacing between grouting holes is 2.5 m. Grout curtain depth varies from 15 to 80 m. The average grout mix consumption in karstified sections was 200 kg/m', and 50 kg/m' in marlstone and chalky limestone. The natural permeability of 16–25 Lugeons respectively was reduced to less than 5 after grouting. The total leakage from the reservoir at its highest water level is about 100 l/s. Tracer tests have shown that the principal seepage filtration is through the limestone and calcareous sandstone formations (Mozafari and Raeisi, 2012).

**Geheyan Dam** on the Qingjiang River (China) consists of a 151 m high concrete gravity arch dam and reservoir with a volume of 3.12 billion m³ (Ruichun and Fuzhang, 2004). The dam site is located on extremely karstified Cambrian limestone, 180 m in thickness, that is overlain and underlain by shale formations.

Geological investigations over more than 30 years included: geological mapping of 905 km²; 380 boreholes (48,675 m); four inspection boreholes with a diameter of one m and depth of 172 m; water pressure tests over 4445 sections; water pumping or infiltration tests – 42 sections; and 418 m of excavations to enlarge caverns for tracing underground streams. Geophysical investigations included cross-hole radio-wave transmission to determine depths of karstification; cross-hole seismic exploration, acoustic logging and down-hole temperature measurements.

More than 600 caverns were detected, including a few conduit systems with a total volume of about 60,000 m$^3$. Seventy-nine caverns, with an aggregate volume of 22,000 m$^3$, were discovered along the grout curtain route. Before grouting, all caverns were plugged. The grout curtain on both sides and beneath the dam was connected to the impervious shale formations. The 1400 m long curtain was installed from four grouting galleries on each side. The total surface area of the curtain is 180,000 m$^2$. Average cement consumption was 83 kg/m'.

Downstream from the dam, the deeply incised Qingjiang River makes a sharp bend. A key question was whether it would be possible for water to open karst channels directly across the neck of the bend due to the great increase in the hydraulic gradient there when the reservoir was filled. However, due to the limestone between the reservoir and the river downstream having little karstification, a significant hydraulic connection in this direction was considered very unlikely (see Fig. 3.12).

A serious problem of groundwater intrusion during excavation of the diversion tunnel (1.67 m$^3$/s) was successfully dealt with. During the operation of the dam no leakage has been detected to date.

**Grančarevo Dam** (Herzegovina) is a double curvature concrete dam, 123 m high, that creates the Bileća Reservoir, V = 1.27 x 10$^9$ m$^3$. The dam and entire reservoir are in karstified Mesozoic limestone and dolomite. Ten ponors were mapped in the reservoir bottom during the investigation phase. The water tightness of the dam site and reservoir is based on the positive (impermeable) hydrogeological role of a large anticlinal structure with a core of grusified dolomite that separates the reservoir from lower terrain beside it. Two rows of grout curtain are spaced at 3.5 m. The maximum curtain depth was 150 m. A clay-based grout mix was used: 1,509 tons of cement, 2452 tons of clay, 46 tons of sodium, 3 tons of sand, and 88 tons of bentonite. Average acceptance of dry material was four times greater in the upstream row than in the downstream row (200 versus 57 kg/m). Flushing of joints and micro-karst features was done for consolidation of the rock in the foundations: cells in three boreholes were used simultaneously-one for air injection, one for water injection, and the third to discharge the flushed water and debris. Flushing was extended to a depth of 15 m. Leakage at the dam site (full reservoir) is less than 150 l/s. One challenging problem was the instability of a downstream section of the left bank where deep excavations for the foundations of the power house endangered the stability of part of the bedrock slope (Fig. 13.3). This part of the rock was stabilized by installing 200 pre-stressed anchors that were 40–60 m length (Stojić, 1966).

**Haditha Dam** (Iraq) is an earth-filled dam with an asphaltic-concrete core, located on the Euphrates River and constructed in 1982. It is 9,064 m in length and 57 m high. The dam site is on Oligocene-Miocene carbonate rocks (limestone and dolomite with intercalations of breccia, conglomerate, shale and clay). Karstic features are well developed at the surface and underground, particularly in the right bank (Fig. 13.4).

More than 120 sinkholes (about 100 of them in the right bank) were found in a broad area around the Haditha site. Sizes of surface openings were up to 150 x 70 m with depths of 3 to 40 m (Kondratiev, 1979). A number of caverns and horizontal

**Figure 13.3.** Grančarevo Dam. Herzegovina.

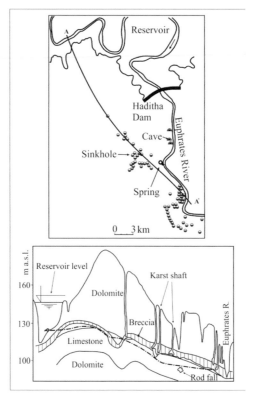

**Figure 13.4.** Haditha Dam, Iraq. Simplified geological layout (upper) and cross-section A–A', (Kondratiev, 1979), from Lykoshin et al., 1992.

solution channels were discovered, mostly a few meters above the water table. The largest caves in the dam site area were 60 m wide, 5m high and 50 m long. Investigation boreholes crossed 50 cavities developed along joints, 0.1–1.7 m in height. The principal breccia layer had a clay-shale matrix and was accepted as a hydrogeologically sufficient structure and the basis for emplacing a watertight structure in the right bank.

**Karacaoren II Dam** (Turkey) is a gravity arch dam, 49 m in height from the foundations. To prevent seepage through karstified Mesozoic limestone, three different protective measures were applied at the surface—an approximately 1 m thick concrete lining close to the dam; 7 cm thick layers of reinforced shotcrete over the limestone, 40 x 650 m; and a clay blanket on the slope). The sealing treatment is successful and justified the investment in protection (Okay and Soidam-Bas, 1999).

**Hoa Binh Dam** (Vietnam) is 128 m high earth-filled structure with a crest length of 734 m. It is one of eight potential dam sites considered along 40 km of the Da River valley. One of the crucial questions at the site was the possibility of leakage towards the adjacent valley of the Ma River. Detailed investigations of the carbonate rock massif were undertaken, including exploratory boreholes down to depths of 550 m. During a first stage, a karst cavern of 134 m$^3$ was filled with concrete to allow the provisional operation of the power plant. During the second stage 17.600 m$^3$ of concrete was used to construct a cut-off wall 10.500 m in length. The grout curtain had to be reinforced by a cut-off, 26 m long, 33 m deep and 6 m wide, in just one intensively tectonized section (Skiba et al., 1992).

**Khao Laem Dam** (Thailand) is a 90 m high rock-filled structure. Karstic features were developed down to 50 m below the river bed. To prevent filtration through the many solution channels (0.2 to 10 m wide) a concrete diaphragm (cut-off) wall was constructed down to a depth of 15–55 m. The diaphragm wall was installed by using overlapping piles (765-mm-diameter piles drilled with 615 mm between their centers, Fig. 6.35). A five level gallery system was used for the cut-off wall. For the grout curtain, 22 km of galleries were excavated in the right bank (Fig. 13.5).

The diaphragm wall is about 15.900 m$^2$ in area. The depth of the conventional grout curtain was between 100 and 150 m. The complete anti-seepage structure is 77,000 m$^2$ in the dam foundations and about 360,000 m$^2$ in the right abutment. Grout mix consumption between depths of 100 and 150 m was 160 kg/m', and below 150 m it was 800 kg/m' (Bergado et al., 1984).

**Normandy Dam** (Tennessee, USA), a 34 m high earth-filled dam, is an example of successful treatment. The dam foundation is on horizontal limestone that includes a weak zone 80 m wide. After detailed investigations, a cut-off wall (overlapping wells) was constructed and reinforced with upstream and downstream grout curtains.

**Oymapinar Dam** (Turkey) is a double curvature arch dam, 185 m high, situated in karstified Permian limestone. The dam was completed in 1983. During the investigation phase there were 37,000 m of core borings, 10 km of exploratory adits, and 400 km of geoelectrical soundings.

The geological and tectonic setup is complex. The dam site is located on Late Paleozoic formations, with karstified Permian limestone in the dam abutments. The

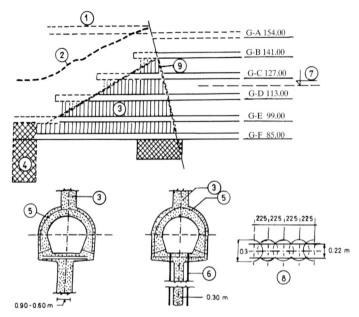

**Figure 13.5.** Khao Laem Dam site: (1) Three Pagodas fault; (2) rockfilled dam; (3) spillway; (4) grout curtain in the valley section; (5) z-axis of the grout curtain and galleries in the right abutment. From Nonveiller E., Grouting, Theory and Practice. Elsevier, Amsterdam, 1989.

grout curtain was tied laterally and under the river bed to an upstream contact between the limestone and impervious schist formations, making a successful bath-tub structure. The total area of the grout curtain is 154.800 m². The total length of grouting boreholes was 286 km and the injected grout mix was 61,000 tons. All important open karstic joints were plugged by concrete. The large Dumanli spring ($Q \approx 20$–100 m³/s), just upstream of the dam, was submerged by 120 m without any influence on the reservoir water tightness (Altug, 1999). Based on the calculations of Günay et al. (1985), groundwater storage in the Oymapinar Reservoir banks range between 35–40% of the volume of the reservoir itself.

**Peruća Dam** (Croatia). This rock-filled dam, 65 m in height from a limestone foundation, was constructed in 1958. Detailed geological investigations and water pressure tests along with 20,000 m of investigation boreholes showed that the permeable karstified limestone in both banks was underlain by impervious rocks. Seepage could occur only through the limestone. The three row grout curtain at the dam site is 200 m deep and 1600 m long. The grout mix (49,000 tons) consisted of 25% cement and 75% of clay, stabilized with some bentonite. The properties of the cement/clay grout mix were verified in a comprehensive laboratory investigation program, in which the strength, erosion resistance, volume stability and durability were tested and documented (Nonveiller, 1989). The injection pressure was up to 40 bars in the upstream and downstream rows, up to 60 bars in the middle row to a depth of 100 m, and then 65 bars to a depth of 200 m. The middle row was grouted after the two external rows were finished. The total surface area of the curtain is 242

122 m². According to Pavlin (1961), after long term monitoring it is estimated that the constant leakage from the reservoir at its full storage level is about 1 m³/s (1.5% of the average river flow).

**Pueblo Viejo Dam** (Guatemala) is a rock filled structure, 130 m in height, located in limestone with both hypogene and epigene karstification operating simultaneously (Fig. 13.6). The principal karst features are created by steeply ascending hot water in several horizons, e.g., between el. 600 and 670 m a.s.l. (Wiesner and Wilhelm, 1985)). It is clear that substantial knowledge of the karst conduits would be necessary for successful waterproofing of the dam foundation. Such knowledge was obtained in good part only during the construction itself.

During the investigation and construction phases thermal water was recorded at springs (37°C) and in exploration adits and galleries (up to 55°C). As a consequence of recent volcanic activity, caverns and conduits were filled with volcanic ash. To seal the dam site a single row grout curtain was constructed 400 m into the right abutment and 900 m in the left abutment. Maximum curtain depth below the valley bottom was 160 m. Drilling and grouting started at a spacing of 12 m and proceeded by systematic reduction of spacing to as little as 0.75 m in some sections. Different grout mixes were applied (V/C between 1.6 and 0.5). Large caverns were filled with mortar with a maximum grain size of 2 mm. Maximum applied grout pressures reached 60 bars. Some partially grouted karstic cavities were treated by jet grouting using pressures up to 300 bars. This particular treatment was halted when grout came back out of neighboring boreholes and the section was then grouted in the standard way. Many of the karst cavities were filled with volcanic ash that was not groutable. According Ewert (1986), in the event of seepage through such ashes, piping and regressive erosion could not be excluded. However, after the reservoir was filled to the maximum level the measured seepage was only 25 l/s.

**Poliphyton Dam** (Greece) is a 105 m high rock-filled structure. The Poliphiton Reservoir is partially in contact with karstified marble and limestone. The large Neraidha Spring (10 m³/s) was submerged by 40 m of water. In spite of intensive karstification, no negative impacts of the spring submergence have been observed.

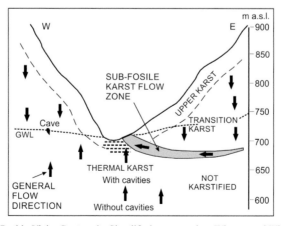

**Figure 13.6.** Pueblo Viejo, Guatemala. Simplified cross-section, Wiesner and Wilhelm (1985).

**Three Gorges Reservoir** (China, volume 39.3 x $10^9$ m³). The dam is in granite in the core of large anticline with limestone on top and on the flanks. The reservoir is partly located in an extremely karstified area: Sanjioba plateau and the area of Wujing and Furongjiang. A number of huge collapses, deep canyons multi-level channels with underground flows and voluminous caverns are registered in this area. One of collapses (Xiaozai Tiankeng) is 650 m deep and 630 m wide. The number of huge underground features is still not completely investigated.

**Wujiangdu Dam** (China, Guizhou). This gravity arch dam in karstified limestone and 165 m in height, was completed in 1980 (Fig. 13.7).

The many karst cavities at the site, particularly in the left bank, have been detected down to depths of about 250 m beneath the river bed. They were filled with soft clay (water content – 56%). To increase the strength of these fillings an unconventional grouting approach was used (Fig. 6.30). The soft clay in the caverns was subject to four successsive processes: hydraulic fracturing, extrusion, consolidation, and chemical hardening. The initial grout mix was cement with a water/cement ratio of 8:1 by weight, and after grouting with this thin mix, it was gradually thickened to 1:0.8–1:0.5 according to local conditions. The grout curtain is 1020 m long and 200 m deep below the river bed. A concrete diaphragm wall, 93 m high, was constructed in the left bank abutment. The total curtain surface is 189,000 m². Grouting pressure ranges between 2 and 60 kg/cm² (Zuomei and Pinshow, 1986).

**Stewartville Dam**, Canada, Ontario is a combined concrete gravity/earth filled dam, 41 m in height, built in 1948 on the Madawaska River, a major south bank tributary of the Ottawa River. The foundations are in horizontal, regularly bedded, platformal limestones of the Simcoe Group (Ordovician) that also underlie the cities of Ottawa and Montreal. The limestones have been heavily glaciated, with resultant post-glacial opening of the shallowest bedding planes and joints during glacial pressure release exfoliation. Scattered shallow karst landforms and small caves are now widespread.

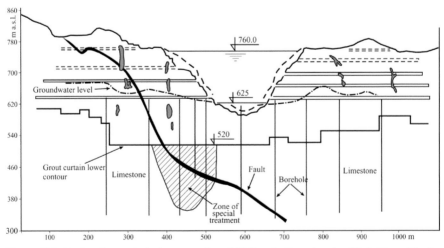

**Figure 13.7.** Engineering geological cross-section of Wujiangdu Dam site, simplified. (From Liu, B., Wujiangdu dam, in *Karst in China,* Lu, Y., Ed., Institute of Hydrogeology and Engineering Geology, Chinese Academy of Geological Sciences, 1986.)

After the reservoir's first filling at Stewartville, there were two significant leakages beneath the dam foundations, one on the north side, the other on the south side, provoking some washout of karst fillings. Conventional grouting with cement-based grout mix was not successful in spite of many years of regrouting efforts. By 1983 the leak rates had increased to 375 and 369 L/s respectively and conventional grout was being washed away by the rapidity of the flow in the conduits. To stop or decrease these high velocities, hot bitumen was injected via one borehole in the south side, followed by a sand-cement slurry through a separate upstream borehole a few minutes later. The reaction was unusually quick (after 4 minutes) and after several hours the southern karst conduit was effectively plugged, its flow reduced to 15 L/s. In 1984 the north channel was sealed almost completely by the same method. It was the first successful use of hot bitumen for karst channel plugging in Canada (ECO Grouting Specialists Inc. 2017). However, "Asphalts are known to have a tendency to creep and shrink. In applications where they are subjected to high pressures, continuing observation is essential" (Jansen, 1988).

## 13.4  CASE STUDIES WITH SUCCESSFUL CORRECTIVE WORKS

The first (experimental) filling of a reservoir always has an important role in the best investigation works because it will reveal the efficiency of the sealing treatment and any deficiencies in it. Because of this the term ***experimental reservoir filling*** should be used instead of the standard term *first reservoir filling.*

In a large number of dams and reservoirs constructed in karst the waterproofing treatment applied during the dam construction was only partially successful or sometimes unsuccessful and declared as a failure. If the leakage was (technically and economically) unacceptably high, then additional extensive waterproofing work had to be carried out. On the basis of the new findings the additional waterproofing work was successful in many cases, i.e., leakage was reduced to an acceptable level, sometimes to a negligible amount. Some examples of successful sealing treatments undertaken after the first experimental filling are presented in this subchapter.

The **Canelles Dam** (Spain) is one of the prominent examples of successful remedial works. The concrete arch structure is 151 m high, with a crest length of 203 m. It is located in the Noguera-Ribagorzana River canyon in karstified Cretaceous limestone (Figs. 13.8 and 13.9). During the first filling there was leakage of approximately 6 m³/s that quickly increased to 8.0 m³/s. Due to the reservoir depth of about 100 m (i.e., very high pressure) the leakage flow rates were very high (over 16 m/s; Weyermann, 1977). Remedial treatment began in spite of these high gradients and rates of flow.

To stop the concentrated flow in karst channel a 300 mm borehole was drilled and a channel was filled with 2130 granite cobbles of ~ 240 mm diameter to form a pre-filter plug. Before final grouting, gravel (25 to 120 mm) was injected over the plug. A new grout curtain below the dam was constructed to a depth of 175 m, the upper 25 m being left free of grouting to protect the dam foundation from possible additional uplift pressure. The remedial grouting included 162.393 meters of boreholes and 51.319 tons of dry components. Average grout mix consumption was 303 kg/m'.

After speleological investigations in the left bank, the karst channel (surface area of 17 m²) was plugged with an 11 m long concrete plug. 1380 tons of dry materials

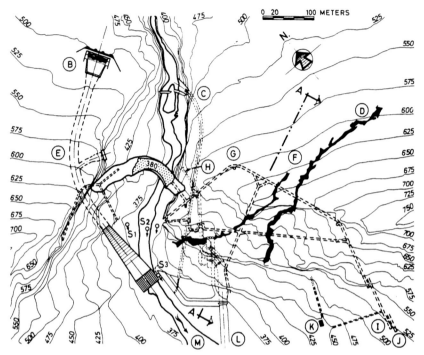

**Figure 13.8.** Layout of the Canelles Dam site (Spain). (A–A) Cross section presented in Figure 13.9. (B) spillway; (C) diversion tunnel; (D) Cueva Negra, a horizontal cave at the dam crest level; (E) inclined spillway tunnel with additional intake for reducing reservoir level control; (F) Cueva de la Tortuga, a deep cave; (G) grouting gallery at an elevation of 380 m; (H) intake; (I) end of grouting gallery at an elevation of 450 m asl; (J) end of gallery at an elevation of 510 m; (K) grouting plan; (L) tailrace: (M) backwater of Santa Ana Reservoir (S1, S2, S3) seepage springs.

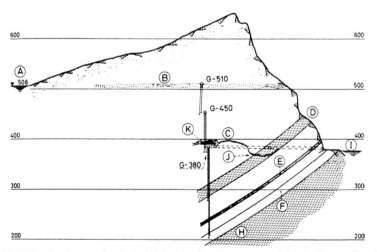

**Figure 13.9.** Canelles Dam section A–A. (A) Projected maximum reservoir level; (B) cave Cava Negra; (C) cave Cueva de la Tortuga; (D) Capa Negra marl beds; (E) lesser marl layer; (F) sandstone with red and yellow veins; (G) grouting galleries; (H) Magres Hondus, deep thick marls; (I) tailrace water level; (J) concrete plug constructed during the first filling of reservoir; (K) plug constructed in accordance with the grout curtain (Weyermann, 1977).

were used to fill other caves. Some of the caverns were sealed through the boreholes only, without direct access into the cavern. By using chemical products (polyurethane foams, acrylic asphalt-resin mixes) the karst porosity and solution channels were sealed down to a depth of 200 m. Finally the leakage was reduced to the negligible quantity of 68 L/s (Weyermann, 1977).

**Dalongdong Reservoir** (Guangxi, China) is constructed in a karst polje in Carboniferous and Permian limestone and dolomite in a region of typical peak cluster topography (Zhuoxin and Guangyan, 1999; Cao et al., 2016). Before dam construction outflow drainage of the polje occurred through the Dalongdong karst system (Fig. 13.10).

 The dam was built in 1957. To prevent leakage solution channels behind it, five sinkholes were plugged with concrete (Fig. 13.11A).

 However, during the reservoir impounding, the karst collapse occurred and 10 new sinkholes were created along the footwall. Seepage through them quickly increased to 15 m³/s. After detailed investigations the leakages were found to be mostly along the three largest faults (F1, F2 and F3). Based on these findings an additional grout curtain was built, 652 m long and 65 m deep (Fig. 13.11B). At present the Dalongdong Reservoir operates without any seepage losses (Cao et al., 2016).

The **Dokan Dam** (Iraq) is a concrete arch dam, 116 m high. The site consists of karstified dolomite overlain by thinly bedded limestone. The single row grout curtain with a total length of 2541 m was extended to a depth of 150–190 m below the design water level. After the reservoir filled to two thirds of its capacity, leakage of 6.0 m³/s occurred, bypassing the end of the curtain through a karstified zone 100 m wide. After grout curtain extension (336 m, to a depth of 150 to 160 m) leakage from the reservoir was eliminated (Perrott et al., 1963). Before grouting the curtain extension, the rapid underground water flow was plugged with a mixture of 50 kg bentonite, 30 L diesel oil, and 0.5 kg cotton flock. For final flow plugging, cement and sand with a calcium chloride mix was applied, using four rigs simultaneously. After the reservoir was filled again no leakage was observed.

The **Douglas Dam** (USA) is a concrete gravity structure, 62 m high, built in 1940–1943. The dam foundation consists of karstified limestone and dolomite to a depth of 18 m. The solution cavities were partially filled with clay and sand. The following procedure was implemented to close the caves in the dam site area:- 22 shot drill holes were drilled from the bedrock surface to the roof of a cave, on 8 m centers. Hoists were set up over the drill holes to remove the muck and excavated material from the cavity, and after the latter had been washed, concrete filling was begun through the shot drill

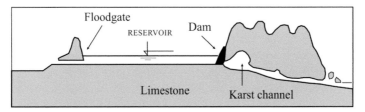

**Figure 13.10.** Dalongdong Reservoir. Simplified cross section.

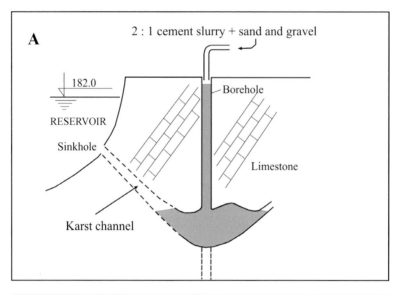

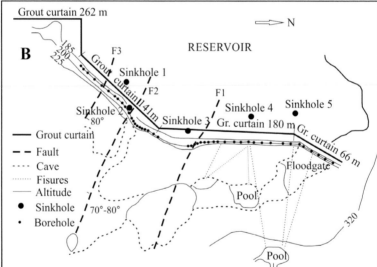

**Figure 13.11.** Dalongdong Reservoir. (A) Design for filling the individual karst channels, (B) Grout curtain route and location of main sinkholes (from Zhuoxin and Guangyan, 1999).

holes. A total of 2479 tons of concrete was required. A large bedding cavity, starting under the end of the southern section of the concrete dam, was discovered and explored by excavation, mining and 106-cm-diameter shot drill holes spaced on 15 m centers and drilled from the ground surface to the floor of the cavity. The height of the cavity ranged from 1 to 3 m, except for one 17-m-long section with a height of 8 m. The cleaning of the cavity and placing of a 4965-ton concrete plug that extended from the floor to the roof of the cavity, was accomplished through a large-diameter shot drill hole. After the concrete hardened, the void between the roof of the cave and top of the concrete plug was grouted through nipples previously placed at approximately 7-m centers along the roof of the cavity. Following completion of the concrete plug, extensive cement grouting was done through holes drilled from the ground surface, thus making the concrete plug a part of the grout curtain extending from the dam into the rim area.

**Walter F. George Dam** (Chattahoochee River between Alabama and Georgia) consists of 500 m of gravity concrete sections, and two embankments with total length of about 4 km. The reservoir, approximately 30 m deep, lies over a limestone formation of almost the same thickness (30 m). During the final stages of construction (1961) two sinkholes developed, including boils, along the downstream toe on both flanks of the dam (Simpson et al., 2006). Remedial works were done in 1968, 1969 and 1970. In 1982 new seepage appeared downstream of the powerhouse. This was plugged by tremie concrete. In 2001 construction of two types of cut-off walls began: two (trench) panel walls and one 410 m long overlapping piles structure. The trench cut-off walls, 66 m deep (0.9 m by 2.7 m slots) was built using hydro-mill technology. The wall on the Alabama side is 100 m long and 55 m on the Georgia side.

The overlapping pile structure (secant cut-off) was constructed under water at a distance of 3.3 m in front of the upstream dam face. The depth of the wall is approximately 65 m below the reservoir level, i.e., approximately 30 m into the karstified limestone (Fig. 13.12).

The large cranes used for cut-off wall construction were mounted on barges anchored in front of the dam. Before wall construction began, some large cavities were discovered along its route and were filled via the boreholes, using a sand, cement, and water mixture. The secant wall consists of 469 overlapping piles 610 mm in diameter. During construction and inspection of works the role of divers was very important (more than 150 dives). After more than 40 years the seepage problems of the Walter F. George Dam were solved.

The **El Cajon Dam** (Honduras) arch Dam (236 m high) is situated in Cretaceous limestone which was karstified before being covered by lava. Thermal water was detected during excavations for the underground power plant. In spite of a bath tub antifiltration structure, during the reservoir filling (1993) seepage of 1.65 $m^3$/s was observed. Voluminous cavities of up to 5000 $m^3$ about 170–180 m below the gallery level had to be grouted. To reach them, boreholes with lengths up to 250 m were used. Total leakage was reduced to a negligible 0.1 $m^3$/s (Guifarro et al., 1996). As a consequence of progressive erosion during dam operation, the leakage had slowly increased to 3 $m^3$/s in 2016. Remedial works are still in progress.

**Great Falls** (Tennessee, USA). Leakage through the karstified left bank of the reservoir increased rapidly from 0.47 $m^3$/s in 1926 to more than 6.6 $m^3$/s in 1939 and to 12.7 $m^3$/s in 1945. Underground leakage flow rates ranged between 60 and 180 cm/s. Combined asphalt and cement grouting was applied. By constructing a grout curtain 1,750 m long to an average depth of 15.25 m, water losses from the reservoir were reduced to only 2% of their maximum.

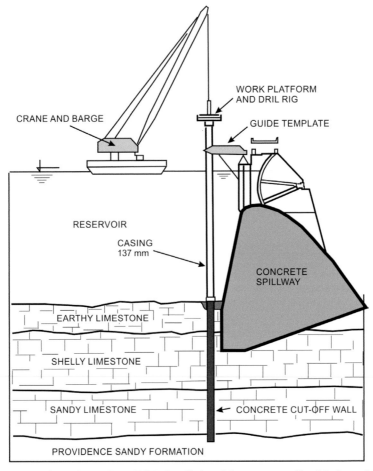

**Figure 13.12.** Walter F. George Dam, U.S.A. Installation of the secant cut-off wall in front of the dam, Simpson et al., 2006.

**Guanting Dam** (Hebei, China) is a 45 m high earth-filled dam. Immediately after construction was completed (1955), collapses and leakage of 1.0 m³/m occurred. Three karst solution channels were discovered. After regrouting, seepage was reduced to 0.1 m³/s (Lu Yaoru, 2012).

**Krupac Reservoir** (Montenegro) is one of three, mutually connected reservoirs situated in Nikšićko Polje. Seepage from the reservoir area occurred through the Krupac ponor and also escaped around both of the dam banks. To prevent the leakage, three grout curtains were constructed. The most important remedial work was the installation of a 247.5 long grout curtain (84.8 m deep on average) that included plugging the solution channel of Krupac ponor, which had a mean flow of 4 m³/s (1957/1959). To prepare the channel zone for plugging, underground flow velocities were reduced by injecting 2280 m³ of a crushed stone, sawdust and grout mix (70% sand and 30% grout) via boreholes. A coagulated (clayey) mixture was applied for the final grouting.

The average grout mix consumption was 1282 kg/m'. One important lesson learned during the process was that delays and breaks during grouting negatively affected the quality of the finished work.

The grout curtain on the left bank is 504 m long and 18.5 m deep on average, and on the right bank, only 159 m long and 37 m deep on average. After these remedial works seepage decreased to negligible amounts.

**Logan Martin Dam** (Alabama U.S.A.) consists of two earthen embankment sections (total length 1,682 m) and a central concrete powerhouse section (187 m), with a maximum height of 30 m. The karstic foundations consist of dolomite, chert and breccias with sandstone and limestone. Large cavities exposed in the underlying bedrock were treated dentally with concrete. During the first filling (1968) seepage was observed downstream from the dam and a collapse sinkhole opened downstream of the crest of the eastern embankment. Remediation grouting started immediately. Besides the original construction grouting (1960–1964) there have been seven successive phases over a duration of more than 50 years. Phase 4 (1991–2006) focused on grouting two cavities with large flows of water below an elevation of 38 m. In the upper cavity, grout mixes containing sand, gravel, sodium silicate and hydrophilic chemical were successfully applied. Burlap strips and polypropylene sacs were added to the grout mix (548 m³). For plugging the lower cavern, a rounded gravel with grout mix was applied (1260 m³). During Phase 5 unconsolidated soil in the overburden at the east end of the concrete spillway was treated by compaction grouting. In Phase 6 grouting continued plus monitoring that included tracer tests, pH measurements and thermal logging. In Phase 7 deep grouting (down to 122 m) was undertaken in some sections. An optical televiewer and calliper logging were used. For the grouting, cement, water, bentonite and a super plasticizer were applied; five thousand cubic meters of grout mix were required (Bruce et al., 2014).

**Maotiohe 4th Cascade** is one of the multi-cascades dam projects that have been built in China, on the Maotiohe and Longjiang Rivers. The dam, 67 m high, is located in karstified limestone. More than 10 large caverns were discovered in the immediate area. The deepest one has a depth more than 100 m (40 m below the river bottom). The volume of largest one is more than 100,000 m³. During reservoir impounding there was seepage of 20 m³/s. After complex and massive treatment, this has been diminished successfully (Yuan, 1981).

**Marun Dam** (Iran) is a rock-filled structure 165 m in height that is located in the karstified Asmari limestone. Seepage occurs through two mutually interconnected karst systems: along the bedding planes and through a thin layer of intensively vuggy limestone in the Middle Asmari. In a few days leakage of only 40 L/min increased to 1.5 m³/s and finally to 10.0 m³/s. The main leaking conduit was investigated and mapped speleologically. After extension of the grout curtain down to a section with shaley interbeds and plugging the main karst channel with concrete, the leakage was reduced to a negligible amount.

**Mujib Dam** (Jordan) is a gravity RCC/CCR structure 67 m in height, with a crest length of 746 m. It was constructed between 1999 and 2003. The foundation consists of Cretaceous limestone with marl intercalations. The grout curtain below the dam is 50 m deep. After very quick filling due to heavy rain, the first impoundment was finished in just one night. Immediate seepage losses of 140 L/s were registered. During the next two weeks these increased to 240 L/s because the abrupt increase of hydraulic pressure was washing out natural infill not removed during the Phase 1 grouting. After a regrouting program had been executed, seepage was practically eliminated – 3 L/s only (Wiesner and Ewert, 2013).

**Times Ford Dam** (Tennessee, U.S.A.), a 460 m long embankment dam, has as its right abutment a ridge largely consisting of karstified limestone (Bruce, 2003). Seepage through the ridge started immediately after reservoir filling (1971) and gradually increased, reaching 0.483 m³/s after 20 years. By constructing a 240 m long and 37 m deep grout curtain the seepage was reduced to an acceptable 0.015 m³/s.

Župica Dam (Bosnia and Herzegovina) is only 24.75 m high (the depth of the water is 23 m). However, due to very deep karstification, the average depth of the grout curtain is 185 m, i.e., eight times greater than the height of the dam that can be seen on the surface. The length of the curtain is 620 m and its surface area is 127,777 m².

## 13.5  DAMS WITH UNSUCCESSFUL OR PARTIALLY SUCCESSFUL REMEDIAL WORKS

**Lar Dam** (Iran) is an earth-filled structure 105 m in height above the foundations. The geological conditions are very complex: lava, tuff, lake deposits over intensively karstified Jurassic and Cretaceous limestones. In natural conditions the groundwater level (water table) was at depths of approximately 200–300 m. Immediately after impounding started (1980) there was seepage at two springs, as well as collapses in the reservoir bottom (see Chapter 10, Fig. 10.3).

The Haraz spring zone is the lowest base level for water in the Lar basin level. This is located eight km downstream and 350 m lower than the Lar Reservoir. The springs have developed in a broad, tectonized and heavily karstified zone. Cavities in limestone cliffs above the spring zone represent earlier stages of the evolution of the Lar aquifer. A few caverns have been discovered at depths of 250 m down to 430 m. The deepest conduits were about 100 m lower than the Haraz spring zone (Mijatović, 1989).

The deepest part of the grout curtain was about 80 m above the water table. The average consumption of grout mix was 44 kg/m (or 13 kg/m²) but in some parts of the deeply karstified zones was between 1,000 and 40,000 kg/m'.

During the first filling of the reservoir (1980) very high leakage (7.7 to 10.8 m³/s) was observed. At the highest water level attained in the reservoir during the past 30 years (24 m lower than the design level) measured discharge was 9.85 m³/s. During the past 30 years massive remedial works have been performed, including intensive grouting and the filling of a very large cavern at a depth of 210 m (volume – 90,000 m³). However, the work has been without success (Djalaly, 1988). The reservoir has never completely filled and the quantity of leakage does not change. Based on analysis of all available reports and findings during reconnaissance visits, the conclusion is that to achieve the required water tightness at Lar Reservoir the only solution is surface treatment by applying compaction and geomembranes to the reservoir floor and shotcrete in the karstified limestone banks (Milanović, 2010).

**Keban Dam** (Turkey) is in two parts: rock-filled and concrete gravity arch sections. The dam height is 211 m and reservoir volume is 30.6 x 10⁹ m³. The dam is situated in heavily karstified marble and limestone. During its construction about 30 caverns were treated, including 'Crab Cavity' which had a volume of 105,000 m³. During the reservoir filling (1976), when the water was 10 m below the design maximum level, a large karst shaft opened near the body of the dam and losses from the reservoir sharply increased to 26 m³/s. This water discharges in the Keban Creek, 2.5 km south of the dam site (Figs. 6.27 and 13.13).

The karst shaft was connected to the huge 'Petek Cavern'. After this was filled with about 605,000 m³ of limestone blocks, gravel, sand and clay, the flow rate in the Keban Creek decreased to 8 to 9 m³/s (see 6.3.17).

**Salakovac Dam** is a gravity concrete structure located in the Neretva River valley, upstream of Mostar (Bosnia-Herzegovina). The dam is 70.0 m high, rising 45 m above the bed of the river. The foundation rock is a tectonized and karstified Cretaceous limestone. About 1.3 km downstream the limestone is dammed by a thick mass of impervious Tertiary sediments. Karstification is intensely developed along the left bank of the reservoir and the dam abutment until the contact with the Tertiary sediments (Fig. 3.13). As a consequence there are a number of springs along the left bank of the

**Figure 13.13.** Keban Dam, Turkey. (A) Seepage section in the left bank; (B) One of the new leakage springs in Keban Creek.

river. The average discharge is estimated at 5 m³/s but it varies from a few hundred l/s in the dry season up to approximately 10 m³/s in periods of high precipitation. After impounding the reservoir, this discharge increased to 15 m³/s. An emergency protective shotcrete blanketing over the reservoir left bank immediately upstream from the dam was unsuccessful.

To intersect the possible deep karst systems and leakage paths, a 400-m-long investigation/grouting gallery was excavated for exploratory drilling, TV logging, dye tests and detailed geological mapping. Many cavities with diameters of 1 to 10 m were discovered along the gallery. Most were filled with very compacted clayey deposits. Only one very large, empty cavern was intersected by the gallery.

**Samanalawewa Reservoir** (Sri Lanka) is 107 m high rock-filled dam with a central clay core. The dam site and reservoir are composed of gneisses with limestone layers. The project area has been subjected to intensive folding, faulting and hydrothermal action. The limestone has been karstified by meteoric solution and hydrothermal action. Large cavities were detected in it. During the initial impounding (1991) seepage through the right bank started when the reservoir reached 16 m. As a remedial measure a grout curtain, 1800 m long and 100 m deep, was constructed along the right bank. In renewed impounding (1992) seepage increased to more than 7 m³/s (Fig. 10.5). Erosion by the leaking water washed away about 25,000 m³ of right bank earth in a period of only 10 hours (Laksiri et al., 2005). Twenty four hours after peak discharge, the flow had decreased and stabilized at about 2 m³/s. The appearance of a karstic channel (ponor) in the reservoir bottom was reported, i.e., there was a concentrated leak point (a solution conduit) in the reservoir.

A second attempted remedial measure was the construction of a wet blanket in 1998. The idea was to dump graded earth into the reservoir to cover the seepage zone. Two 125 m³ bottom-opening barges were used (Fig. 13.14). However, in spite of dumping about 500,000 m³ of graded earth material, the leakage did not change.

**Figure 13.14.** Samanalawewa Dam, Sri Lanka. Bottom opening barges.

In 2006, after heavy monsoonal rains, the leakage increased again to approximately 5 m³/s and then suddenly stabilized at 2.2–2.5 m³/s. Over the past few years there have been a number of collapses in the reservoir area, but relatively far from the seepage area in the right bank. The relationship between the collapse area and the right bank sinking zone is now a matter of analysis. At present Samanalawewa Dam operates with more or less constant leakage of 1.8 m³/s.

**Višegrad Dam** (Bosnia and Herzegovina) is a gravity arch concrete dam 50 m in height (70 m from foundations) built across the Drina River. The foundations are limestone and dolomite karstified by both hydrothermal and meteoric water (hypogene and epigene karstification; Fig. 13.15). Hypogene features suggest the possibility of deep karstification and play an important role in the Višegrad Dam leakage problems. The deepest karst features detected are 283 m below the river bed.

After excavation of the surface foundation (20,000 m²) 140 seepage points were recorded, but totalling only 0.36 m³/s. Water temperatures were mostly greater than 19°C. Grout curtain depth below the dam foundation was 60 m. During the first filling of the reservoir (1986) under-seepage below the suspended grout curtain began and rose to 1.4 m³/s. It was followed by a massive washing out of clay from caverns situated below the curtain and muddy water appeared at springs downstream. During a major regrouting of the main seepage zone beneath the middle section of dam foundations, 47 flow zones were reconsolidated out of 52 recorded. To fill cavernous spaces, 3–10 mm and 8–16 aggregate was applied together with the grout mix. The trend of rapidly increasing leakage was temporarily stabilized at 6–7 m³/s.

In spite of intensive regrouting and deepening the local curtain down to 110 m, the leakage increased steadily from 6.5 m³/s in 1996, to 9.4 m³/s in 2003 and 13.92 m³/s in 2008. Karst channels filled with clay at depths of more than 130 m below the dam foundations were being reactivated.

A study of geological data from historical records, particularly the geological mapping data of the surface of the foundations, led to an intense site investigation program via boreholes (including TV logging), tracer testing and monitoring of leak points downstream from the dam. Particularly successful was the exploration of

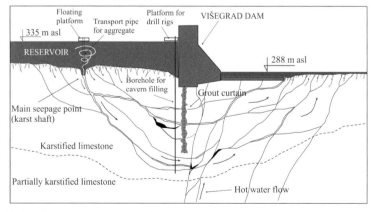

**Figure 13.15.** Višegrad Dam, Bosnia and Herzegovina. Simplified cross-section of the main seepage flows, including the general concept of remedial works (Milanović, S. 2015).

the bottom of the reservoir by divers. During investigations in 2009 and 2010 they discovered a large karst opening (ponor – diameter 3.5 m) with massive water inflow, approximately 160 m upstream from the dam.

An important restriction on the work was that the reservoir could not be drawn down to a low level to minimize seepage and so permit treatment of the deep seepage zone under lower pressure. To organize intensive treatment of the incipient karst features (the ponor at the bottom and large karst channels beneath the dam) two floating structures were constructed: (i) a structure to plug the ponor, and (ii) a floating platform near the upstream dam face for deep grouting boreholes (Ø 122.7) to fill caverns at depths 100–150 m below the reservoir bottom (Fig. 13.16).

The ponor in the reservoir bottom was filled with 1286 m³ of 1–4 mm aggregate. A crushed stone aggregate with a composition of 4–8 mm = 20%; 8–16 mm = 30% and 16–32 mm = 50% was introduced into the empty cavernous space. After 38000 m³ of aggregate had been inserted the leakage was recorded as being reduced to about 4.5 m³/s. The treatment is still not completed, however.

**Center Hill Dam** (Tennessee) is an 80 m high concrete-earthen embankment dam founded directly on highly karstified limestone (2008–2012). Foundation treatment was insufficient. After filling, sinkholes appeared in the abutment and leakage of approximately 10 m³/s was recorded. To prevent this seepage a concrete cut-off (80 cm thick) was constructed. The problem is still pending?

**Salanfe Reservoir** (Switzerland) is retained by a 52 m high gravity dam. It was impounded in 1952. The dam and reservoir area are on quartzite, dolomites and limestone. One year later (Sept 1953) thermal springs at 30°C appeared in the vicinity of the village of Val d'Illez about 9 km away. At the same time significant leakage losses were noted in the reservoir, always beginning when reservoir levels rose up to 1890–1895 m a.s.l. (Roth, 1994). Because of high seepage (1150 L/min) through the left bank of the reservoir during many years of operation, only 50% of its volume was

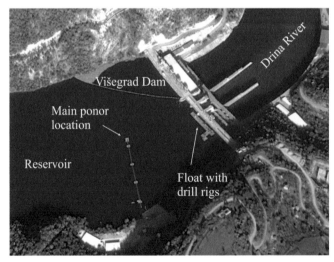

**Figure 13.16.** Višegrad Dam, Bosnia and Herzegovina. Facilities constructed for remedial works. From Google.

being used. When the reservoir was filling, a number of low intensity earthquakes were recorded. Monitoring of these events indicate a strong relationship between the impoundment of the reservoir, the appearance of the thermal springs and the earthquakes. Investigations and analysis suggest that a long and deep seepage flow system has been created. The seepage water needs a few months in deep circulation to become thermal and to discharge at the new springs. Remedial works (1992–1994) included construction of an additional grouting gallery (600 m long) and installation of a one row grout curtain to reduce the water losses (Bianchetti et al., 1992; Dawans et al., 1993).

**Slano Reservoir** (Montenegro) is situated in the extremely karstified and seasonally flooded Nikšić karst polje. The minimum groundwater level is about 40 m beneath the polje. In natural conditions, depending on flood levels, seepage ranges between 5 and 12 m³/s (Fig. 13.17, Graph). During maximum floods the registered losses were 34 m³/s. The reservoir depth is 12 m. By construction of grout curtains along the reservoir rim and beneath the dam, with a total length of 7.011 m and average depth of 65 m, the losses have been reduced to the acceptable amount of 3.5 m³/s. In some sections the depth of the curtain is approximately 100 m. The main problem was to seal the active karst channels. To seal one cave stream channel alone 1.073 tons of dry component was used. Total grout curtain surface area is 404.224 m². The grouting works took 11 years (1961–1971). Seepage was reduced from 8 m³/s to 3.5 m³/s.

However, during the reservoir operation from 1971 to 2001 the grout curtain became severely degraded and new ponors opened between the reservoir banks and the grout curtain (Fig. 13.17, Photo). Leakage increased to 7 m³/s, mostly through the uppermost part of the curtain (Vlahović, 2005). To prevent groundwater flow through a cavern detected at a depth of 90–100 m, plugging was attempted over a period of four years and five months, but without success. 4,124 m³ of gravel and 4,500 tons of cement were injected. The reservoir still operates, but with leakage losses that cannot be accepted.

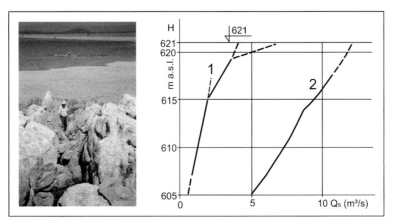

**Figure 13.17.** Slano Reservoir, Montenegro. Photo: Swallow-holes created due to degradation of the grout curtain, Milanović, 2003; Graph: reservoir water seepage versus reservoir water level. (1) Reservoir water seepage after sealing works. (2) Reservoir water seepage before sealing the reservoir banks (Guzina, 1997).

**Tichy Haf Dam,** Algeria, is founded on limestone. There was a water outburst on the left bank. There was no safety problem but the leak was important in size. Attempts to close it by dumping materials upstream were a failure.

**Perdikas Reservoir** (Greece) is 40 m high earth-filled dam. The groundwater level was 70 m below the reservoir bottom, in karstified Upper Cretaceous limestone. During the first filling a number of subsidence sinkholes (ponors) occurred inside the reservoir. An additional loam blanket and directing streams of suspended silt into the leakage area proved to be ineffective. As a consequence, only the lower section of the reservoir can be filled.

**May Reservoir** (Turkey) has a rock-filled dam 28 m in height, built for flood control and irrigation. The reservoir bottom is alluvium 15–20 m in thickness, deposited over karstified limestone, conglomerate and marl. As a consequence of the first filling, a number of subsidence (ponors) were created in the reservoir floor as well as along the right bank close to the dam. The reservoir has never filled completely. A lack of careful geological and hydrogeological explorations in the planning stage of the May Reservoir was responsible for this surprising outcome (Yevjevich, 1981).

**Beni Haroun** in Algeria is 107 m high RCC dam founded on karstified limestone. At a hydraulic head of about 90 m, leakage close to 1 $m^3$/s occurs. In spite of extensive investigation and analysis, the location of the underground flow was not precisely detected. There are no large karst channels; the leaking water most probably flows through diffuse karst porosity. Extensive grouting was only partially successful. Despite very extensive research with the latest technology, it was not possible to locate the groundwater inlets under and in the banks of the dam. The left bank was grouted under a partial water head, with good efficiency, but it required a very significant quantity of cement.

**Kalecik Dam** (Turkey) is 77 m high. The right bank of the dam foundation consists partly of karstified Paleocene limestone. The grout curtain on the right abutment, 200 m long and 60 m deep, was not sufficiently successful to prevent leakage to downstream springs. In spite of three phases of intensive remedial works, including new deep grouting, seepage from the reservoir was not eliminated (Turkmen, 2003).

## 13.6  REDESIGNED ANTI-LEAKAGE STRUCTURES

**Salman Farsi** (Iran). The concrete gravity arch dam is 125 m in height above the foundations (Fig. 13.18). It is located at the entrance of the Karzin Gorge on the Gareh-Agaj River, south of Shiraz. Reservoir volume is 1,400 million $m^3$ and when full (normal water level of 855.00 m asl) the large karst Yarg Spring will be submerged by 27 m of water.

The dam foundation area is composed of heavily karstified Asmari limestone. The springs at the dam site indicate two groundwater systems. One is a shallow, phreatic aquifer with a temperature of 28°C, and the other is a deeper hydrothermal groundwater system with temperatures up to 42°C. Obviously, in the case of the Salman Farsi dam site, hypogene karst processes must play an important role. The deep-sourced upward flow of water through karstified joints creates an additional problem for achieving the

**Figure 13.18.** Salman Farsi Dam, Iran.

required water tightness beneath the dam foundation. A system of sub-vertical faults in the left bank at the dam is one of the main transmission structures for upward flow.

The grout curtain option that was selected and applied on the both sides of the dam makes a closed connection by tying downstream extensions of the curtain tightly into the core of the impermeable Changal Anticline. By this means a 'positive cut-off' structure was put in place. The grout curtain was constructed from galleries at five levels, with the vertical distances between galleries varying from 18 to 36 m in order to intersect as many of the karst caverns and conduits as possible (Fig. 13.19).

A one-row grout curtain was installed. Curtain depth below the dam foundation is 140 m. Total length of grouting galleries is 5,817 m. The grout curtain itself was constructed along a total of 3,725 m. The length of executed (as built) grouting sections is 1.055 m longer than in the tender design. To construct this curtain with its surface area of 260,992 m², 620,217 meters of boreholes were drilled. Total grout mix consumption was 49,117 tons, with an average consumption of 77 kg/m'.

The largest cavern, 'Golshan's' was situated in the course of the grout curtain in the right bank. Its volume was about 150,000 m³. Following detailed speleological investigations this problem was solved by a modification of the curtain, changing its alignment to bypass Golshan's on the upstream side (Figs. 4.21 and 6.2).

For plugging the six largest karst systems discovered along the grout curtain route, 3,125 m³/7,500 tons of self-compacting concrete were used (Dolder at al., 2001). Estimated seepage is now less than 0.5 m³/s.

**Freeman Dam** (Kentucky, USA). Originally this was a clay embankment dam, 15.25 m high, constructed in the mid-1960s. Large leakage (~ 10 m³/s) through big solution features below the dam was reported in 1997. Sinkholes in the abutment were created by piping process. The entire dam was removed and a positive cutoff wall, rock treatment, and filters were included in the redesigned structure. However, almost same amount of seepage loss still occurs.

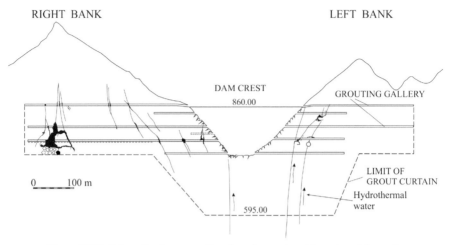

**Figure 13.19.** Salman Farsi, cross-section along the grout curtain route and dam axis.

**Sklope Dam** (Croatia) is a 74 m high rock-filled structure. The dam foundation consists of stratified, highly karstified Cretaceous limestone. During construction of a grout curtain large caverns were discovered in both abutments. The volume of caverns inside the left bank was about 25,000 m³. After detailed speleological investigations the grout curtain route was redesigned. The new route made a detour around a left bank cavern on the downstream side and around a cavern in the right bank from the upstream side (Fig. 13.20). Leakage from the reservoir (when full) is estimated at about 1 m³/s.

**Seymareh Dam** (Iran) is a concrete arch dam 135 m in height (185 m from the foundation) situated on karstified limestone. It is located in the northern limb of the regional Ravandi Anticline. The reservoir volume is 3.205 x 10⁹ m³. A huge land-slide that occurred downstream approximately 15,700 years ago has had a crucial influence on the intensity and depth of karst evolution inside the large mass of Asmari limestone

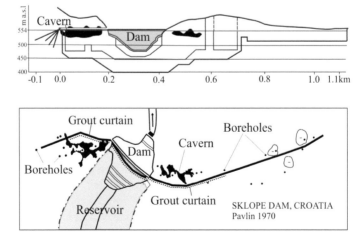

**Figure 13.20.** Sklope Dam, Croatia. Cross-section (above) and layout of the dam site (below) with the grout curtain alignment.

between the reservoir and the river channel downstream where the power plant is built. The local erosion base level was 40 to 50 m lower than at present (~ 600 m a.s.l.). A number of caverns have been developed along faults parallel to the anticline axis (Fig. 13.21). Clearly, the entire fault zone is weak and prone to erosion.

Evaporite rocks (the Gachsaran Formation), in the left reservoir bank just upstream of the dam site, consisting of folded marls and gypsum beds ~ 20 m thick. From the karstification and seepage viewpoints, the gypsum can be declared as a weak zone with considerable leakage potential.

An inverse thermal gradient that was discovered in some boreholes in the left abutment indicate that there is active groundwater flow 10 to 20 m lower than the present river bed. Further, cavities filled with clayey-silty material and in some cases with sand, were detected in some boreholes at depths approximately 100 m lower than the river, i.e., about 50 m lower than the elevation of the fossil river bed; during the karst evolution of this part of the aquifer there were siphon (phreatic looping) conditions.

During grout curtain construction, due to the findings underground, the curtain was redesigned in direction and length. Another problem appeared during excavation of the foundation pit for the Power Plant, which was in the large spring zone. Excavation of the Plant access tunnel was in trouble due to high amounts of groundwater intrusion (from 1.2 $m^3$/s in the dry season up to 2.0 $m^3$/s in the wet season): the downstream section of the Head Race Tunnel and the Power Plant foundation pit were re-designed, and a massive anti-seepage treatment was applied.

The impounding of the reservoir displayed all the complexity of dam sites in karstified rocks. For example, impounding began before the watertight sealings works were completed. A key seepage zone ('hydrogeological window' of about 10,000 $m^2$) at the end of the left bank grout curtain was not completed. Grout mix consumption in the boreholes already grouted was between 1.0 and 30 tons/m'. Immediately after impounding began, the leakage increased extremely rapidly. After the reservoir level had risen up to half of the design depth, leakage had increased from 4 $m^3$/s in 2013 to more than 8 $m^3$/s (Fig. 3.15).

**Figure 13.21.** Seymareh Dam, Iran. Grouting gallery, Cavern developed along the fault parallel to the Ravand Anticline, and gypsum at left bank close upstream from dam site.

**Clarence Cannon Dam** (Missouri, USA), a 46 m high and 700 m long concrete/earth filled structure, is located on karstified Carboniferous limestone with a sequence of shale/siltstone layers 25 m in thickness close to the river level. The original design included deep excavation down to the shale/siltstone, a shallow cut-off beneath the dam body and a short, single row grout curtain. After large cavities were discovered during the construction, the original consolidation and waterproofing plans were modified. The grouting program was completely redesigned. The new design includes a concrete cut-off wall in the left embankment, consolidation grouting and extended triple-line and single-line grout curtains (Jahren et al., 1981).

**Chichik Dam** (Uzbekistan), a 168 m high earth-filled structure, is located in karstified Lower Carboniferous limestone. Numerous caverns 10s of meters in size were detected, including large ones at depths more than 100 below the river bed. Karstification had occurred in various stages, the deepest due to thermal water rising along tectonized zones. During the first filling of the reservoir the grout curtain structure was modified, i.e., adapted to the existing hydrogeological conditions by extension and locally deepening (Kagan and Krivonogova, 1999).

**Darwin Dam** (Tasmania, Australia) is a 21 m high and 400 m long gravel filled embankment dam, built on gravels, sandstone and silts overlying karstified limestone. Empty and silt-filled cavities were encountered 30 m below the top of the limestone. In a first design a positive cut-off would be constructed through the gravels and silts and into the limestone to prevent seepage. On the basis of detailed geological investigations, however, it was concluded that construction of a reliable cut-off was too problematic and risky. The project was redesigned and the dam axis shifted to a new site where the limestone was deepest underground. There was no cut-off structure. Waterproofing included a conservative embankment section and an upstream surface foundation treatment. After filling the reservoir only a small change in the piezometric levels was detected, including minor seepage downstream of the dam (Guidici, 1999).

**Akkopru Dam** (Turkey) is a rock-filled structure 110.5 m in height, built on an impermeable peridodite-serpentine rock mass (Şekercioğlu and Özgüler, 1999). A small part of the reservoir, 2 km upstream of the dam, is in direct contact with karstified Aktaş limestone, however (Fig. 6.47). After a thick alluvial cover had been removed, a few 10s of large diameter karst shafts were discovered in an area of 250,000 $m^2$, in both the reservoir bottom and its banks. The groundwater level is very deep, 100–116 m below the reservoir floor (close to the sea level). Water that sinks in this area discharges at the Innica Springs at an elevation of about 20 m on the sea coast. To prevent leakage a few different solutions were considered. The final decision (DSI, Ankara) was to construct a thick reinforced-concrete slab on the reservoir floor and over the left bank up to 100 m above the bottom (Figs. 6.48 and 6.77).

To prevent leakage below the concrete slab, a cut off wall (concrete piles, 800 m long and 40 m deep) was constructed. It crosses permeable alluvial sediments and terrace conglomerates to be tied into impermeable-to-weakly-permeable flysch, (0–20 Lu; Günay and Milanović, 2005).

## 13.7  DAMS AND RESERVOIRS WITH ACCEPTABLE SEEPAGE LOSSES

**The Adam Beck PGS Reservoir** (Ontario, Canada, 1956) consists of a 7 km long, oval clay- and rock-filled dike. The reservoir has an elliptical shape, 3.2 km long and 1.2 km wide, that is situated on the crest of the Niagara Escarpment. It is built on top of 10–40 m of glacial silts, sands and till over weakly karstified dolomite and limestone (Ford and Worthington, 2012). The surface of an old creek bed in the reservoir bottom was protected by a clay blanket (Fig. 13.22).

Many leakage problems have been detected in the reservoir and dike but all remain minor more than 50 years after containment and storage there. After a small embankment failure in 1998 there were a number of investigations including boreholes, geological mapping and a sonar survey (bathymetry). Suffosion features (sinkholes) were detected by the sonar. A synthetic liner, grouting and repair sinkhole treatments were applied.

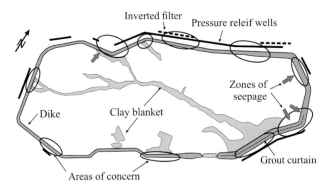

**Figure 13.22.** Adam Beck Reservoir, Canada. Simplified sketch of the reservoir with potential weak zones along the embankment (Ford and Worthington, 2012, courtesy Hatch Energy and Ontario Power Generation).

**Hutovo Reservoir** (Herzegovina) is situated at the downstream end of the highly karstified Popovo Polje. In natural conditions leakage through the 75 registered ponors in the reservoir bottom ranged between 7.0 and 10.0 $m^3/s$. In the dry season the groundwater level is 100 m below the surface, but in wet periods of the year the rock becomes completely saturated and floods rise up to 40 m above the polje floor. The designed depth of the reservoir is only 10 m (Fig. 13.23).

Several sealing methods were applied to achieve the required impermeability. Exposed limestones along the reservoir banks were protected by shotcrete. The alluvial bottom was compacted and ponor zones were infilled (reverse filter) and blanketed with plastic foil.

After reservoir filling (1975), new ponors and fissures 5–30 cm wide appeared and there was leakage of 3.0 $m^3/s$ as a consequence (Figs. 6.51, 6.52 and 13.24). In some places the foil exploded due to strong air pressure from below (Fig. 6.54). To prevent further destruction of the foil, an aeration pipe was installed through 10 m of alluvium and down into a karst channel in the paleorelief (Fig. 6.55). During three

**Figure 13.23.** Hutovo Reservoir, Popovo Polje, Herzegovina.

**Figure 13.24.** Hutovo Reservoir, Herzegovina. Cracks and collapses occurred during the first reservoir filling.

cycles of reservoir filling and emptying more than 120 ponors and 1300 m' of fissures were sealed with clay-cement mixture. Finally, the leakage rate was reduced to an acceptable 1.0 m³/s (Milanović, 2004).

**Kowsar (Tang-e-Duk) Dam** (Iran) is a concrete gravity arch dam rising 125 m above river in a deep and narrow canyon (Fig. 13.25).

The Asmari limestone at the dam site is selectively karstified, with syngenetic (vuggy) porosity plus fracture-guided karstification developed during the rapid fluvial entrenchment. Karstified joints are 10–15 cm wide locally. The depth of the suspended

**Figure 13.25.** The Kowsar Dam site, Iran.

grout curtain beneath the dam foundation is 120–180 m, and 50 to 70 m at the ends of the lateral sections. The reservoir operates with constant leakage of approximately 3 m³/s. The leaking water flows through the karstified zone in the right bank and discharges at a large spring just downstream of the dam.

**Karun I (Shahid Abbaspoor) Dam** (Iran) is a double-arch concrete dam, 200 m in height. Asmari limestone at the dam site is karstified. Leakage through the most critical, clay-filled, fault zone in the right bank is prevented by a concrete cut-off wall 100 m high and 30 m wide (Fig. 6.38). However, discharge of the 'Big Spring' in the right abutment close to the powerhouse increased from 3–15 m³/s in natural conditions to 10–16 m³/s after reservoir filling and during its later operation.

**Karun 4** is one of the other dams on the Karun River, Iran. It is a double arch concrete structure, 230 m in height. The foundations are at 802 m asl and crest elevation at 1032 m a.s.l. Reservoir volume is 2.3 billion m³ at the normal water level of 1028 m a.s.l. The grout curtain surface area is 271,000 m².

The dam site is in the karstified Asmari Limestone, with the impermeable Pabdeh Formation (marlstone and marly limestone) upstream of it.

In the left bank a cavern was encountered (by borehole) at a depth between 157.35 (el. 858.83 m) and 161.00 (el. 855.18 m); i.e., the aperture of the open void (cavern) was 3.65 m. At other depths maximum rod drops were 0.7 m. In the same section, between 193 and 200 m (the approximate base of the foundation level) a Lugeon test showed undefined high permeability (Lu > 100). A large cavern (Fig. 13.26) developed along Fault F4, was only discovered during excavation of grouting gallery GL-4 at 890 m a.s.l.

Five hundred meters downstream of the dam site two different spring waters discharge from the 14 km long Monj Fault: springs with fresh water and nearby springs with a sulfuric water discharge. Very specific forms in the cavern walls support the possibility of hypogene karstification and sponge-like karst features with vugs of different sizes on the surface of the left abutment are more likely consequences of hypogene than of epigene karst processes.

The grout curtain drainage system consists of two drainage galleries with drain holes. Part of the drainage curtain is nearly parallel with the grout curtain at a distance of 30 to 40 m. The last section of the grout curtain is extended into the mountain.

During reservoir filling, leakage in the lower drainage gallery (el. 860–876 m) increased between 500–600 L/s (corresponds to the reservoir level rising from 1000 to 1025 m a.s.l.) in the form of a large eruption (Fig. 6.38). It is clear that a drainage hole had intercepted a solution channel under pressure that is hydraulically connected to the karst system upstream of the grout curtain.

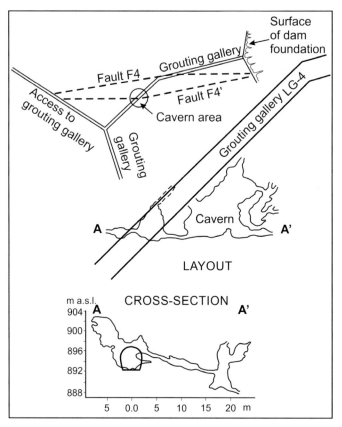

**Figure 13.26.** Karun 4, Iran. Cavern discovered during grout curtain construction in the left bank.

All drainage holes were then closed for one month and pressure was monitored for any significant changes in piezometers downstream of the grout curtain and in the drainage holes. The effect of the closure was negligible, ~ 0.2–0.5 bars. There was practically no increase of uplift pressure on the dam foundation.

**Buško Blato Reservoir** (Bosnia and Herzegovina) is located in the highly karstified Livanjsko Polje (Cretaceous limestone). The reservoir area is 57 km$^2$, volume 800 hm$^3$ and average depth is 14 m. There are a number of ponors along the southwestern perimeter of the polje. In the dry season groundwater level below these sinkhole zones is at depths of 80–100 m, rising to about 30 m below in the wet season. Along the eastern perimeter of the polje the water table rises to the surface (estavelle zone) in the wet season. To prevent leakage through the ponors and estavelles, which have an aggregate swallowing capacity of 35 m$^3$/s, two embankment dams (length 3000 + 600 m) and a discontinuous grout curtain 6206 m in length were constructed along the perimeter of the reservoir (Fig. 6.40). The larger karst channels are closed by fitting vertical shafts from the surface. The loss of water from the reservoir (seepage + evaporation) ranges between 2 and 6 m$^3$/s. Average annual inflow into it is 11 m$^3$/s (Nikolić et al., 1976; Nonveiller, 1989).

**Camarassa Dam** (Spain) is a gravity-arch structure 92 m in height, constructed between 1927 and 1931. The foundations consist of karstified Jurassic dolomites and Cretaceous limestones. During the first reservoir filling seepage increased from 3 to 11.26 m$^3$/s. A grout curtain 1400 m in length and varying in depth from 112 to 394 m in order to reach an underlying impermeable formation, was installed. After a few additional regrouting steps, leakage was reduced to 2.6 m$^3$/s.

**Špilje Dam** (FYUR Macedonia) is an earth-filled dam, 101 m high. During the filling of the reservoir (1969/70) seepage of about 2 m$^3$/s occurred immediately downstream of the dam. The discharge from the largest single spring was 350 L/s. The deepest underground flows are estimated to be at a depth of 250 m below the body of the dam. From long-time monitoring these deep flows do not appear to be producing any progressive erosion. The possibility that there is hypogene karstification at this site cannot be excluded.

**Ataturk Dam** (Turkey) is an earth-filled dam with a crest length of 1800 m and 179 m in height. The reservoir volume is 49 billion m$^3$. The dam foundations are in karstified limestone and dolomitic limestone extending to a depth of about 600 m. The karst conduits range from 0.1 and 1.5 m in diameter, plus a few caverns of 1000 m$^3$. The grout curtain is 5.5 km long and up to 300 m deep, with a surface area of about 1,2 x 10$^6$ m$^2$—the world's largest grout curtain. During reservoir filling (1990) leakage began through the dam foundations and from grouting galleries. The discharge was 2.5 m$^3$/s at very low reservoir level, rising to about 14 m$^3$/s at a level 15 m lower than the operational level. These losses are considered tolerable because the average discharge of the River Euphrates is 850 m$^3$/s (Riemer et al., 1997).

**Tang Ab Dam** (Iran) is a 60 m high rockfilled structure with a clay core. The reservoir elevation is 1447 m a.s.l. The thickness of alluvium at the dam site is 30–35 m. The ground water level is 40–60 m below the reservoir. During first impounding, seepage of 3.5 m$^3$/s was measured and three sinkholes were detected in the reservoir bottom. According to borehole data, the depth of the base of karstification is between 220–450 m below the river level. Underground connections between the reservoir and the Atashkadeh Spring, 5 km away near Ardeshire Babakan Castle, were confirmed by a tracer test.

**Alakir Dam** (Turkey) is constructed on karstified limestone. Because the dam is for flood control only, no protective measures were taken to prevent leakage.

**Mornos Dam** (Greece) is a 135 m high rock-filled structure. The groundwater level is about 200 m below the reservoir level. The seepage protection measures (rockfill along the reservoir bank and an asphalt blanket on the lower parts of its flanks) are not fully succesful. There is still some leakage ($\approx 0.5$ m$^3$/s; Pantzartzis et al., 1993).

**Gorica Dam** (Herzegovina) is a 35.5 m high concrete gravity dam built on karstified limestone (Fig. 13.27).

By intention, only a partial grout curtain was constructed (below the dam and for 200 m in the right bank). There was no grout curtain built in the left bank. The reason was to allow some free by-pass groundwater flow through the right bank and to keep downstream flow constant (Fig. 13.28). A large part of the seepage water just by-passes the dam and discharges into the Trebišnjica River a short distance downstream of the dam. However, some proportion of the sinking water flows off in a different direction through the karst aquifer to reach springs on the sea coast, a distance of approximately 25 km. The largest concentrated sinking zone is an estavelle located 600 m upstream of the dam.

From the stability viewpoint the dam was not endangered by the groundwater flow around it. Due to progressive washing out of clayey deposits from caverns and joints in the flowpaths, the rate of seepage has increased gradually: from 1.5 m$^3$/s in 1965 to 3.0 m$^3$/s in 1983, and up to 4.0 m$^3$/s in 2003.

**Figure 13.27.** Gorica Dam Herzegovina.

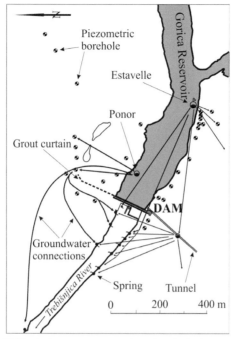

**Figure 13.28.** Gorica Dam, Herzegovina. The principal seepage paths.

**Piva (Mratinje) Dam** (Montenegro) is a concrete arch dam 220 m in height. The reservoir volume is 790 million m³. The dam is situated in a 1500 m deep canyon in massive but karstified Triassic limestones. The cavities and solution-enlarged discontinuities are mostly filled with clayey-sandy sediments and limestone debris. The largest cavern discovered (100 x 100 m) is filled with partially lithified clay: it was left in its natural condition—not treated (Božović, 1985). The depth of the three-row grout curtain below the river bed is 250 m. The designed length of the right bank curtain was 1.8 km but, in actuality, it was stopped after 400 m because after impounding and analysis of the leakage, it was found unnecessary to extend it.

Grout pressures in the curtain ranged from 3 to 40 bars, with a linear increase of 0.6–0.7 bars per meter of the borehole. For the final design pressure, a saturation criterion of 10 L/m' in 10 minutes was adopted. The boreholes in the middle row were grouted completely, as one section. The cavities were sealed using the following grout mix:

- components were 65% sand, 34% cement, and 1% bentonite.
- proportion of dry weight to water was 1:0.4.
- in the left bank 28 cavities were sealed using 5 to 75 tons per cavity.
- in the right bank 38 cavities were sealed using 5 to 98 tons per cavity.

Depending on water levels in the reservoir leakage varies from 1.0 to 2.0 m³/s. This leakage is considered to be acceptable.

**Karun 3 Dam** (Iran) is situated on the middle section of the Karun River in Khouzestan Province. It is a concrete-arch dam, 200 m high and with a crest length of 388 m (Fig. 13.29). The reservoir volume is 2750 x $10^6$ $m^3$. The dam site is on moderately karstified Asmari limestone.

Evidence of karst features has been found in boreholes down to at least 200 m below river level. There was higher grout mix take at this level, indicating that the base of karstification is further below. However, at the surface and during underground excavation, no large karst features (caverns of large volume) were detected. In some sections the Lugeon test results ranged from a few to more than 100 Lu (= undefined high permeability). These results indicate that the rock mass is riddled with solution-enlarged fractures, a very porous rock mass.

A total of 457,000 m of grouting hole have been drilled along the 2530 m of galleries to construct the grout curtain of 200,000 $m^2$ (Fig. 13.30). Grout mix consumption was 89 kg/m' of solids.

The drainage system to reduce uplift pressure consists of nine drainage galleries constructed behind the curtain at three levels. A total of 28,700 m of drainage holes was drilled. During reservoir impounding the drainage system discharged a huge amount of water (Fig. 13.31). For a reservoir water level of 810 m inflow into the drainage galleries reached 1.2 $m^3$/s, reservoir water that had penetrated through the grout curtain.

Between the dam site and the river downstream (Dehzir Valley) seepage occurred through bedding planes and the anticline limb (Fig. 3.6) that is not treated by the existing grout curtain. After impounding, the leakage is reported to be about 0.5 $m^3$/s. During the same period of time the discharge of the Abolghasem Spring downstream has increased from 0.3 $m^3$/s to, roughly, 1.0 $m^3$/s.

**Figure 13.29.** Karun 3 Dam Iran. The photo was taken before the dam was completed.

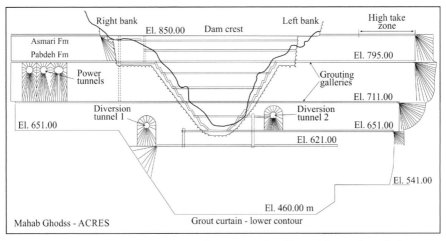

**Figure 13.30.** Karun 3, Iran. Grout curtain contours and positions of the grouting galleries.

**Figure 13.31.** Karun 3 Dam. Water discharge in a drainage gallery.

## 13.8 DAMS THAT WERE DISPLACED, ABANDONED OR HAVE LIMITED USE

**Montejaque Dam** (Andalusia, Spain). This concrete arch dam, 72 m in height, was constructed on the Gaduares River in the 1920s. The average flow at the dam is 4 m³/s. It is located in a karstified limestone gorge just upstream of the entrance to the eight km long Hundindero-Gato Cave system (Fig. 13.32). Due to the highly karstified limestone at the dam site and upstream the reservoir was never filled, despite the use of 1942 tons of cement for grouting. Due to its inacceptable leakage of 4 m³/s, the Montejaque Dam was abandoned in the 1950s.

**Figure 13.32.** Montejaque Dam, Spain. (A) Dam, (B) Entrance to the underground karst system downstream from the dam.

**Hales Bar Dam** (Tennessee, U.S.A.), 34 m high and 706 m long. The dam was built in the period, 1905–1913. Leakage started during the first impounding. Corrective works also started immediately and continued until 1944, using cement and hot asphalt and sometimes filling entire caverns with concrete. This treatment was partially effective in stopping leakage. Wherever asphalt was used it was followed immediately by cement grout to confine and consolidate the asphalt, and adjacent holes were cement-grouted reinforcing the effect. Leakage continuously increased, however. In 1960 an enormous leakage rate of 57 m³/s was measured. The dam was abandoned in 1968 and replaced by building the Nickajack Dam 9.7 km downstream (Tennessee Valley Authority, 1972).

**Wolf Creek Dam** (Kentucky, U.S.A.) consists of a combined earth and concrete gravity structure, 1748 m long and 79 m high (Fetzer, 1979). The dam area consists of nearly horizontal argillaceous limestone interbedded with thin, well-cemented, calcareous shale (Catheys Formation, about 30 m thick). The overlying Leipers Formation (thin-bedded argillaceous limestone and calcareous shale) is also 30 m thick. There is a network of solution channels in the horizontal bedding planes and adjoining vertical joints in the Leipers strata; a number of them were encountered during the excavation of a 15 m deep cut-off trench. The caverns were cleaned and filled with hand-tamped impervious soil. A one-row grout curtain was installed at a depth of 15 m from the bottom of the cut-off trench.

The dam construction took from 1941 to 1951. Some indication of the problems was observed in 1967 when a small sinkhole was found near the downstream toe. New subsidence features (ponors) formed in 1968 and there were seepage water discharges downstream of the dam. New collapses developed in the toe of the earth embankment. The depth of karstification was estimated to be 80 to 100 m beneath the dam body. Emergency grouting started in 1968.

Over 300 piezometers have been installed to monitor the seepage since 1971. The emergency grouting brought the seepage temporarily under control but did not solve the problem. A concrete cut-off wall was selected as the most feasible and least costly of several alternatives. The question of the safety of the dam is still discussed frequently, with much difference of opinion concerning its viability.

**Liverovići Dam** (Montenegro) is built on Triassic dolomites and limestone. The concrete arch dam, 45.5 m high, was constructed in 1955/57. The reservoir volume is ~ 9.5 million cubic meters. Under natural conditions, and after reservoir impoundment, the groundwater level was 35–50 m below the reservoir bottom (Vlahović, 1991). Grout curtains were constructed along the both sides of river (reservoir) banks for a length of 714 m. A few known ponors were also grouted but not all of them. Immediately after impounding a large new ponor opened in the right bank of the reservoir, and leakage between 0.5 and 2.0 m³/s was detected. The underground leakage is very deep. Additional remedial works were never undertaken.

**Apa Reservoir** (Turkey) was developed for irrigation of the Konya – Cumra region. The reservoir leaks water into limestone underlying the upstream portion of the reservoir. According to some estimates, about 20–25% of the total annual inflow into the reservoir is leaking away into the karstified limestone.

**Cevisli Dam** (Turkey) is 20 m high rock-filled structure built to store water for irrigation and flood control. The dam itself is built on impermeable rocks but upstream areas of the reservoir are in karstified limestone with abundant typical karst features (Yevjevich, 1981). When the reservoir was first filled (1979/80) large swallow holes (ponors) were created in the limestone bottom of the upstream section of the reservoir (Fig. 13.33).

Some of the ponors had diameters of one to several meters. Several branches from the ponors converged into a deep karst channel that had been masked by alluvial sediments under natural conditions.

**Vrtac Reservoir** (Nikšić Polje, Montenegro) is retained by an earth-filled dam 16.5 m in height. In natural conditions groundwater recharge from the reservoir area was very large, up to 27 m³/s. Groundwater levels fluctuated between 10 and 20 m beneath the karst polje bottom. Making the Vrtac Reservoir watertight was based on the concept of isolating the principal ponors and estavelles with cylindrical dams, constructing non-return valves in the estavelles, and covering the highly karstified limestone banks with concrete blankets (1958–1962). The two largest estavelles, Opačica and Misor, are isolated by cylindrical dams (Figs. 6.37D and 13.34). After the first filling more

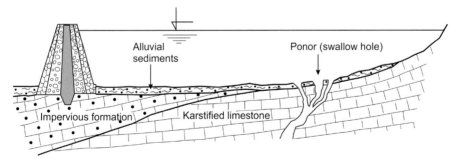

**Figure 13.33.** Cevisli Dam and reservoir, Turkey. Simplified longitudinal cross-section (Yevjevich, 1981).

**Figure 13.34.** Vrtac Retention, Montenegro. Cylindrical dam on the Misor estavelle.

than one hundred new ponors were formed and new leakage began at a rate of > 2.0 m³/s (Fig. 6.38).

The solution to this new problem was to add a large grout curtain. The designed length was 1500 m beneath the dam and 4500 m along the southern bank. The criterion for grouting depth was that it should be 20 m below the average minimum groundwater level. The highest permeability and caverns are mostly concentrated in this zone. The grout curtain consists of one row with boreholes spaced at 2 m. Along one 400 m section of the curtain (depth 30 to 130 m) 32 caverns were intercepted. Locally, grouting holes reached a depth of 106 m. Consumption of the grout mix generally ranged between 206 and 720 kg/m', on average 505 kg/m'. In the cavernous zones, however, grout consumption exceeded 4000 kg/m'. In spite of such extensive and long-lasting grouting efforts, leakage remained at the same levels as before the remedial works. When reservoir levels reached their maximum, the estimated leakage was approximately 20 m³/s (Radulović, 2000).

Finally, Vrtac Reservoir was abandoned as a reservoir and redesigned for temporary flood retention storage.

**Perdikkas Dam** (Greece, 1961) is an earth-filled structure 40 m in height, built on unconsolidated Plio-Pleistocene sediments (clayey silts, silty sand, coarse sand and gravel) deposited over fractured and karstified Upper Cretaceous limestone. The thickness of the cover sediments ranges from zero to over 90 m. The groundwater level was about 70 m below the reservoir bottom. During the first filling, numerous ponors, cracks and subsidences occurred inside the reservoir. Loam blankets were applied to all leakage areas as rehabilitation measures. This clogging process was effective at low reservoir levels but new ponors developed in response to higher storage levels.

Streams of suspended silts were directed from the reservoir banks as reinforcements but were ineffective. Eventually the reservoir was abandoned because the cost of full rehabilitation would be much higher than the benefits gained (Pantzartzis et al., 1993).

Other examples of dams that have been abandoned or operated with limited use are: **Lone Pine** (USA) 31 m high; and **Onac** Dam, Turkey.

In the case of **Havasan Dam** (Iran) the standard grouting techniques were not considered reliable enough for the first site investigated. The dam location was moved a few kilometers upstream.

## 13.9 RESERVOIR FAILURES AFTER MANY YEARS OF OPERATION

**Mavrovo Reservoir** (FYUR Macedonia) is partly located in karstified marble. At the time of first filling (1960), two large and a few small collapse ponors provoked leakage of about 9 to 12 m³/s. The leakage area was successfully protected with an impervious blanket. During its operation the top five meters of the reservoir could never be filled, however. After 25 years of operation several new subsidence (swallow holes) appeared over a 10 day period following intensive rain and the rapid rise of groundwaters to the maximum operational level (Fig. 13.35). A local road and a number of houses were damaged. The remedy was to isolate (amputate) the leaking end section of the reservoir by constructing a shallow dike.

**Hammam Grouz** (Algeria) is a concrete gravity dam, 49.5 m in height from the foundation, which is limestone karstified by hydrothermal water. To achieve the necessary water tightness of the reservoir bottom an overlaying protective impervious clay blanket (surface 860,000, thick 1.8 m, Ip-43%) was constructed. A drainage system (~ 2,200 m') was installed beneath the clay blanket to drain the thermal springs. The

**Figure 13.35.** Mavrovo Reservoir, FYRM Macedonia. Collapses at reservoir bank after extreme rain (1986).

limestone right bank was protected by a shallow cut-off and a reinforced shotcrete blanket (thick 0.17 m). A drainage curtain 15 m deep was constructed below the central part of dam foundations. Average spacing of drainage boreholes was 3.5 m. Reservoir filling started in 1987/88.

In 2003, after 17 years of operation, significant leakage from the reservoir was detected (Fig. 13.36). After heavy precipitation and an abrupt increase in the reservoir water level, this seepage gradually increased. Two large ponors are created at the perimeter of reservoir.

Flow in the drainage galleries increased to 1.2 m³/s and about 40 tons of clay, sand and pebbles were washed out of the karstified rock mass and transported downstream.

**Figure 13.36.** Hammam Grouz, empty reservoir. Shallow sinks in clay blanket.

To halt this loss a grout curtain was constructed to a depth ~ 50 m in the right bank in 2006. To plug some large open cavities, 8000 kg of sawdust, 1624 kg of artificial sponge, 5262 tons of pebbles and 150 plastic bottles filled with sand were used, together with a standard grout mix. However, the reservoir dried up completely in 2007 (see Chapter 10, Fig. 10.4). One possible solution that was suggested was to plug the leaking channel, place a reinforced concrete slab in the reservoir bottom to cover the plug and then put a compacted blanket over the slab (Milanović et al., 2007). Due to the effects of cold (meteoric) water leaking from the reservoir, the temperature of the thermal water in a nearby spa decreased considerably. According Benfetta et al. (2017) new leaking water was recorded in 2015, mostly as overflow from boreholes downstream of the dam that, until then, had been dry for decades.

## 13.10 DAMS UNDER DESIGN OR CONSTRUCTION

In spite of the karst is not friendly and safe environment for dams and reservoirs number of new dams are under design or construction in countries with large karstic areas. Particularly intensive is the building of the large size dams and reservoirs in countries with huge water potential in karst regions: China, Turkey, Iran and many other countries with karst areas rich with high water potential. Some of the dams at a feasibility stage and construction are: Bakhtiary, Kavar, Khersan III, Kuhrang, Beheshtabad, Abolabas, Nargesi Dam, Iran; Wudongde Dam, China; Pošćenje and Rilja dams and reservoirs (Herzegovina) and Nenggiri Saddle dam (Malaysia). At any of those projects engineering karstology plays an important role. A few of the above listed examples are presented below.

**Bakhtiary Dam** in Iran is one of the most challenging dam projects in limestone at the present time. The height of this concrete arch dam is still under discussion, with proposals up to about 300 m. The dam site is located in an extremely folded and faulted formation, as can be seen in Fig. 13.37.

Limestone is the dominant rock in the region, particularly in the dam site area. However, it is frequently shaly and intercalated with beds of marly limestone, laminated marlstone, plus shale and siliceous limestone that are either insoluble or only weakly soluble. Most of these intercalation are continuous layers, 1 cm to 30 cm in thickness and with a frequency of four, five or more per meter. Only the silica layers are sometimes discontinuous. In spite of the large percentage of limestone in it, the entire formation of this rock mass is not prone to major karstification.

From the hydrogeological viewpoint two systems of open fractures at the site are important: pressure release and interbedding fractures. Release fractures are mostly developed in the left bank and are open for free groundwater circulation. In the case of interbedding fractures, areas of sharp kinks in the folding are important from the perspective of water circulation. During the process of folding, relative movement

**Figure 13.37.** The proposed Bakhtiary Dam site, Iran, showing investigation works on the right bank.

between beds created empty spaces in the kink area (along the axial plane). Free water flow can occur through this space in the synclinal sections of the folding. Due to the frequency of very sharply folded structures, a substantial number of such interbedding 'channels' along the kinks are connected to the release fractures, creating a well-developed network, i.e., high permeable zones. Water pressure tests show surprisingly high permeability and predominantly turbulent flow in these networks below the existing river level. A summary diagram of the Lugeon values v elevation is presented in Fig. 5.6.

**Wudongde Dam** (China, Jinshajiang River) is a double arch concrete structure, 270 m high, with a reservoir volume of 7.43 billion $m^3$. The site is located in Middle Proterozoic limestone and marble. It occupies a deep river canyon created by fluvial entrenchment responding to strong uplift of the Yunnan-Guizhou plateau, with karst features developed along sub-vertical structures (mostly bedding planes). Probably karstification in deeper sections, particularly below the river bed, was influenced by rising thermal water. During the drilling of 154 boreholes only nine of them encountered small karst features. Exploration adits were more efficient (Figs. 5.7 and 5.8). Ninety karst features were detected in 77 of the 186 adits that were excavated (total length 16.5 km). Most caverns were small, 0.2–0.5 m with 13 attaining widths of 1–2 m. However, within right bank a huge cavern, K25, with a volume of 369,000 $m^3$ was discovered. It is filled with limestone blocks, lesser breakdown and silty sandy deposits (Fig. 13.38). Locally thick, horizontally bedded, silty clay layers indicate the periodic cessation of vigorous fluvial erosion at their level in the canyon, with underground flooding and transport of fines. The origin of this cavern is, most probably, a consequence of intensive tectonic uplift coupled with the solution and erosion processes.

The geotechnical design calls for all naturally deposited debris in the cavern that is below the elevation of the crest of the dam to be removed and replaced with concrete.

**Kavar Dam** (Iran) according to the design, should be a 60 m high, rock-filled for irrigation. Dam site located in marl (bottom and left bank), and limestone (right bank) formations. The bottom of reservoir area consists of more than 100 m thick lake deposits. In marlstone formation (Razak) is present gypsum in the form of veins, laminations and locally, disconnected lenses. At the dam site area the Razak formation is about 70 m lower than river bed. The number of permanent springs discharges along the reservoir bank, 5–15 m above reservoir bottom (Fig. 13.39). All these springs will be submerged 30 to 40 m after reservoir impounding.

After many years of investigations the project was substantially redesigned, particularly the sealing technology—a watertight surface blanket on the right bank close to the dam.

**Abolabas dam site** is situated at the folded belt of the Zagros Mountain, tectonic province of southern Iran. The dam site (deep and narrow gorge) consists of karstified Asmari limestone (Fig. 13.40).

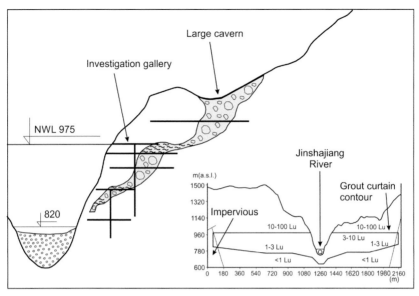

**Figure 13.38.** Wudongde, China. Simplified cross-section of dam site showing a section through the large cavern section and contours the grout curtain design. Source: Presentation of 'Study of Karst Problems at the dam site of the Wudongde project on the Jinshajiang River', Changjiang Survey, Planning, Design and Research Co., LTD. 2011/2012.

Lugeon tests and core logs indicate low permeability. Undefined high permeability (Lu > 100) and cavities are registered at the upper few meters, however groundwater table is 60 m below the gorge bottom.

**Figure 13.39.** Kavar dam site. (A) Downstream view on the Kavars dam site; and (B) One of the springs discharges above river level.

**Nargesi Dam** (Iran), earth filled with clay core, is 77.5 m high from the river bed. The Saddle dam at right bank is 16 m high with a crest length of 88 m. The saddle dam foundation is situated at Gachsaran Formation (Lower Miocene). This formation consists of limestone (basal unit), marl, limestone, gypsum/anhydrite (middle unit) and red marlstone, alternating with gypsum/anhydrite and marl (upper unit). A cut-off wall intersects the river bottom and grouting gallery is being excavated beneath the dam site bottom (Fig. 13.41A). The crucial questions are: (1) how to prevent seepage through the soluble evaporates at saddle dam area; (2) The solution process of evaporates can provoke collapses and endanger right dam abutment; and (3) There is possibly considerable deterioration of water quality. Number of sinkholes and shafts in gypsum are present at the saddle dam area (Fig. 13.41B).

Presently plugged outlets in the creek downstream from the dam site belongs to the separated karst system in evaporates, most probably are not connected with reservoir area.

To prevent solution problems at the saddle dam area the cut-off wall structure is suggested. The gallery downstream and parallel to the cut-off wall for investigation, monitoring and maintenance is suggested as the acceptable solution.

## 13.11  ABANDONED SITES, REDESIGNED OR TEMPORARY FROZEN PROJECTS

In some cases, after many years of detailed geological investigations, some dam projects proposed on karstified rocks have been temporarily halted or outrightly abandoned. In these examples it has usually been found that the necessary sealing works are too expensive and/or the expected results are too problematic, and consequently the entire project is too risky. In some particularly risky and hazardous cases, there have been 30–40 or more years of investigations before the final decision and selection of

**Figure 13.40.**  Abolabas dam site, Iran.

**Figure 13.41.** Nargesi dam site. (A) Foundation area; (B) Solution cavities in gypsum; (C) Outlet channel downstream from dam area presently filled with clayey sediments.

proper dam type and proper anti-seepage works. Some of the abandoned or temporarily halted dams and reservoirs are: **Steno-Calaritico Project** (Greece); **Mangum dam site** (Oklahoma, USA); **Taka Lake** (Peloponnesus, Greece); **Nevesinje** and **Cernica Reservoirs** and **Šnjetica dam site** (Herzegovina); **Beheshtabad, Sazbon, Paalam** and **Kuhrang 3 dam sites** (Iran); **Bogovina dam site** (Serbia); **Lower Gordon** dam site (Tasmania) and **Subansiri dam site** (India).

**Nevesinje Dam** and **Reservoir** are situated at Nevesinjsko Polje at an elevation of 800–870 m. The large part of polje consists of Eocene/Oligocene conglomerates with maximum thickness of about 800 m. Layers and lenses of shale, locally deposited between conglomerates, were detected by boreholes. Due to the matrix of conglomerate is predominantly carbonate, karst features (caves and shafts) are frequently developed. Conglomerates are completely surrounded by karstified Cretaceous limestone (Fig. 13.42). During the rainy period two separated areas of water storage are formed, one in the northern part of polje and second (much larger) in the southern part of polje in the area of Biograd Ponor. Biograd Ponor is terminus of Zalomka River with maximum sinking capacity of 110 m³/s. The river flow is active 212 days per year in average. Average annual river flow is 11.2 m³/s. Maximal registered flow is 440 m³/s. Maximum measured depth of flood water is 38.30 m. All water of the Nevesinjsko Polje catchment area discharges at Buna and Bunica springs at an elevation of 36 m a.s.l. (Fig. 13.43). In periods of fully saturated aquifer velocity of underground flows between ponors situated at perimeter of polje and springs is between 5.82 and 33.67 cm/s.

The hydropower reservoir was designed during the early 60s. By construction the dam (38.0 m high) in front of the large Biograd Ponor and the reservoir of volume V = 145 x 10⁶ m³ and surface of 15 km² should be created. Detailed, comprehensive and long lasting investigations were done to confirm feasibility of the project:

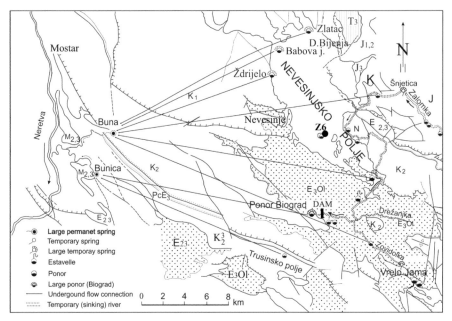

**Figure 13.42.** Nevesinjsko Polje, Herzegovina. Simplified hydrogeological map. Z6–deep piezometer with GWL fluctuation 315 m.

detailed geological mapping, quantitative geomorphological analysis, drilling the deep piezomeric boreholes, groundwater level monitoring, measurements of hundreds of geoelectrical sounds, radioactive borehole logging, large scale tracer tests using dye and radioactive tracers, simultaneous hydrological measurements, speleological investigations and excavation of swallow hole and estavelle areas. A number of ponors, estavelles and deep flows beneath the reservoir have been detected.

Based on more than 50 years of investigations, construction of the large reservoir was determined as too risky. Accordingly new data project was redesigned a few times. To avoid most critical areas the reservoir volume was reduced to $61.8 \times 10^6$ m³ and finally to $59.3 \times 10^6$ m³. Even, with such a reduced reservoir, it is not possible to avoid all potential seepage localities and areas. The reservoir is under construction. Very complex prevention measures are foreseen to achieve required water tightness.

**Taka Lake** temporary flooded karst polje situated at Arkadia plateau (Peloponnesus, Greece). Maximal floods reach 7 m. During humid years the flood of polje last all year. The surface of temporary flooded areas is 7.6 km² and catchment area is about 81 km². Polje dewatering occurred through the two ponor zones. The largest ponor, in the south-west side of the lake, is protected by stone-concrete wall. Along the north perimeter of polje, in Olinos–Pindos limestone are developed in the second ponor zone. This zone consists of a few collapse sinkholes in alluvium and some openings in the limestone ridge of polje, Milanović, 1986. After detailed geological and geophysical investigations including a number of investigation boreholes, it was concluded that Taka karst polje is too risky a space for reservoir construction. Finally,

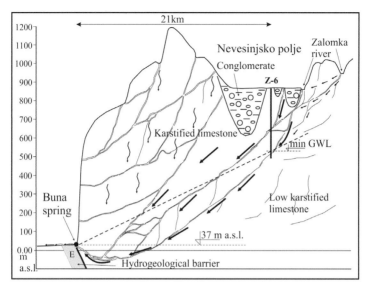

**Figure 13.43.** Nevesinjsko Polje–Buna Spring. Layout sketch of main underground flows.

instead of the reservoir, a pond surrounded by impervious dike has been constructed at the polje bottom.

**Cerničko Polje** (eastern Herzegovina) at elevation of 850 m is one of stepwise closed karst poljes in Dinaric karst (Figs. 13.44 and 13.45). The polje was selected for the reservoir as part of Hydropower System Trebišnjica. The polje was developed along the overthrust discontinuity (Figs. 3.5 and 13.44). Karstified Cretaceous limestone was overthrust on Eocene flysch. Flysch sediments, 200 m deep, acts as a hanging hydrogeological barrier.

The main inflow source is Vilina Pećina, at northern perimeter of the polje, with maximum discharge capacity of more than 50 m³/s (Fig. 13.46).

Along the contact flysch/limestone (southern perimeter of the polje) the number of ponors is developed. Sinking capacity of the largest one is about 20 m³/s (Fig. 2.34). As a result the total sinking capacity of all ponors is much less than inflow in maximum polje flooding. Sinking water discharges at 350 m lower in the Obod Spring in Fatničko polje. Investigations started during the early 60s: geological mapping, borehole drilling, tracer tests, geophysical (geoelectrical) investigations, speleological investigations, hydrological measurements and groundwater level monitoring. Finally, the idea to construct reservoir in Cerničko Polje with volume of 100 million cubic meters, by plugging all ponors and sinking areas was abandoned as too risky.

However, recently, due to knowledge of engineering karstology and technology of prevention and remediation is at a much higher level than 50 years ago, the idea to analyze, once more, the possibility for construction of reservoir in Cerničko Polje has reappeared.

**Bogovina Dam** and **Reservoir** (Serbia) is one of the frozen projects due to negative influence on the Bogovina cave, one of the largest caves in Serbia. On the base of detailed hydrogeological and speleological investigations, it was concluded that the

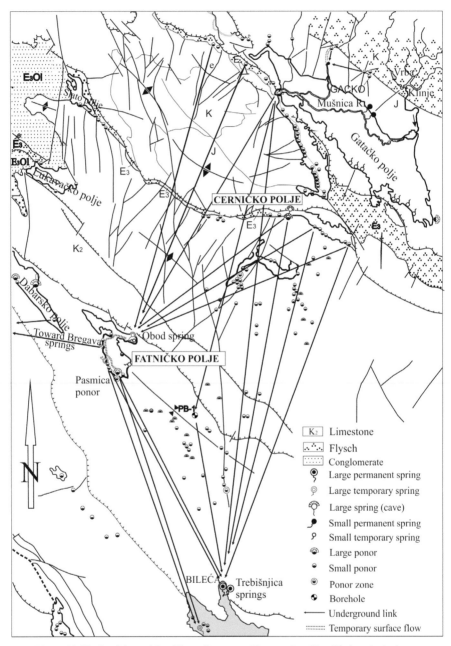

**Figure 13.44.** Cerničko and Fatničko polje, eastern Herzegovina. Simplified geological map.

suggested reservoir level is too high and will provoke filtration in the upper part of karstified rock mass (Fig. 13.47).

Groundwater flows currently orientated toward the future reservoir would reactivate currently through unsaturated pathways, and form a reverse discharge

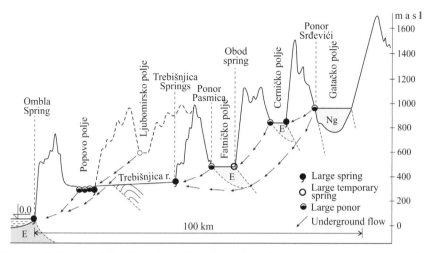

**Figure 13.45.** Position of Cerničko and Fatničko polje at schematic cross-section of eastern Herzegovina.

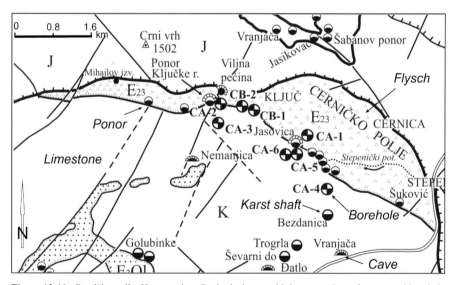

**Figure 13.46.** Cerničko polje, Herzegovina. Geological map with important karst features and borehole locations.

outside of the reservoir area. On the base of these findings the redesign of project is suggested—the dam height should be reduced by 9 m, Milanović et al., 2010.

**Šnjetica dam site** at Zalomka River valley (Herzegovina) was abandoned after detailed investigations of dam site and reservoir area. The large sinking section along the river bed close upstream from the dam site has been detected and investigated. The dam site is shifted 7 km upstream, see Chapter 3, Fig. 3.14.

Due to intensive karstification of the dam site and left reservoir abutment the **Havasan** dam site, in Iran, was abandoned and the dam site shifted a few kilometers upstream.

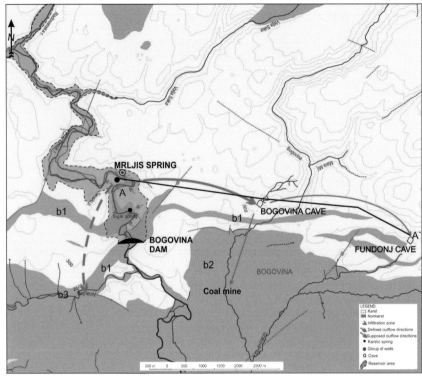

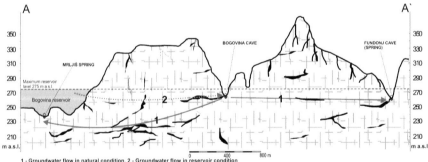

1 - Groundwater flow in natural condition, 2 - Groundwater flow in reservoir condition

**Figure 13.47.** Bogovina. Schematic hydrogeological map. Layout, non-karst: b1 Aptian Orbitoline sandstones; b2 Oligocene deposits of Bogovina basin; b3 Upper Cretaceous volcanic rocks. Cross-section along karstic spring Mrljiš, Bogovina and Fundonj Cave, (1) groundwater flow in natural condition, (2) groundwater flow in reservoir condition, Milanović S. et al., 2010.

**Lower Gordon** dam site (Tasmania) would have flooded a large karst area containing caves of great archeological importance. The project was abandoned for environmental and legal reasons (Kiernian, 1988).

In the case of the **Cuyaguateje Reservoir (Cuba)** the project was abandoned after the investigation stage. Geological mapping, borehole drilling and tracer tests show highly karstified limestone and dolomite and too great a risk to construct dams in such an environment.

Recently, environmental problems have become more complex than karstification of the dam site and seepage from reservoirs. Due to this reason (**Messochora Dam**, Greece, **Ombla underground dam,** Croatia) were temporarily halted.

## 13.12  TAILINGS DAMS AND PONDS IN KARST

Tailings dams are more vulnerable than other types. In the event of failure, the environmental impact is catastrophic and long-lasting. Because tailings ponds usually contain high concentrations of different noxious chemicals they pose a major potential threat of environmental contaminant. In some cases the tailings are contaminated by extremely dangerous chemicals such as heavy metals or cyanides.

Tailings dam failures have occurred for many different reasons but mostly after heavy rainfall caused overtopping, seepage and/or foundation failure, or there was dam wall failure or liquefaction during an earthquake. Poor management and inadequate dam construction methods are also causes of failure.

Tailings ponds within karstified rocks at high elevations are particularly dangerous because any failure can affect the water quality in springs and surface streams at many places at lower elevations. The risk of seepage losses from the tailings impoundment via karstic conduits is a real possibility during the initial stage of mine life. As a consequence of the nature of karst, solution-enlarged fractures and conduits usually do not appear to have been encountered by early investigations (drilling failed to identify them). However, it is important to start with preventive measures during the initial stages. It is particular hazardous when the tailings lie above permanent or temporary karst springs. In that case complicated and expensive measures are required.

The general hydrogeological rules for construction on karst are the same for tailings containment in karstified rocks—the key prerequisite is to understand the local evolution of the karst. If this is clearly understood, the correct design to protect tailings from leakage should be apparent. Two examples are the Cerro Corona copper/gold mine in Peru and the Gacko coal mine in eastern Herzegovina, B&H.

The Cerro Corona tailings are located within karstified carbonate formations at an elevation of about 3800 m (Fig. 13.48). To achieve successful leakage prevention and protection of local springs, there was a very detailed and complex program of geological and hydrogeological investigations and analysis and massive geotechnical measures were undertaken: construction of large compaction blankets; installation of a variety of geomembranes; construction of multi-row grout curtains; and construction of complex drainage systems. The re-tapping of an existing spring to convey potable water beyond the influence of any leakage from the tailings was particularly complicated and expensive.

In the example of the Gacko coal mine (elevation about 950 m), there were comprehensive hydrogeological investigations to select a favorable location around 1100 m asl for the deposition of highly polluted ash and slag. Due to the rapid channeled underground flow common in karst, the eventual failure of tailings containment there

**Figure 13.48.** Cerro Corona, Peru. Tailings pond located within karstified limestone at an elevation of 3800 m.

could have catastrophic implications. If it penetrates underground, the polluted water is beyond any possible control and in a only few days important sources for water supply would be seriously endangered. Protective measures underground (a grout curtain) and an impervious membrane on the surface were applied in order to achieve the required water tightness. Monitoring of surface and groundwater water quality over the entire region is undertaken on a regular basis, with extra sampling at any time if required, particularly in emergency situations. After the tailings pond was completely filled, a layer of productive clean soil was deposited over the ash and slag and the site is now used for a fruit plantation.

## 13.13  DAMS FOR SUBMERGED SPRING TAPPING

Tapping of submerged springs, particularly submarine springs is a challenging task from ancient times. Some very special structures are needed in the case of submarine and sublacustrine spring tapping. One not frequently used concept is damming. Two examples are presented below: Kiveri-Anavalos (Greece) and Bolje sestre (Montenegro).

## 13.13.1 Kiveri-Anavalos Marine Dam

Kiveri-Anavlos submarine spring at Peloponnesus (Greece) is one of two concentrated discharge points of the large Tripolis aquifer. The main structure of intake system is a semicircular concrete gravity dam with its foundation at the sea bottom (Fig. 13.49).

The dam was designed by the German geologist Dr. Staender and built in 1973 to prevent seawater intrusion and pollution of coastal part of aquifer. The purpose of this pilot structure was to provide 35 cm of overpressure (with respect to sea level) in the space surrounded by the dam reducing salinity to 190 mg/L Cl. There is no grouting treatment at the contact between the dam body and the foundation rock. Because of this, a part of aquifer still discharges beneath the dam into the sea, Milanović (1986).

**Figure 13.49.** Kiveri marine dam, Peloponnesus. In corner: sketch, dam layout.

## 13.13.2 Bolje Sestre Spring

Lacustrine spring zone Bolje Sestre (Montenego) is submerged by lake water that fluctuate between 4.6 to 9 m a.s.l. The main discharge zone (Fig. 13.50 zone 2) is submerged between 3 m and 5 m depending on the lake level. The tapping structure consists of elliptical concrete dam structure gated with a rubber gate. The dam height is between 5.1 and 7.0 m and thickness 0.6 m. To achieve proper tightness between the dam body and limestone nearly 2 m of unconsolidated clayey/sandy sediments are removed along the foundation surface.

The large diameter vertical shaft for water pumping, protected by still casing, is situated in karstified rock mass behind the submerged spring zone. The system operates perfectly with possible discharge more than two m³/s, Stevanović (2010).

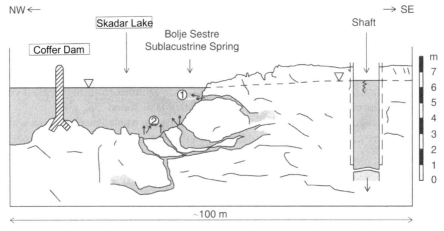

**Figure 13.50.** Bolje Sestre tapping structure (Montenegro), Stevanović, 2010.

# REFERENCES

Abkemeier, T.J. and Stephenson, R.W. 2005. Remediation of a sinkhole induced by quarrying. pp. 605–614. *In*: B.F. Back (ed.). Sinkholes and The Engineering and Environmental Impacts of Karst. ASCE Geotechnical Special Publication No. 122.

Adamo, N., Al-Ansari, N., Issa, E.I., Sissakian, V.K. and Knutson, S. 2015. Mystery of mosul dam the most dangerous dam in the world: Maintenance grouting. Journal of Earth Sciences and Geotechnical Engineering, Vol. 5 No. 3, Scienpress Ltd., pp. 47–58 and 71–77.

Ai, W. 1982. A preliminary study on the air explosion in karst passageways underground. Selected papers for the Second All-China Symposium on Karst. Academic Press of China, pp. 43–49.

Altug, S. 1999. Oymapinar arch dam, Turkey: Foundation treatment in karstic limestone and reservoir curtains. International Commission on Large Dams (ICOLD), Antalia, pp. 193–212.

Altug, S. and Saticioglu, Z. 2001. Berke arch dam, Turkey: Hydrogeology, karstification and treatment of limestone foundation. Proceedings of the 6th International Symposium and Field Seminar Present state and future trends of Karst studies. G. Gunay, K.S. Johnson, D. Ford and I.A. Johnson (eds.). Technical Documents in Hydrology/No. 49, Vol. I, UNESCO, Paris, pp. 315–323.

Anagnosti, P. 1987. Prediction and control of seepage in soluble grounds. IX European Conference, International Society of Soil Mechanics and Foundation Engineerig, Dublin.

Aquilina, L., Ladouche, B., Doerfliger, N. and Bakalowicz, M. 2003. Deep Water Circulation, Residence time, and Chemistry in a Karst Complex. Groundwater – Wiley Online Library.

Arandjelović, D. 1976. Geophysics in the Karst. Geozavod, Geophysical Institute, Vol. 17 (special ed.), Beograd.

Atkinson, T.C. 1977. Diffuse flow and conduit flow in limestone terrain in the Mendip Hils, Somerset (Great Britan). Journal of Hydrology, 35(1-2): 93–110.

Audra, P., Macochian, L., Bigot, J-Y. and Nobecourt, J-C. 2009. Hypogene cave patterns. Ukrainian Institute of Speleology and Karstology, Special paper.

Avdagić, I., Preka, N. and Preka, N.L. 1976. Sodium Chloride Tracing of the Pivka River undergriund flow. 3th International Symposium of underground water tracing (3. SUWT), Ljubljana-Bled, pp. 9–16.

Avdagić, I. 1987. Properties of water regime in karst hydrological systems. Faculty for civil engineering, Sarajevo. Scientific publication (In serbo-Croatian).

Bakalowicz, M. et al. 2007. Hydrogeological settings of submarine springs and aquifers of the Levantine coast (Syria, Lebanon). Toward their sustainable exploration. pp. 721–732. *In*: A. Pulido Bosch, Lopez Getta and Ramos Gonzales (eds.). TIAC'o7. Coastal Aquifers: Challenges and Solutions, IGME, Almeira Springs, 23.

Bakalowicz, M. 2014. Karst at depth below the sea level around the Mediterranean due to the Messinian crisis of salinity. Hydrogeological consequences and issues. Geologica Belgica, 17/1, pp. 96–101.

Balif, P. 1896. Wasserbauten in Bosnia and Hercegovina. Meliorationarbeiten und cisternen im karstgebiete. Wien.

Balissat, M. and Spinnler, U. 1987. Abdichtungsssnahmen bei grossen talsperren im karst. MITTEILUNGEN der Schweizerischen Gesellschaft für Boden-und Felsmechanik. Bienne.

Barjasteh, A. 2012. Salt tectonics impact on dam construction in Khuzestan Province of Iran. International Symposium on Dams for a changing world. Kyoto, Japan.

Benfetta, H., Ouadja, A., Hocini, N., Boudina, S. and Moulla, A. 2017. Updating and Improving the Study of Water Leaks in the Algerian Dam of Foum El Gherza. Manuscript.

Bergado, T.D., Areepitak, C. and Prinzl, F. 1984. Foundation problems on karstic limestone formation in western Thailand: A case of Khao Laem Dam. pp. 397–401. *In*: B.F. Back (ed.). Proceedings of the First Multidisciplinary Conference on Sinkholes, Orlando.

Bianchetti, G. Roth, P., Vuataz, F.D. and Vergain, J. 1992. Deep groundwater circulation in the Alps: d'Illiez, Wallis, Switzerland. Ecologae Geol. Helv., 85(2): 291–305.

Biondić, B., Biondić, R. and Dukarić, F. 1998. Protection of karst aquifers in the Dinarides in Croatia. Environ. Geol., 34(4): 309–319.

Bögli, A. 1960. Kalklosung und Karrenbildung. Zeitschrift für Geomorphologie, Suplement-band, pp. 4-21.

Bonacci, O. 1987. Karst Hydrology, With Special References on Dinaric Karst. Springer-Verlag, Berlin, Heidelberg, New York.

Bonacci, O. 2016. Hydrological analysis of turbidity at karst springs: interpretation of data measured at Ombla Spring. In Croatian language. Hrvatske vode, No. 95, Zagreb, pp. 311–321.

Borghi, A., Renard, P. and Janni, S. 2010. How to model realistic 3D karst reservoir using a pseudo-genetic methodology – Example of two case studies, Advance in Research in Karst Media, Springer.

Borić, M. 1980. The use of temperature changes in locating storage leakages in karst areas. 6th Yugoslav Symposium, Hydrogeology and Engineering Geology, Portorož, Yugoslavia, p. 179.

Božičević, S. 1971. Application of speleology in grouting of karst terranes, in 1st Yugoslav Symposium on Hydrogeology and Engineering Geology, Hercegnovi, Yugoslavia.

Božović, A., Badanur, H., Nonveiller, E. and Pavlin, B. 1981. The Keban Dam foundation on karstified limestone: A case history, Bull. Int. Assoc.Eng. Geol, Geol., 24, 45.

Božović, A. 1985. Foundation Treatment for Control of Seepage. Question 58. Commission Internationale des Grandes Barrages Quinzieme Congres des Grandes Barrages. Lausanne.

Bruce, D.A. 2003. Sealing of massive water inflows through karst by grouting: Principles and practice, pp. 615–626. *In*: B.F. Back (ed.). Sinkholes and The Engineering and Environmental Impacts of Karst. ASCE Geotechnical Special Publication No. 122.

Bruce, D.A., Green, B.H., Williams, B.E. and Williams, H.J. 2014. Evolution of grouting in karst at Logan Martin Dam, Alabama. The Journal of Dam Safety, 12(4): 21–35.

Brune, G. 1965. Anhidrite and gypsum problems in engineering geology. Engineering Geology. Sacramento, USA, 2: 26–33.

Burdon, D.J. and Papakis, N. 1963. Handbook of Karst Hydrogeology, with Special Reference to The Carbonate Aquifers of the Mediterranean Region. United Nations Special Fund, Karst Groundwater Investigations, Athens, Greece.

Burdon, D.J. and Safadi, C. 1963. Ras-el-Ain: The Great Karst Spring of Mesopotamia. Journal of Hydrology, 1: 81–137.

Butscher, C. and Huggenberger, P. 2007. Implications for karst hydrology from 3D geological modelling using the aquifer base gradient approach. J. Hydrology, 342: 184–198.

Castany, G. 1963. Traité pratique des eaux souterraines, Chapitre 22,DUNOD, Paris, pp. 552–583.

Cao, J., Jiang, Y. and Milanović, P. 2016. Hydrogeological settings for underground dam construction – Four case studies from southwest karst area of China. *In*: Z. Stevanović, N. Krešić and N. Kukurić (eds.). Karst without Boundaries. CRC Press, Taylor & Francis Group, Boka Raton.

Chen, J. 1986. Percolation and collapse of reservoirs in karst regions. Carsologica Sinica. Vol. 5. No. 6. Institute of Karst Geology, Guilin, Guangxi, China. In Chinese.

Chen, J., Thomas, G. and Jones. I. 1995. Mechanical impacts of air and water compression in karst conduits, *In*: A.A. Beck (ed.). Karst GeoHazards, Balkema, Rotterdam.

Chen, W. 1988. The study of Disu Underground River system. Du an County. In Karst of China, Yuan D. 1991. Geological Publishing House, Beijing China. Monograph.

Chengjie, X. 1988. A study of geothermal field and karstic leakage in karstic area. Proceedings of the IAH 21st Congress, Geological Publishing House, Beijing, China, p. 1127.

Chengije, Z. 1994. Karst Engineering Geology for water resources and Hydropower Development. Press of Water Resources and Hydropower, Beijing. In Chinese.

Coffin, D.L. 1970. A preliminary evaluation of bank storage associated with Libby Reservoir in Northvest Montana, USGS Water Supply Paper 1899-L, p. 25.

Cooper, A.H. and Calow, R.C. 1998. Avoiding gypsum geohazards: Guidance for planning and construction. British geological survey. Technical report WC/98/5 overseas geological series.

Cvijanović, D. 1980. Precipitation and earthquakes. 5th Symposium of Yugoslav Society for Rock Mechanic and Underground Works. Split, pp. 355–360.

Cvijić, J. 1893. Das Karstphänomenen. Versuch einer geographischen Morphologie. Feogr. Abhandlungen herausgegeben von A. Penck, V, 3.

Cvijić, J. 1900. Karstna polja zapadne Bosne i Herzegovine. In Serbian. Bull. De l'Acc. De Science de l'Art de Serbie, LIX, Beograd.

Cvijić, J. 1918. Hidrographie souterraine et evolution morphologique du karst. Grenoble.

Cvijić, J. 1925. Types morpholigiques des terrains calcaires. Comptes Rendus, Academie des Sciences, Paris.

Dai, J. 1984. The development of underground streams and its problems in engineering geology in Guizhou karst region. Carsologica Sinica. Vol. 3 No. 2, Institute of Karst Geology, Guilin, Guangxi, China. In Chinese.

Dawans, P., Gandais, M., Schneider, T.R. and Waldmeyer, J.P. 1993. Sealing of the Salanfe reservoir (Switzerland)-grout curtain. pp. 259–268. *In*: Widmann (ed.). Grouting in Rock and Concrete. Balkema, Rotterdam.

Deer, D.U. 1982. Cement-bentonite grouting. Grouting in Geotechnical engineering. Proceedings of Conference sponsored by the Geotechnical Engineering Division, ASCE, New Orleans, pp. 279–300.

Djalaly, H. 1988. Remedial and Watertightening Works of Lar Dam. Paper presented at Sezieme Congress das Grandes Barages. San Francisco.

Djigaury, G.M., Mastickiy, A.K., Kereselidze, S.B., Neyshtad, L.I. and Okudzava, V.I. 1980. Inguri Dam on Inguri River. Geology and dams (in Russian). Institut Gidroproekt, Moskva.

Dobrin, E.Z., Molokov, L.A. and Skiba, S.I. 1985. Basic engineering-geological questions of Hoabin Dam at Da River, Vietnam. In Russian. Hydrotechnical Structures, No. 2. Moskow.

Dolder, T., Kreuzer, H. and Milanović, P. 2001. Salman Farsi Dam Project: Final Report Phase 2, Electrowatt-Econo, unpublished, Zurich, 2001.

Dreybrodt, W. 1988. Processes in Karst Systems: Physics, Chemistry, and Geology. Springer-Verlag. New York, N.Y.

Dreybrodt, W., Romanov, D. and Gabrovšek, F. 2001. Karstification below dam sites: A model of increasing leakage from reservoirs. 8th Multidisciplinary Conference on Sinkholes and the Engineering and Environmental Impacts of Karst. Luisville, Kentucky, U.S.A.

Dreybrodt, W., Gabrovšek, F. and Romanov, D. 2005. Processes of Speleogeneze: A Modeling Approach, Chapter 6, Karstification Belowe Dam Sites. Karst Research Institute at ZRC SAZU, Postojna – Ljubljana.

Drew, D.P. and Smith, D.I. 1969. Techniques for the tracing of subterranean water. British Geomorphological Research Group, Technical Bulletin, p. 36

Drogue, C. 1980. Essai d'identification d'un type de structure de nagasins carbonates fisures. Application à l'interpretation de certains aspects du fonctionnement gydrogeologique. Mem. H. ser. Soc. Geol. De France, Paris, pp. 101–108.

Drogue, C. 1985. Geothermal gradients and groundwater circulation in fissured and karstic rocks. J. Geodin., 4.

Dublyansky, V. and Dublyanski, Y. 2000. The role of condensation in Karst Hydrogeology and Speleogenesis. *In*: A. Klimchouk, D. Ford and A. Palmer (eds.). From, Speleogenesis, Evolution of Karst Aquifrs. Dreibrodt. National Speleological Society, Huntsvill, Alabama, U.S.A.

Ewert, F-K, 1985. Rock grouting, Springer Verlag.

Ewert, F-K. 1986. The hydraulic situation of the subsoil at the Pueblo Vijeo Dam (Guatemala). 5th International IAEG Congress, Buenos Aires.

Ewert, F-K. 1992. Evaluation and interpretation of water pressure test. Grouting in ground. Proceedings of the conference organized by the Institute of Civil Engineers. A.L. Bell (ed.). London.

Ewert, F.K. 2005. Hydrofracturing of latent discontinuities in rock and implications for successful and economical execution of grouting. Dam Engineering, Wilmington Publishing. London.

Fetzer, C.A. 1979. Wolf Creek Dam – Remedial Work, Engineering concepts, Actions and Results. Commission Internationale des Grandes Barrages, Q.59, R.5. New Delhi.

Filipponi, M. and Jannin, P.Y. 2008. Possibilities and limits to predict the 3D geometry of karst systems with interception horizon hypothesis, geophysical research abstracts, Vol. 10. EGU General Assembly 2008, EGU2008-A-02825.

Filipponi, M., Jeannin, P.-Y. and Tacker, L. 2009. Evidence of inception horizons in karst conduit networks. Geomorphology 106, Elsevier, B.V.

Filippov, A.G. and Shkolnik, O.A. 1988. Geology of new gypsum caves of the Eastern Sibiria. *In*: Peshchery (Caves). Perm University, Perm pp. 52–64. In Russian.

Fleury, P. 2013. Sources sous-marines et aquifers karstiques cotiers Mediterranees. Fonctionnement et characterisation. Environmental Sciences. Universite Pierre et Marie Curie – Paris VI.

Fleury, P., Bakalowicz, M. and de Marsily, G. 2007. Submarine springs and coastal karst aquifers: a review. Journal of Hydrogeology, 339: 79–92.

Fleury, P., Bakalowicz, M., de Marsely, G. and Cortes, J.M. 2008. Functioning of a coastal karstic system with a submarine outlet, in southern Spain. Hydrogeology Journal, 16. I, pp. 75–85.

Ford, D. 1968. Features of cavern development in central Mendip. Transactions of the Cave Research Group of Great Britain.

Ford, D. 1979. A review of alpine karst in the southern Rocky Mountains of Canada. Bulletin of the National Speleological Society, 41: 53–65.

Ford, D.C. 1980. Threshold and limit effects in karst geomorphology. pp. 345–62. *In*: D.L. Coates and J.D. Vitek (eds.). Thresholds in Geomorphology George Allen - Unwin, London.

Ford, D. 2009. Carbonate-hosted massive sulfide deposits and hypogene speleogenesis: A case study from Nanisvik zinc/led mine, Baffin Island, Canada. pp. 136–148. *In*: K.W. Staford, L. Land and G.Veni (eds.). Advances in Hypogene Karst Studies. National Cave and Karst Research Institute, Symposium 1.

Ford, D. and Williams, P. 1989. Karst Geomorphology and Hydrology. Unwin Hyman Ltd London UK.

Ford, D. and Williams, P. 2007. Karst Hydrogeology and Geomorphology. John Wiley & Sons, Chichester, West Sussex, England.

Ford, D. and Worthington, S. 2012. Notes for the Mid-Congress Karst Field Excursion. 39th IAH Congress, Niagara Falls, Canada.

Ford, D. 2015. The science of caves and karst: From the beginning of the Geological Society of America to ca. 1960. The Geological Society of America, Special Paper 516.

Forkasiewicz, J. and Paloc, H. 1965. Le regime de tarissement de la Foux de la Vis. Colloque sur l'hydrologie des roches fissurees. Dubrovnik.

Frumkin, A. and Shirmon, A. 2006. Tinnel engineerig in the Iron Age: geoArcheological Science 33: 227/237.

Galli, M. 1999. Timavo, Espllozione e studi. Suplemento no. 23 di Atti e Memorie della Commissione Grotte Eugenio Boeagan, Trieste.

Gams, I. 1974. Karst, Monograph Publisher Slovenska Matica, Ljubljana.

Gorbunova, K.A., Maximovich, N.G. Kostarev, V.P. and Andrechuk, V.N. 1991. Technogenic impact on the karst in Perm region. Geology, Climate, Hydrology and Karst formation: Project IGCP 299, Newsletter - Guilin, China, pp. 85–90.

Gospodarič, R. 1984. Hydrogeological features of some karst parts of Slovenia. *In*: B. Mijatovic (ed.). Hydrogeology of the Dinaric Karst. International Association of Hydrogeologists, Volume 4. Verlag Heinz Heise, Hannover.

Grund, A. 1903. Die Karsthydrographie: Studien aus Westbosnien. Geographyscen Abhandlungen, Band VII, Heft 3, von Penck, 7, Leipzig, pp. 103–200.

Grund, A. 1914. Der geographische Zukulus im Karst. Gesellschaft für Erdkunde. Wina.

Guidici, S. 1999. Darwin Dam design and behaviour of an embankment on karstic foundations. International Committee of Large Dams (ICOLD), Antalia, pp. 619–698.

Guifarro, R., Flores, J. and Kreuzer, H. 1996. Francisco Morozan Dam, Honduras: the successful extension of a grout curtain in karstic limestone. Reprint from Int. J. Hydropower and Dams, 3(5): 1–6.

Guun, J. 2003. Encyclopedia of Caves and Karst Science. Fitzroy Dearborn, London.

Günay, G. and Karanjac, J. 1980. Dumanly Spring discharge measured by fluorometric techniques. Proceedings, International Seminar on karst hydrogeology, Ankara, pp. 193–206.

Günay, G., Arikan , A., Bayari, S. and Ekmekci, M. 1986. Quantitative determination of bank storage in reservoirs constructed in karstic areas: case study of Oymopinar Dam. Kars Water Resources Ed. by G. Günay and I. Johnson. IAH Publication No. 161. Printed in Ankara.

Günay, G. and Milanović, P. 2005. Karst engineering studies at the Akkoprü Reservoir area, SW of Turkey. *In*: Z. Stevanovic and P. Milanovic (eds.). Water Resources and Environmental Problems in Karst. National Committee of the International Association of Hydrogeologists (IAH) of Serbia and Montenegro, Belgrade.

Gutierrez, F., Desir, M. and Gutierrez, M. 2003. Causes of the catastrophic failure of an earth dam built on gypsiferous alluvium and dispersive clays (Altorricon, Huesca Province, NE Spain). Environmental Geology, 43: 842–851.

Gutierrez, F., Mozafari, M., Carbonel, D., Gomez, R. and Raeisi, E. 2015. Leakage problems built on evaporates. The case of La Loteta Dam (NE Spain), a reservoir in a large karstic depression generated by interstratal salt dissolution. Engineering Geology, 185: 139–154.

Guzina, B. 1970. Some methods and problems pertaining to the investigation of permeability of karst reservoirs. Dixieme Congres des Grandes Barages, Q.37, R.14, Montreal.

Guzina, B., Sarić, M. and Petrović, N. 1991. Seepage and disolution at foundations of a dam during the first impounding of the reservoir. Congres des Grandes Barages, Q66 Vienne, Austria, p. 1459..

Guzina, B. 1997. Impact of Impounding on a Karst Reservoir Zield. Congress des Grandes Barrages, Florence, Italy, p. 489.

Gvozdecki, N.A. 1981. Karst. Publisher "Misly", Moskva.

Hajdin, G. and Ivetić, M. 1976. An attempt to explain hydraulic conditions in underground karst courses by observation of piezometric levels and outflow. In Serbian. XIII Yugoslav Congress on Rational and Applied Mechanics, Sarajevo, Yugoslavia.

Herak, M. 1977. Tectogenetic approach to the classification of karst terrains. Carsus Iugoslavie 9/4, Zagreb.

Hidrotehnika. 2012. Belgrade. Web site: http://www.hidrotehnika.rs/alzir/brana-ourkiss/.

Hocini, N. and Moulla, A.S. 2005. Detection of water leakage in Foum el-Gherza Dam (Algeria). Ninth International Water Technology Conference, IWTC9, Sharm El-Sheikh, Egypt, pp. 581–589.

Houlsby, A.C. 1977. Engineering of grout curtains to standards. Journal Geot. Div., ASCE, 103: 953–970.

Housby, A.C. 1992. Construction and Design of Cement Grouting. John Wiley & Sons Inc., N.Y./Chichester.

ICOLD. 1985. Question 58 Foundation treatment for control of seepage.

Imamura, A. 1937. Theoretical and Applied Seismology. Tokyo.

Isailović, D. 1976. Optimal operation of physically coupled surface and underground storage capacities. Hydrology Papers G 83, Colorado State University, Fort Collins, Colorado.

Ivanov, G. 1976. Some views about testing of water permeability, one of mostly used methods at exploration for grout curtain design. (In Serbian). Working Papers of the First Yugoslav Symposium for the Soil Consolidation. Zagreb.

Jaćimović, N., Dašić, T., Stanić, M., Đorđević, B., Milanović, P., Sudar, N. and Savić, S. 2015. Distributed Hydrologica Model for runoff Simulation in Karst Basin. (In Serbian) Vodoprivreda Vol. 47, Belgrade, pp. 29–40.

Jahren, N., Rockaway, J. and Easterly, M. 1981. Foundation improvement for prevention seepage through limestone bedrock at Clarence Cannon Dam, Missouri, USA. International Engineering Geology Symposium: Engineering Geological Problems of Construction on Soluble Rocks, Istanbul.

James, A.N. and Lupton, A.R.R. 1978. Gypsum and anhydrite in foundations of hydraulic structures. Geotechnique, 3: 249–272.

James, A.N. and Kirkpatrick, I.M. 1980. Design of foundations of dams containing soluble rocks and soil. Quarterly Journal of Engineering Geology, London, 13: 189–198.

Jansen, R.B. 1988. Advanced dam engineering for design, construction, and rehabilitation. Van Nostrand, New York.

Jarvis, T. 2003. The Money Pit: Karst Failure of Anchor Dam, Wyoming. pp. 271–278. *In*: K.S. Johnson and J.T. Neal (eds.). Evaporite Karst and Engineering/Environmental Problems in the United States: Oklahama Geological Survey Circular 109.

Jeannin, P.Y., Groves, C. and Häuselmann, P. 2007. Speleological investigations. pp. 25–44. *In*: N. Goldscheider and D. Drew (eds.). Methods in Karst Hydrogeology, vol. 26. International Contribution to Hydrogeology, IAH, Taylor and Francis/Balkema, London.

Jennings, J.N. 1971. Karst. Australian National University Press, Canberra.

Jian, C. 1986. Percolation and collapse of reservoirs in karst regions. Cartologica Sinica. No. 5, 2. Guilin. pp. 106–111.

Jianghua, C., Yuchi, J. and Milanović, P. 2016. Hydrogeological settings for underground dam construction – Four case studies from southwest karst area of China. *In*: Z. Stevanović, N. Krešić and N. Kukurić (eds.). Karst without Boundaries. Chapter 21, CRC Press, Taylor & Francis Group. Published by CRC Pres/Balkema, EH Leiden, The Netherlands.

Johnson, K.S. 2003. Gypsum karst as a major factor in the design or the proposed Lower Magnum Dam in southwestern Oklahoma//Evaporate karst and engineering/environmental problems in the United States/Edits. K.S. Johanson, J.T. Neal. Oklahoma Geological Survey Circular 109.

Johnson, K.S. 2004. Problems of dam construction in areas of gypsum karst. Karstology – XXI century: theoretical and practical significance. Proceedings of the International Symposium. Perm, Russia.

Johnson, K.S. 2008. Gypsum-karst problems in constructing dams in the USA. Environmental Geology, 53: 945–950.

Johnson, K.S. and Wilkerson, J.M. 2013. Gypsum karst causes relocation of proposed Ceder Ridge Ram, Throckmorton County, Texas. *In*: L. Land, L.D.H. Doctor and J.B. Stephanson (eds.). Proceedings of the Thirteenth Multidisciplinary Conference. National Cave and Karst Research Institute, Karlsbad, New Mexico.

Kagan, A. and Krivonogova, N. 1999. Impact of engineering geological conditions on decision making during the design and construction of hydraulic structures in areas of karst development. International Commission on Large Dams (ICOLD), Antalya, pp. 673–678.

Kang, Y. and Zhang, B. 2002. Karst and engineering handling to the karst in Wulichong Reservoir, Yunnan Province. Carstologica Sinica, Vol. 21, No. 2. Institute of Karst Geology, Guilin, Guangxi, China.

Karamehmedović, E. 1965. Construction of grout curtain and consolidation works at dam site. Conference: Construction of Hydropower System Trebišnjica. Trebinje, Yugoslavia, pp. 153–163.

Käss, W. 1976. 100 Jahre Uranin. 3th International Symposium of Underground Water Tracing (3. SUWT), Ljubljana, pp. 113–122.

Katzer, F. 1909. Karst and Karsthydrographie. Zur kunde der Balkanhalbinsel. Sarajevo.

Keys, S.W. and McCary, L.M. 1971. Application of Borehole Geophysics to Water-Resources Investigation. US Department of Interior, Washington.

Kiernian, K. 1988. Human impacts and management responses in the karst of Tasmania. Resources management in limestone landscape. Department of geography and oceanography, University College. The Australian Defence Force Academy, Canberra (Special publication no. 2), pp. 69–92.

Klimchouk, A. 2000. Speleogenesis under deep-seated and confined settings, *In*: A. Klimchouk, D. Ford and A. Palmer (eds.). Speleogeneses – Evolution of Karst Aquifers. Dreibrodt. National Speleological Siciety, Huntsvill, Alabama, U.S.A.

Kondratyev, N.N. 1979. Characterisation of carbonate karst development at the area of Hadita Dam construction. Scientific Bulletin of Gidroproject, Moskov (In Russian).

Kovacs, A. 2003. Geometry and hydraulic parameters of karst aquifers—a hydrodynamic modelling approach. Ph.D. thesis, La Faculte des sciences de I'Universite de Neuchatel, Suisse, p. 131.

Kovačina, N., Skopljak, E. and Kesić, P. 1978. Pressure micropulzation in piezometer and possibility for estimation of groundwater fluctuation (in Serbian language). Civil Engineering Faculty, Sarajevo.

Kozyreva, E.A. and Trzhtsinsky, Yu.B. 2004. Karst and its correlation with other geological processes (with reference to the zone of influence of Bratsk Reservoir). F. Guo and D.Tang (eds.). World Correlation of Karst Ecosystem, Newsletter, Karst Dynamic Laboratory, Guilin, China.

Kranjc, A. 1997. Tracer Hydrology. Proceeding book of 7th International Simposium on Water Tracing, A.A. Balkema, Rotterdam.

Krasnopoljsky, A.A. 1947. Underground and artesian water, Upper journal I, II and III. Moscow.

Krašovec, M. 1989. Results of speleological investigations and cave diving in Ombla Spring. Speleological report. Not published.

Krešić, N. 2010. Modelling, Chapter 5 in Groundwater Hydrology of Springs, Engineering, theory, management, and sustainability, N. Krešić and Z. Stevanović (eds.). Elsevier, Burlington, MA, USA.

Kukurić, N. 2015. Transboundary Aquifers in Karst. Karst Aquifers – Characterization and Engineering. Ed. Stavanović, Z., Chapter 17.6. Springer International Publishing Switzerland.

Kutepov, V.M., Parabuchev, I.A. and Kalin, Y.A. 2004. Karst and its influence on territory development. pp. 192–198. *In*: Proceedings of the International symposium, Karstology—XXI century: theoretical and practical significance. Perm.

Laksiri, K., Gunathilake, J. and Iwao, Y. 2005. Engineering geology evaluation of reservoir leakage problem. 73rd Annual Meeting of ICOLD, Tehran, Iran, Paper No.: 151-S3

Lauritzen, S.E. 1988. A geomorphological approach to engineering in karst. IAH 21st Congress, Karst Hydrogeology and karst environment protection, Guilin, China.

Lauritzen, S.E. 2000. Solution and erosional morphology of caves. *In*: Klitchouk et al. (eds.). Speleogenesis – Evolution of Karst Aquifers, National Speleological Society, Huntsville, U.S.A.

Legget, R.F. 1939. Geology and engineering, New York and London.

LeGrand, H.E. and LaMoreaux, P.E. 1975. Hydrogeology and hydrology of karst. *In*: A. Burger and L. Dubertret (eds.). Hydrogeology of Karst Terrains. International Association of Hydrogeoloists, Paris.

Lehman, O. 1932. Die Hidrographie des Karstes. Enzyclopedie des Erkunde 6. Leipzig und Wien, p. 212.

Lombardi, G. and Deer, D. 1993. Grouting design and control using the GIN principle. Water Power and Dam Construction.

Lombardi, G. 1996. Selecting the grouting intensity. Hydropower and Dams, Issue 4.

Lombardi, G. 2003. Grouting of Rock Masses. 3rd International Conference on Grouting and Grout treatment, New Orleans.

Londe, P. 1970. General Report, Q 37. Dixieme Congres des Grandes Barrages, Montreal.

Lu, Y. 1972. The development of karst in China and some hydrogeological and engineering geological conditions. Beijing.

Lu, Y. 1986. Some problems of subsurface reservoirs constructed in karst regions of China. Insitute of Hydrogeology and Engineering Geology. Ministry of Geology and Mineral Resources, China.

Lu, Y. and Cooper, A.H. 1997. Gypsum karst geohazards in China. pp. 117–126. *In*: Beck, F.B. and Stephenson, J.B (eds.). The Engineering Geology and Hydrogeology of Karst Terranes. Proceedings of the Sixth Multidisciplinary Conference on Sinkholes and the Engineering and Environmental impacts of Karst, Springfield/Missouri.

Lu, Y. and Zhang, F. 2006. Sulphate rock karst and sulphate-carbonate rocks' compound karst. Hogh Education Press, Beijing.

Lu, Y. 2012. Karst in China – A World of Improbable Peaks and Wonderful Caves. Ministry of Land and Resources, P.R.C. and China Geological Survey. China.

Lu, Y. and Duan, G. 1997. Artificially induced hydrogeological effects and their impact of environments on karst of North and South China. J. Fei and Krothe (eds.). Proc. 30th Int'l. Geol. Congr., pp. 113–120.

Lu, Y., Zhang, F., Liu, C., Tong, G. and Zhang, Y. 2006. Groundwater Systems and Eco-hydrological Features in the Main Karst Regions of China. Acta Geologica Sinica, 80(5): 743–753.

Lugeon , M. 1933. Barages at geologie, Dunod, Paris.

Lykoshin, A.G., Molokov, L.A. and Parabutchev, I.A. 1992. Karst and Dam Engineering. In Russian. "Gidroproekt", Moskva. (In Russian).

Lykoshin, A.G. 1968. Karst and construction of hydrostructures, Moscow.

Lykoshin, A.G. 1968. Discussions of the filtration deformations and different permeabilty of rock masses. Bulletin of Gidroproect, No. 48. Moscow.

Mahjoob, D.F., Sadatifard, A., Hassani, H. and Zia, A. 2014. Upper Gotvand Dam and Hydro Power Plant Dealing With Salinity in Reservoir. Challenges, Remedies and Evaluations. International Symposium on Dams in a Global Environmental Challenges, Bali, Indonesia.

Mancebo Piquera, J.A., Sanz Perez, E. and Menendez-Pidal, I. 2012. Water seepage beneath dams on soluble evaporate deposits: a laboratory and field study (Caspe Dam, Spain). Bulletin Eng. Geol. Environ., 71: 201–215.

Mangin, A. 1969. Etude hydraulique du mechanisme d'intermitence de Fontestorbes (Blesta-Ariege). Annales et Speleogie.

Mangin, A. 1975. Contribution à la l'étude hydrodinamique des aquifers karstiques. Laoratorie souterain du C.N.R.S., Moulis, Ann. Spéléol. Moulis, pp. 21–124.

Marković, M. 1973. Geomorphological Evolution and Neotectonics of the Orjen Mountain. PhD thesis (In Serbian), Belgrade University, Belgrade.

Martel, E.A. 1894. Les abimes, Delagrave, Paris.

Maximovich, N.G. 2006. Safety of dams on soluble rock (The Kama hydroelectric power station as an example). Book 212 pages. Publisher "Garmonia", Perm Russia.

Mijatović, B. 1968. Investigation methods of karst aquifer hydrodinamque regime by analysis of recession curve. In Serbian. Bulletin of Geological Survey (Geozavod), Volume VIII, B Belgarde, pp. 43–82.

Mijatović, B. 1989. Report on watertightening problems of Lar Dam and Reservoir, unpublished, UNDP Project, Tehran, Iran.

Mijatović, B. 1990. Hydrogeology of karstic aquifers. Geozavod, Belgrade. Book.

Mijatović, B. 2005. Jovan Cvijić - the precursor and founder of the modern karst hydrogeology. In: Cvijić and Karst. Ed. by Z. Stevanović and B. Mijatović. Serbian Academy of Science and Arts, Board on Karst and Speleology, Belgrade. Serbian Academy of Science and Arts, Board on Karst and Speleology.

Mikszewski, A. and Krešić, N. 2015. Mathematical Modeling of Karst Aquifers. pp. 283–298. *In*: Z. Stevanović (ed.). Karst Aquifers – Characteristion and Engineering. Springer International Publishing Switzerland.

Milanović, P. 1977. Hydrogeology of the Ombla Spring drainage area. pp. 187–255. *In*: Serbian. Herald Geological, Sarajevo.

Milanović, P. 1979. The Keban Dam leakage. Report on the Consultant's Mission in Turkey, unpublished, UNESCO

Milanović, P. 1981. Karst Hydrogeology. Water Resources Publications. Littleton, Colorado, U.S.

Milanović, P. 1984. Mission Report, Steno-Kalaritiko dam site, Elati Basin, and Koromilia Project. Public Power Corporation, Not published, Athens.

Milanović, P. 1984. Some methods of hydrogeological exploration and water regulation in the Dinaric karst with special reference to their aplication in Eastern Herzegovina. *In*: B.F. Mijatović (ed.).

Hydrogeology of the Dinaric Karst. International Contributions to Hydrogeology, Volume 4. General editors: G. Castany, E. Groba, E. Romjin. Verlag Heinz Heise, Hannover.

Milanović, P. 1986a. Influence of the karst spring submergence on the karst aquifer regime. Journal of Hydrology, 84, Elsevier Science Publishers B.V., Amsterdam - Printed in The Netherlands, pp. 141–156.

Milanović, P. 1986b. Report on UNDP Consultancy Mission, GRE/85/001.

Milanović, P. 1988. Artificial underground reservoirs in the karst—experimental and project examples. IAH 21st Congress, Guilin, China, pp. 76–87.

Milanović, P. 1989. Power Plant Ombla: Geological Part. Feasibility Study, unpublished, Energoproject, Beograd, Yugoslavia.

Milanović, P. 1990. Influence of construction on hydrogeological and environmental conditions in the karst region, Eastern Herzegovina. Environmental Geology Water Sci., Vol. 15, No.1. Springer-Verlag New York Inc.

Milanović, P. 1995. Criteria for groundwater protection in karst regions. Ecologica No. 7. Society of Engineers of Yugoslaviae, Belgrade.

Milanović, P. 2002. The environmental impacts of human activities and engineering constructions in karst regions. Episodes, Journal of International Geoscience. International Union of Geological Sciences.

Milanović, P. 2003. Prevention and remediation in karst engineering. Proceedings of the ninth multidisciplinary conference, Sinkholes and the Engineering and Environmental Impacts of Karst. Ed. Back B.F. Geotechnical special publication No. 122, ASCE.

Milanović, P. 2004–2010. Gotvand Dam Project. Watertightness of reservoir and dam site. Mission Reports. Mahab Ghodss Consulting Engineers, Teheran.

Milanović, P. 2004. Water Resources Engineering in Karst. CRC Press, Boca Raton.

Milanović, P. 2006. Karst of eastern Herzegovina and Dubrovnik littoral. Publisher ASOS, Beograd.

Milanović, S. 2007. Hydrogeological characteristics of some deep siphonal springs in Serbia and Montenegro karst. Environmental Geology. Vol. 51 No. 5, Springer Berlin/Heidelberg, pp. 755–759.

Milanović, P. 2010. Lar Dam Rebilitation Project. Mission Report, Tehran, Iran.

Milanović, P. 2015. Catalog of engineering works in karst and they effects. *In*: Z. Stevanović (ed.). Karst Aquifers – Characterisation and Engineering, Chapter 13, Springer International Publishing Switzerland.

Milanović, S. 2012. Speleology and speleo-diving in karst hydrogeology. In Serbian. Faculty for Mining and Geology, Hydrogeology Department, Belgrade University

Milanović, P. and Aghili, B. 1990. Hydrogeological characteristics and groundwater mismanagement of Kazerun karstic aquifer, Zagros, Iran. pp. 163–171. *In*: G. Günay, I.A. Johnson and W. Back (eds.). Hydrogeological Processe in Karst Terranes. IAHS Press Walingford, UK.

Milanović, P. and Milanović, S. 2001. Exploitation potential and conditions for underground storage at Perućac Spring. Technical Documents in Hydrology/No. 49, Vol. I, UNESCO, Paris.

Milanović, P., Stevanović, Z. and Beličević, V. 2007. Dams and Reservoirs: Hammam Grouz, Saf-Saf and Ourkis. Consultancy Report. Belgrade/Algeri. Springer.

Milanović, S., Stevanović, Z. and Jemcov, I. 2010. Water losses risk assessment: An example from Carpathian Karst. Environ. Earth Sci., 60(4): 817–827.

Milićević, M. 1987. Karst environment and influence of hydropower structures on karst water regime. Scientific project, Civil engineering faculty, Sarajevo.

Milanović, S., Dragišić, Radulović, M.M. and Stevanović, Z. 2015. Prevent leakage and mixture of karst groundwater. *In*: Z. Stavanović (ed.). Karst Aquifers – Characterization and Engineering. Chapter 16. Springer.

Miller, S.A. 2008. Note on rain-triggered earthquakes and their dependence on karst geology. Geophysical Journal International 173: 334–338.

Mingzhang, W. 2006. Exploitation and utilization of underground water resources in Guizhou karst area. *In*: China Geological Survey, Institute of Karst Geology, CAGS China Geological Publishing House (in Chinese).

Misiewicz, J. 1988. The geology and metallogeny of the Otavi mountain land, Damara origen, Swa/Namibia, with particular reference to the Berg Auks Zn-Pb-V deposits - a model of ore genesis. Master these at Rodes University, Grahamstown.

Mozafari, M. and Raeisi, E. 2012. Water leakage paths in the Doosti Dam, Turkmenistan and Iran. Environmental Earth Science, 65, Springer, pp. 103–117.

Nikolić, R., Raljević, B., Franić, M. and Zidar, M. 1976. Possibility for solving of underground areas gaps and cavities in the area of storage "Buško Jezero", in Working papers of the 1st Yugoslav Symposium for Soil Consolidation, JUSIK, Yagreb, p. 91.

Nonveiller, E. 1989. Grouting—Theory and Practice. Elsevier, Amsterdam.

Okay, G. and Soidam, B.A. 1999. Experience on two karstic sites. Int. Comm. of Large Dams (ICOLD), pp. 709–722.

Osipov, V.I. 1981. Dissoluble salts in clays and their properties dependence upon the forming. International Engineering Geology Symposium: Engineering Geological Problems of Construction on Soluble Rocks. Istanbul.

Öziş, Ü., Ünal, E. and Benzeden, E. and Harmancioğlu, N. 1981. Underground storage capacity of dams in karst regions, estimated through recession hydrograph. International Engineering Geology Symposium: Engineering Geological Problems of Construction on Soluble Rocks. Istanbul.

Palmer, A.N. 1991. Origin of limestone caves: Geological society of America Bulletin. V. 103(1): 1–21.

Palmer, A. and Palmer, M. 2009. Caves and Karst of the USA. National Speleological Society, Huntsville, Alabama, U.S.A.

Pantzartzis, P., Emmanuilidis, G., Krapp, L. and Milanović, P. 1993. Karst phenomena and dam construction in Greece. pp. 65–74. *In*: G. Gunay I. Johnson and B. Back (eds.). Hydrogeological Processes in Karst Trains. IAHS 207.

Parizek, R.R. 1976. On the nature and significance of fracture traces and lineaments in carbonate and other terraines. Proceeding book "Karst Hydrology and Water Resources". W.R.P., Colorado, U.S.A.

Paviša, T. 2017. The Ombla Spring. Excursion Guidebook, 44th Annual Congress of the IAH, Dubrovnik, Croatia.

Pavlin, B. 1961. Reservoir Peruća in Dinaric karst - successfully constructed. V Congress of Yugoslav Committee for Large Dams, Belgrade.

Pavlin, B. 1973. Establishment of subsurface dams and utilization of natural subsurface barriers for realization of underground storages in the coastal karst spring zones and their protection against sea-water intrusion. Ozieme Congres des Grandes Barages, Madrid.

Payton, C.C. and Hansen, M.N. 2003. Gypsum karst in southwestern Utah: Failure and reconstruction of Quail Creek Dike/Evaporite karst and engineering/environmental problems in the United States. K.S. Johnson and J.T. Neal (eds.). Oklahoma Geological Survey Circular 109.

Pearson, R. 1999. Geology and Safety of Dams Case Histories in Gypsum Karst for Horsehooth Dam and Reservoir and Carter Lake Dam No. 2, Colorado Big Thomson Project, Ft. Collins and Loveland, Colorado. U.S. Department of the Interior. Bureau of Reclamation. USBR Technical Service Center Center D-8321, Denver.

Pechorkin, I.A. and Pechorkin, A.I. 1979. Theoretical aspects and engineering-geological forecast on the on the shores of karst reservoirs. Bull. Int. Assoc. Eng. Geol., 20.

Peric, B., Gabrovšek, F. and Pipan, T. 2006. Establishment of Monitoring of the Reka River subterranean course. Published by Park Škocjanske jame, Slovenia.

Perić, J. 1963. Underground reservoirs as structures for artificial water retardation behind karst springs. Bulletin, Volume III B, Engineering geology and hydrogeology. Institute for geology and geophysics. Belgrade.

Perrott, W.E. and Lancaster-Jones, P.F.F. 1963. Case Records of Cement Grouting. From "Grout and Drilling Mud in Engineering Practice". Institute of Civil Engineers, London.

Petrović, B. 1965. Experimental plugging of Obod estavelle in Fatničko polje (in Serbian language). Energoinvest, Sarajevo.

Plotnikov, N.A. 1959. Estimation of groundwater storage. in Russian. Gosgeoltehizdat, Moscow.

Pokrajčić, B. 1976. Epidemcs at karst region of Yugoslavia due to spring contamination. Karst Hydrology and Water Resources. Sarajevo.

Potie, L., Ricour, J. and Tardieu, B. 2005. Port-Miou and Bestouan freshwater submarine springs (Cassis – France) Investigation and works (1964–1978). Water resources and environmental problems in karst. Z. Stevanović and P. Milanović (eds.). Proceedings of the International Conference and Field Seminars Belgrade & Kotor. Special edition from the Institute for Hydrogeology & Faculty of Mining and Geology, Begrade, pp. 267–274.

Preka, N. and Preka, L.N. 1976. Investigation of self-purification capabilities of karst underground flows. Karst Hydrology and Water resources. In Serbo-croatian. Faculty for Civil Engineering, Sarajevo. pp. 577–584.

Quinlan, J.F. 1976. New fluorescent direct tracer suitable for tracing groundwater and direction with cotton. 3rd International Symposium of underground water tracing, Ljubljana, Yugoslavia, pp. 257–262.

Quinlan, J.F. 1978. Types of karst, with emphasis on cover beds in their classification and development. Univ. of Texas at Austin PhD. thesis.

Radulović, M. 2000. Karst Hydrogeology of Montenegro. In Serbian. Geological Institute of Montenegro, Podgorica.

Ravbar, N. and Goldscheider, N. 2007. Proposed Methodology of Vulnerability and Contamination Risk Mapping for the Protection of Karst Aquifers in Slovenia. Acta Carsologica, 36/3. Ljubljana, pp. 397–411.

Ravnik, D. and Rajver, D. 1998. The use of inverse geotherms for determining underground water flow at the Ombla karst spring near Dubrovnik, Croatia. Journal of applied Geophysics. Elsevier. Printed in Netherlands.

Riemer, W., Gavard, M. and Turfan, M. 1995. Ataeturk dam: Hydrogeological and Hydrochemical Monitoring of Grout curtain in Karstic Rock, ASCE Committee on Grouting, ASCE Convention, San Diego, CA.

Riemer, W., Gavard, M., Soubrier, G. and Turfan, M. 1997. The Seepage at the Ataturk Fill Dam. Commission Internationale Des Grandes Barrages (ICOLD), Florence, Italy, p. 613.

Riemer, W. 2015. Investigation and Treatment of Problematic Foundations for Storage Dams: Some Experiance. Engineering Geology for Society and Territory – Volume 6. Springer International Publishing, Switzerland.

Rižikov, V.D. 1954. Nature of karst and key rules of its development (In Russian) Academy of sciences of USSR, Moscow.

Roglić, J. 1964. Les poljes du karst Dinarique et les modifications climatices du Quartenaire. Rev. Belge Geographique, 80(1/2).

Roglić, J. 1972. Historical review of morphologic concepts. *In*: M. Herak and V.T. Stringfield (eds.). Karst, Important Karst Regions of the Noerthern Hemisphere. Elsevier, Amsterdam.

Roksandić, M. 1970. Effects of the Load by a Reservoir on Seismic Activity, Proceedings of the second Congress of International Society for Rock Mechanics. Belgrade.

Romanov, D., Gabrovšek, F. and Dreibrodt, W. 2005. Leakage below dam sites in limestone terrains by enhanced karstification: A modelling approach. *In*: Water Resources and Environmental Problems in Karst. pp. 645–650. Z. Stevanović and P. Milanović (eds.). Proceedings of the International Conference and Field Seminars Belgrade & Kotor. Special edition from the Institute for Hydrogeology & Faculty of Mining and Geology, Begrade.

Roth, J. 1999. Threats to endemic cave species. American Caves.

Roth, P. 1994. Reservoir Induced Earthquakrs and Thermal Spring in Valais. PROSEIS AG. P 4.

Rubinić, J. and Fistanić, I. 2005. Application of time series modelling in karst water management. pp. 417–422. Z. Stevanović and P. Milanović (eds.). Proceedings of the International conference and field seminars Belgrade & Kotor. Special Edition From the Institute for Hydrogeology & Faculty of Mining and Geology, Begrade.

Ruichun, X. and Fuzhang Yan. 2004. Karst geology and engineering treatment in Geheyan Project on the Qingjiang River, China. Engineering Geology 76, Elsevier B.V., www.sciencedirect.com, pp. 155–164.

Sari, S. 2013. Détection des fuites d'eau dans le barrage de Joumine et étude de la sédimentation dans le barrage de Ghezela par la méthode nucléaire. Rapport interne INIS-TN--193–Université de Jendouba, Tunisie, pp. 20–40.

Satoshi, I., Motoi, K., Eiichi, A., Fazal, M.A., Takeo, T. and Masayuki, I. 2003. Construction of subsurface dams and their impact on the environment. Groundwater and Geological Engineering, Materials and Geoinvironment, Volume 50, Ljubljana.

Schoeller, H. 1962. Les eaux souterraines. Masson, Paris.

Sebev, 1970. Giant Caves in Rhodopes. Rodopski Peschernjak (Chepelare) 50 in Bulgarian.

Şekercioğlu, E. and Özgüler, E. 1999. Geotechnical investigations on leakage problem of Akköprü Dam, Turkey. International Commission on Large Dams (ICOLD), Antaliya, pp. 497–506.

Selimović, M. 1977. Analysis of design criteria of Yugoslav grout curtains in the Karst. M.Sc. Dissertation, Facultz of Civil Engineering, University of Zagreb.

Semenov, M.P. et al. 1962. Consolidation of massive limestone - Gergebelska Dam. From Lykoshin et al. Karst and dam engineering, 1992, "Gidroproekt" Moscow.

Sikošek, B. 1954. Tectonics of Bileća - Trebinje region. Collection of Paper of Geological Institute "Jovan Žujović" No. 7, Beograd, Yugoslavia.

Simpson, D., Phipps, M. and Ressi di Cervia, A.L. 2006. Constructing a Cutoff Wall in Front of Walter F. George Dam in 100 Feet of Water. Hydro Review, March, HCI Publications.

Sissakian, V., Al-Ansari, N., Issa, I.E., Adamo, N. and Knutsson, S. 2015. Mystery of Mosul Dam the most dangerous dam in the world (five volumes). Journal of Earth Sciences and Geotechnical Engineering, vol. 5, no. 3.

Sket, B. 2003. Cave fauna – The particular case of Vjetrenica. In "Vjetrenica Cave", I. Lučić & and Sket B.eds Zagreb – Ravno.

Skiba S.I., Molokov, L.A. and Dobrin, E.Z. 1992. Hoa Binh Dam on Da River (Vietnam). Geology and Dams. M. Energoatonizdat, v. XII, pp. 101–110.

Skopljak, E. and Kovačina, N. 1978. Pressure pulsation in piezometric borehole, possibility to estimate groundwater level fluctuation - Phase I. In Serbian., Sarajevo.

Smart, P.L. 1976. The use of optical brighteners for water tracing. Trans. British Cave Research Assoc., 3(2): 62–76.

Smart, C.C. 1983. The hydrology of the Castelguard Karst, Columbia Icefields, Alberta, Canada. Artic and Alpine Research, 15,4, pp. 71–86.

Sokolov, D.S. 1962. Fundamental conditions for the development of karst. In Russian. Gosgeoltekhizdat, Moscow.

Solgi, K. 2010. Personal communication. Photograph taken during excavation at the Seymareh dam site, Iran, 2009.

Song, L.H. 1979. Geochemistry and karstification in Huanghou underground drainage basin, Dushau County, Guizhou Province. The Geomorphology Department of the Institute of Geography, the Chinese Academy of Science, Peking, China.

Springer, G. 2004. A pipe-based, first approach to modelling closed conduit flow in caves. J. Hydrogeology, 289: 178–189.

Stanić, M. and Dašić, T. 2005. Modeling the groundwater regime in karst. In Serbian. Vodoprivreda No. 37. Belgrade, pp. 83–93.

Stevanović, Z., Dragišić, V. and Špadijer, S. 1998. Application of geophysical logging in Hydrogeological investigations: Eastern Kučaj, in 13th Yugoslav Geological Congress, Vol. 5, Hercegnovi, Yugoslavia, p. 499.

Stevanović, Z. 2005. Jovan Cvijić studies of eastern Serbian karst—a foundation of karst hydrogeology. *In*: Z. Stevanovic and B. Mijatovic (eds.). Cvijić and Karst, Serbian Academy of Sciences and Art, Board of Karst and Speleology. Belgrade.

Stevanović, Z. 2010. Case study: Intake of the Bolje Sestre karst Spring for the regional water supply of the Montenegro coastal area. *In*: N. Krešić and Y. Stevanović (eds.). Groundwater Hydrogeology of Springs, Chapter 10.5, Elsevier, Burlington, MA, USA.

Stevanović, Z. 2015. Engineering regulation of karstic springflow to improve water sources in critical dry period. pp. 490–523. Z. Stevanović (ed.). In Karst Aquifers – Characterisation and Engineering, Springer.

Stojić, P. 1966. Bearing capacitz of abutment and improvement of stabilitz of left slope of Grančarevo Dam. VII Congress of the Yugoslav Committee for Large Dams. Sarajevo.

Stojić, P. 1980. Effects of Reservoir in Karst Areas on Earthquakes. Hydrology papers No. 99. Colorado State Universitry. Fort Collins, Colorado.

Sweeting, M.M. 1972. Karst Landforms. Macmillan Press, London.

Šumarac, V. 2008. Application of geomembrane to prevent water seepage from Ourkis Reservoir, Algeria. In Serbian. First Congress of Serbian Commitee for Large Dams. Bajna Bašta, Serbia.

Team of Survey and Exploration. 1979. On the Problems of Air-liquid Pressure in Conduit flow of karstic water. In Chinese. Hydroelectric Survey and Project Institute of Hunan Province.

TVA-Tennessee Valley Authority projects. 1949. Geology and foundation treatment, Technical report No. 22, Knoxville, Tennessee, U.S.A.

TVA-Tenneesse Valley Authority. 1972. Technical Report 16. U.S.A.

Timeus, G. 1928. Nei misteri del mondo sotterraneo. Resultati delle ricearche idrogeologiche sul Timavo 1895–1914, 1918–1927 (Alpi Giulie, v. 29): Atti e Menorie della Commissione Grotte Eugenio Boegan, 22: 117–133.

Tolmachev, v. and Leonenko, M. 2011. Experience in Collapse Risk Assessment of Building on Covered Karst Landscapes in Russia. Karst Management, Ed. by P.E. van Beynen, Springer, Heidelberg London New Yourk.

Turkmen, S. 2003. Treatment of the seepage problems at the Kalacik Dam (Turkey). Eng. Geol., 68(3-4): 156–169.

Uljarević, M., Jokanović, V. and Traparić, R. 2003. Report on investigation results including concept of remedial works to prevent seepage at Gorica dam site (in Serbian). Hydropower Trebišnjca Co, Trebinje, Herzegovina.

U.S. Dept. of Interior. 1965. Geology Groundwater and Bank Storage, Lake Powell: Bureau of Reclamation, Regional Office 4, Salt Lake City, December, p. 10.

U.S. Department of the Interior, 1975. Annual Report: 1974 Operation of the Colorado River Basin and 1975 Project Operations. Bureau of Reclamation, January, p. 34.

Valvasor, J.W. 1689. Die Ehre des Herzogsthums Krain. Endeter, Nürenberg.

Vasić, Lj. 2017. Genesis and Circulation of Groundwater of Complex Karst Systems of Kučaj-Beljanica Massif. PhD. Disertation. University of Belgrade, Mining-Geology Faculty, Serbia.

Vlahović, M. 2005. Hydrogeological properties for construction of reservoirs in karst environment. Case study Nikšićko Polje. MS thesis, University of Belgrade - Faculty of Mining and Geology, Belgrade.

Vlahović, V. 1981. Karstic reservoir Slano, (special ed.), Vol. 14, Montenegro Academy of Science and Arts, Podgorica.

Vlahović, V. 1991. Remediation of rock mass at dam site Liverovići (In Serbian). Montenegro Geological Survey. Harald Geological, Vol. XIV, pp. 155–165.

Vučković, D. and Milanović, S. 2001. Salman Farsi Dam Project, Speleological Investigation, Mission Report. Not published. Belgrad.

Xie, Y., and Li, Y., 2001. Karst Geology, Geomorphology and Ecosystems of Shilin, Yunan. Guidebook for Ecosystems of Semiarid Karst in North China and Subtropical Karst in Southwest China. IGCP 448 China National Working Group. Yuan et al. (eds.).

White, W.B. 1977. The role of solution kinetics in the development of karst aquifers. *In*: J.S. Tolson and F.L. Doyle (eds.). Karst Hydrogeology, Memoir 12, International Association of Hydrogeologists, p. 503.

Wiesner, E. and Wilhelm, R. 1985. Injektionen zur Unterground-Abdichtung in tiefverkarstetem Fels einer Sperrstelle in Guatemala. Geotechnic, Sonderausgabe, pp. 155–163.

Wiesner, E. and Ewert, K-F. 2013. Resolving serious seepage through karstified limestone at the Mujib Dam, Jordan. Bulletin of Engineering Geology and the Environment, pp. 149–162.

Weyermann, W.J. 1977. The karstic rock mass of Canelles Dam. pp. 16–23. *In*: Rock conditions improved through pressure grouting, RODIO in collaboration with The Institute for Engineering Research, Zurich.

Willems, L. and Ek, C. 2011. Field trip on Underground System of Wamme/Lomme rivers in Rochefort. Abstract book. Karst Research, Chalanges for the XXIst century. Brussels.

Wittke, W. 1968. Zur von Injectionen in klüftigen Fels. Fels Mechanic & Ing. Geology., Suppl. IV, Springer Verlag., Wien – New York, pp. 12–78.

Worthington, S.R.H. 2011. Management of carbonate aquifers. pp. 243–262. *In*: Van P.E. Beynen (ed.). Karst Management Springer, Heidelberg London New York.

Worthington, S.R.H. Smart, C.C. and Ruland, W. 2012. Effective porosity of a carbonate aquifer with bacterial contamination: Journal of Hydrology.Walkerton, Ontario, Canada.

Wuzhou, H. 1988. A study on the formation of Triassic "Gupsum-disolved-strata" in Guizhou Province and the seepage prevention for reservoirs. pp. 1117–1126. *In*: Proceedings of IAH 21st Congress, Geological Publishing House, Beijing.

Yaoru, Lu. 1985. The karst geomorphological mechanisms and types in China. Institute of Hydrogeology and Engineering Geology, Ministry of Geology and Mineral Resources, China.

Yevjevich, V. 1981. Karst Waters of Southern Turkey. Final Technical Reporet of UNDP/DSI Project TUR/77/015.

Yoshikawa, M. and Shokohifard, G. 1993. Underground dams: A new technology for groundwater resources development. Proceedings of Articles of International Karst Symposium, Shiraz, Iran, p. 205.

Yu, W. 2008. Water resources exploration from big karst spring by flow narrowing-water pressure adjusting water table rising—A case in Pijiazhi spring, Luxi county, Yunnan. Carstologica Sinica, 17(1): 1–6 (in Chinese with English Abstract).

Yuan, D. 1981. A brief introduction to China's research in karst. The Institute of Karst Geology, Guilin, Guangxi, China.

Yuan, D. 1983. Problems of environmental protection of karst areas. Institute of Karst Geology, Ministry of Geology and Mineral Resources, Guilin, Guangxi, China.

Yuan, D. 1990. The construction of underground dams on subterranean streams in South China Karst. Institute of Karst Geology, Guilin, pp. 62–72.

Yuan, D. 1991. Karst of China. Geological publishing house, Beijing.

Zaisheng, H. 2006. Transboundary Aquifers in Asia With Special Emphases to China. UNESCO Publication.

Zhang, B. and Wu, M. 2000. Seepage control treatment of blind valley reservoir, Gao Yaoji, China, Hydroelectric Power Construction Company, Yunnan Province.

Zhang, X. and Huo, P. 1982. Grouting of karstic caves with clay fillings. Proceedings of the conference on Grouting in Geotechnical Engineering. New Orleans, Louisiana. Published by the American Society of Civil Engineers.

Zhuoxin, W. and Guangyan, S. 1999. Treatments against the seepage of Dalongdong Karst Reservoir in Shanglin County, Guangxi. Pearl River 48–52 (in Chinese).

Zhi Ping, C. 1979. Prediction of the distribution of the underground drainage system in karst terrain. The Geomorphology Department of the institute of Geography, the Chinese Academy of science, Beijing, China.

Zoumei, Z. and Pinshow, H. 1986. Grouting of the karst caves with clay fillings. *In*: Proceedings of the Conference on Grouting in Geotechnical Engineering. Published by the American Society of Civil Engineers, New Orleans, pp. 92–104.

Zhuoxin, W. and Guangyan, S. 1999. Treatment against the seepage of Dalongdong Karst Reservoir in Shanglin County, Guangxi. Pearl River, pp. 48–52 (in Chinese).

Zogović, D. 1990. Storage watertightness problems and technical solutions in the high developed karst area of the Nikšićko Polje. Hydrogeological Processes in Karst Terranes. G. Günay and A.I. Johnosn (eds.). IAHS Publication No. 207. IAHS PressWallingford, UK.

Zwahlen, F. 2004. Vulnerability and risk mapping for protection of carbonate (karstic) aquifers. Final Report COST action 620. European Commission, Directorate-Generalfor Research, Brüssels.

Živanović, V. 2015. Delineation of Karst Groundwater Protection Zones. *In*: Z. Stevanović (ed.). Karst Aquifers-Characterization and Engineering. Springer International Publishing Switzerland, pp. 625–642.

# INDEX

# ABOUT THE AUTHOR

Petar Milanović is Professor Emeritus, President of the Serbian Chapter of International Association of Hydrogeologists and member of Governing Board of International Karst Institute founded by Chinese Academy of Sciences and UNESCO. Since 1973 he has been a member of the IAH Karst Commision and was instrumental in the organization of a few Karst Commision Meetings. He has more than 50 years of experience in investigation, analysis and design related to hydrogeological and geotechnical issues for different civil structures. He has specialized in problems related to karst rock masses. He has been a visiting professor and distinguished lecturer on karst theory and engineering karstology at universities across the world, and countries with construction problems in karst.

Dr. Milanović is the author of several books on hydrogeology and engineering in karst: Karst Hydrogeology, 1981, WRP Littleton, Colorado; Water Resources Engineering in Karst, 2004, CRC Press, Boca Raton; Karst of Eastern Herzegovina and Dubrovnik Littoral, 2006, ASOS, Belgrade. He wrote more than 100 professional and scientific papers and has contributed to numerous other publications on this subject. As a consultant and member of boards of experts, he participated in investigations, design and construction of more than 80 dams, reservoirs, tunnels and tailings in different karst regions worldwide.

He has received many awards for his work, including: Mostar University Award 1987; the Yugoslav National Committee for Hydrogeology and Engineering Geology Award 1987; Hacettepe University Award, Turkey 2000; Karst China 2001 Award, Beijing. In 2011 he was awarded Honorary Membership of the US National Speleological Society and in 2017 President's Award of International Association of Hydrogeologists.